COMPREHENSIVE SERIES IN PHOTOSCIENCES

Series Editors
Donat-P. Häder
Professor of Botany

and

Giulio Jori
Professor of Chemistry

European Society for Photobiology

COMPREHENSIVE SERIES IN PHOTOSCIENCES

Series Editors: Donat-P. Häder and Giulio Jori

Titles in this Series

COMPREHENSIVE SERIES IN PHOTOSCIENCES – VOLUME 2

PHOTODYNAMIC THERAPY AND FLUORESCENCE DIAGNOSIS IN DERMATOLOGY

Editors

Piergiacomo Calzavara-Pinton

Department of Dermatology, Azienda Spedali Civili di Brescia, Brescia, Italy

Rolf-Markus Szeimies

Department of Dermatology, University of Regensburg, Regensburg, Germany

Bernhard Ortel

Wellman Laboratories of Photomedicine, Massachusetts General Hospital, Harvard Medical School, Boston, MA, USA

2001

ELSEVIER

AMSTERDAM – LONDON – NEW YORK – OXFORD – PARIS – SHANNON – TOKYO

ELSEVIER SCIENCE B.V.
Sara Burgerhartstraat 25
P.O. Box 211, 1000 AE Amsterdam, The Netherlands

First edition 2001

Library of Congress Cataloging in Publication Data
A catalog record from the Library of Congress has been applied for.

ISBN: 0-444-50828-7
ISSN: 1568-461X

♾ The paper used in this publication meets the requirements of ANSI/NISO Z39.48-1992 (Permanence of Paper).
Printed in The Netherlands.

SERIES EDITORS' PREFACE

"It's not the substance, it's the dose which makes something poisonous!" When Paracelsius, a German physician of the 14th century made this statement he probably did not think about light as one of the most obvious environmental stress factors. But his statement applies as well to light. While we need light for example for vitamin D production too much light might cause skin cancer. The dose makes the difference. These diverse findings of light effects attracted the attention of scientists for centuries. The photosciences represent a dynamic multidisciplinary field which includes such diverse subjects as behavioral responses of single cells, cures for certain types of cancer and protective potential of tanning lotions. It includes photobiology and photochemistry, photomedicine as well as the technology for light production, filtering and measurement. Light is a common theme in all these areas. In the last decades a more molecular centered approach changed both, the depth and the quality of the theoretical as well as the experimental foundation of photosciences.

An example for the relationship between global environment and the biosphere is the recent discovery of ozone depletion and the resulting increase in high energy ultraviolet radiation. The hazardous effects of high energy ultraviolet radiation on all living systems is now well established. This discovery of the result of ozone depletion put photosciences in the center of public interest with the result that in an unparalleled effort scientists and politicians worked closely together to come to international agreements to stop the pollution of the atmosphere.

The changed recreational behavior and the correlation with several diseases in which sunlight or artificial light sources play a major role in the causation of clinical conditions (e.g. porphyrias, polymorphic photodermatoses, Xeroderma pigmentosum and skin cancers) have been well documented. As a result in some countries (i.e. Australia) public services inform people about the potential risk of extended periods of sun exposure for every day. The problems are often aggravated by the phototoxic or photoallergic reactions produced by a variety of environmental pollutants, food additives or therapeutic and cosmetic drugs. On the other hand, if properly used, light-stimulated processes can induce important beneficial effects in biological systems, such as the elucidation of several aspects of cell structure and function. Novel developments are centered around photodiagnostic and phototherapeutic modalities for the treatment of cancer, artherosclerosis, several autoimmune diseases, neonatal jaundice and others. In addition, classic research areas like vision and photosynthesis are still very active. Some out of these developments are unique to photobiology, since the peculiar physico-chemical properties of electronically excited biomolecules often lead to the promotion of reactions which are characterized by high levels of selectivity in space and time.

Besides the biologically centered areas, technical developments have paved the way for the harnessing of solar energy to produce warm water and electricity or the development of environmentally friendly techniques for addressing problems of large social impact (e.g. the decontamination of polluted waters). While also in use in Western countries, these techniques are of great interest for developing countries.

The European Society for Photobiology (ESP) is an organization for developing and coordinating the very different fields of photosciences in terms of public knowledge and scientific interests. Due to the ever increasing demand for a comprehensive overview over the photosciences the ESP decided to initiate an encyclopedic series, the 'Comprehensive Series in Photosciences'. This series is intended to give an in-depth coverage over all the very different fields related to light effects. It will allow investigators, physicians, students, industry and laypersons to obtain an updated record of the state-of-the-art in specific fields, including a ready access to the recent literature. Most importantly, such reviews give a critical evaluation of the directions that the field is taking, outline hotly debated or innovative topics and even suggest a redirection if appropriate. It is our intention to produce the monographs at a sufficiently high rate to generate a timely coverage of both well established and emerging topics. As a rule, the individual volumes are commissioned; however, comments, suggestions or proposals for new subjects are welcome.

Donat-P. Häder and Giulio Jori
Summer 2000

INTRODUCTION

Photodynamic therapy has been widely investigated over the past two decades and is emerging as a promising therapeutic modality for skin cancers and several inflammatory diseases. This growing interest is based on the availability of a new simple, effective and safe regimen using the topical application of a pro-drug, 5-aminolevulinic acid, as well as on the development of new "second generation" photosensitizers, namely 5-aminolevulinic acid-esters, phthalocyanines, chlorins, porphycenes and hypericin. In contrast to hematoporphyrin derivatives, these compounds are characterized by short-lasting generalized skin photosensitivity. These dyes are available for either topical or systemic delivery and are well characterized. The U.S. FDA has recently approved a topical formulation of 5-aminolevulinic acid (Levulan Kerastick®) plus irradiation with blue light for the treatment of actinic keratoses. Meanwhile, several clinical phase II/III studies are investigating the use of new "second generation" photosensitizers for non-melanoma skin cancers, Kaposi's sarcoma, psoriasis. In addition, pilot studies have found that PDT is an effective treatment modality for a wide range of neoplastic, inflammatory and infectious skin diseases. These indications include very different diseases such as Kaposi's sarcoma, cutaneous T cell lymphoma, acne vulgaris, hypertrichosis, palmo-plantar warts, HPV-induced condylomata and others.

Besides its usefulness in PDT, 5-aminolevulinic acid also exhibits a unique feature for diagnostic purposes. After topical or systemic application, porphyrines are induced in epithelial tumors rather selectively with a high ratio of tumor to surrounding tissue. Tumour-localised porphyrin fluorescence can be viewed after excitation with ultraviolet A or blue light. By using a CCD-camera system and digital image processing, the contrast of the acquired fluorescence images can be significantly enhanced. This diagnostic procedure provides structural information which can be utilized either for site-directed biopsy or for preoperative planning.

The basic principles of PDT is more complex than chemotherapy or other pharmacological modalities. PDT involves not only a drug but an otherwise harmless compound that is activated by visible light. The interaction of these two treatment components is PDT. The variability of these both components results in a complexity of the treatment that may disorient the clinician who does not have specific experience in this field. This book aims to focus experimental and clinical findings on PDT in order to attract and direct the attention of a growing number of dermatologists who want to apply this novel therapeutic approach for the benefit of their patients.

Piergiacomo Calzavara-Pinton
Rolf-Markus Szeimies
Bernhard Ortel

THE EDITORS

Piergiacomo Calzavara-Pinton attended the Medical School in Milan, Italy, and completed his clinical training in Dermatology at the Department of Dermatology I of the University of Milan. Since 1984 he is a member of the staff of the Department of Dermatology of the Spedali Civili of Brescia and, since 1998, he is the head of the Photobiology and Phototherapy Unit. His main fields of investigative and clinical investigation are phototherapy, photochemotherapy, photodynamic therapy and connective tissue diseases. He is involved in clinical and experimental trials for the development of new "second generation" photosensitisers for photodynamic therapy.

Rolf-Markus Szeimies studied at the University of Munich, Germany, finishing with a doctoral thesis in 1989. From 1989 to 1991 he was a resident physician at the Department of Dermatology of the University of Munich. Since 1991 he has been at the Department of Dermatology of the University of Regensburg where he now holds a position as Assistant Professor.

Apart from Photodermatology, his research interests include Photodynamic Therapy, Oncology and Laser Therapy.

Bernhard Ortel is a native of Vienna, Austria, where he went to Medical School and completed his clinical training in Dermatology at the Department of Dermatology I. Early on during his residency his interest focused on photosensitivity diseases and phototherapies. During his training and his time as fulltime staff and Dozent at the Division of Special and Environmental Dermatology at the University of Vienna he pursued both the clinical and basic research sides of photosensitization. In 1994 he assumed a position at the Wellman Laboratories of Photomedicine at Massachusetts General Hospital in Boston, where he spends most of his time investigating photodynamic therapy using aminolevulinic acid-induced porphyrins. He currently holds the position of Assistant Professor of Dermatology at Harvard Medical School.

CONTRIBUTORS

Christoph Abels
Department of Dermatology
University of Regensburg
Franz-Josef-Strauss-Allee 11
93042 Regensburg
Germany

Günther Ackermann
Department of Dermatology
University of Regensburg
Franz-Josef-Strauss-Allee 11
93042 Regensburg
Germany

Claudia Alge
Department of Dermatology
University of Regensburg
Franz-Josef-Strauss-Allee 11
93042 Regensburg
Germany

Béatrice M. Aveline
Wellmann Laboratories of Photomedicine
Massachusetts General Hospital
Harvard Medical School
55 Fruit Street
Boston, MA 02114
USA

Wolfgang Bäumler
Department of Dermatology
University of Regensburg
Franz-Josef-Strauss-Allee 11
93042 Regensburg
Germany

Kristian Berg
Department of Biophysics
Institute for Cancer Research
The Norwegian Radium Hospital
Montebello, 0310 Oslo
Norway

Robert Bissonnette
Division of Dermatology
University of Montreal Hospital Centre
Notre-Dame Hospital
1560 Sherbrooke Street East, Rm K-5201
Montreal, Quebec
Canada H2L 4M1

Wolf-Henning Boehncke
Department of Dermatology
University of Frankfurt
Theodor-Stern-Kai 7
60590 Frankfurt
Germany

Piergiacomo Calzavara-Pinton
Dermatology Department
Azienda Spedali Civili di Brescia
Piazza Spedali Civili 1
25123 Brescia
Italy

P. Mark Curry
QLT Inc.
887 Great Northern Way
Vancouver, BC
Canada V5T 4T5

Giuseppe De Panfilis
Dermatology Department
Azienda Spedali Civili di Brescia
Piazza Spedali Civili 1
25123 Brescia
Italy

Christine C. Dierickx
Department of Dermatology
University Hospital Ghent
De Pintelaan 185
9000 Gent
Belgium

Julia Dräger
Department of Dermatology
University of Regensburg
Franz-Josef-Strauss-Allee 11
93042 Regensburg
Germany

Regina Fink-Puches
Department of Dermatology
University of Graz
Auenbruggerplatz 8
8036 Graz
Austria

Michael R. Hamblin
Wellman Laboratories of Photomedicine
Department of Dermatology
Massachusetts General Hospital
Harvard Medical School
55 Fruit Street
WEL 224
Boston, MA 02114
USA

David W.C. Hunt
QLT Inc.
887 Great Northern Way
Vancouver, BC
Canada V5T 4T5

Sigrid Karrer
Department of Dermatology
University of Regensburg
Franz-Josef-Strauss-Allee 11
93042 Regensburg
Germany

Michael Landthaler
Department of Dermatology
University of Regensburg
Franz-Josef-Strauss-Allee 11
93042 Regensburg
Germany

Olle Larkö
Department of Dermatology
Sahlgrenska University Hospital
41345 Göteborg
Sweden

Harvey Lui
Division of Dermatology
University of British Columbia
835 West 10th Avenue
Vancouver, BC
Canada V5Z 4E8

Anne C.E. Moor
Medac Gmbh
Theaterstrasse 6
22880 Wedel
Germany

Colin A. Morton
Department of Dermatology
Falkirk Royal Infirmary
Major's Loan, Falkirk
Scotland FK1 5QE
U.K.

John R. North
QLT Inc.
887 Great Northern Way
Vancouver, BC
Canada V5T 4T5

Bernhard Ortel
Wellmann Laboratories of Photomedicine
Massachusetts General Hospital
Harvard Medical School
55 Fruit Street
Boston, MA 02114
USA

Qian Peng
PDT Laboratories
Department of Pathology
Institute for Cancer Research
University of Oslo
Montebello
0310 Oslo
Norway

Alexis Sidoroff
Department of Dermatology and
   Venereology
University of Innsbruck
Anichstrasse 35
6020 Innsbruck
Austria

Charles R. Taylor
Massachusetts General Hospital
Harvard Medical School
55 Fruit Street
Boston, MA 02114
USA

Manju Trehan
Massachusetts General Hospital
Harvard Medical School
55 Fruit Street
Boston, MA 02114
USA

Ann-Marie Wennberg
Department of Dermatology
Sahlgrenska University Hospital
41345 Göteborg
Sweden

Peter Wolf
Department of Dermatology
University of Graz
Auenbruggerplatz 8
8036 Graz
Austria

Cristina Zane
Dermatology Department
Azienda Spedali Civili di Brescia
Piazza Spedali Civili 1
25123 Brescia
Italy

TABLE OF CONTENTS

Part I:

History and Basics

Chapter 1

History of photodynamic therapy in dermatology

Rolf-Markus Szeimies, Julia Dräger, Christoph Abels and Michael Landthaler

Table of contents

Abstract

A "photodynamic reaction" describes a photochemical process involving the absorption of light by a photosensitizer and the subsequent generation of reactive oxygen species. Hermann von Tappeiner coined the term "photodynamic" after numerous experiments in 1904 in order to distinguish this photooxidative process from the sensitization during photography. Already at this early stage patients with dermatological conditions like lupus vulgaris or basal cell carcinoma were treated by von Tappeiner in cooperation with the dermatologist A. Jesionek with PDT. However, it took over 90 years until photosensitizers were approved, first in disciplines like pulmonology or gastro-enterology. Although dermatology was the discipline were the very first patients were treated, PDT was not approved until 1999, when in the US 5-aminolevulinic acid for PDT of actinic keratoses was registered.

1.1 The role of light in medical therapy

The importance of sunlight and the roots of the therapeutic use of light can be found already in ancient Egypt medical reports (1,300 b.c.). At that time it was known that on one hand certain drugs were negatively influenced in their potency by sunlight and on the other hand sunlight was necessary for the mechanism of action of several drugs.

More detailed reports on the physiologic effects of sunlight on the human body can be found in the Corpus Hippocraticum (460–375 b.c.) [1]. First observations with the combination of photosensitizing drugs (probably psoralens) and consecutive irradiation with sunlight were reported in the scripts by Abn Mohamed Abdullah Ben Ahmed ("Ebn Baithar") (about 1,200 a.c.) from Malaga [2]. Friedrich Wilhelm Herschel discovered the infrared spectrum of sunlight in 1800, ultraviolet light was described by Johann Wilhelm Ritter in 1806. Already in the beginning of the 19th century it was supposed that the ultraviolet spectrum of light was responsible for reddening of the skin, sunburn and tanning, but it took until 1889 when the Swedish ophthalmologist Erik Johann Widmark proved this in an experimental setting. However, already in 1877 Arthur Henry Downes and Thomas Porter Blunt discovered the bactericidal effects of ultraviolet light.

The widespread use of light for elimination of bacteria is based on the fundamental work of Niels Ryberg Finsen, first in patients with smallpox, later with tuberculosis of the skin. In November 1895 he started irradiation sessions in a patient with lupus vulgaris using carbon arc light resulting in complete remission in January 1896. In April of that year he founded the Finsen Light Institute which was closed 100 years later due to financial problems. Until 1903 he performed phototherapy on 800 patients (published by Finsen and his colleague Forchhammer) and received the Nobel prize the same year. Unfortunately, he died one year later due to Pick's disease [3].

1.2 Early observations of photoreactions

Besides the positive reactions of light, already in the 19th century several reports dealt with the observation that the intake of drugs or plants is harmless even in high quantities unless patients or animals are consecutively exposed to sunlight. Baumstark described in 1874 the influence of light in the spectrum of clinical symptoms in acute intermittent porphyria. In 1892 Charles Darwin and Karl Dammann reported the presence of exanthematic reactions in animals which received buckwheat. The exanthemata were more severe in animals with fair fur and those which were exposed to extensive sunlight [4]. The same observations have also been reported in sheep and cattle for St. John's worth [5].

In 1900 a French neurologist, Jean Prime observed swellings of the fingers, ulcerations, blisters and nail loss in epileptics who received eosin dye solution systemically. Those skin changes were only present in light exposed areas. Eosin was at that time of interest since it contained bromide as a potential antiepileptic drug [6].

However, the first attempt to use this amplifying effect of light was carried out by Oscar Raab, a student. Raab enrolled in the Medical School at the University of Munich in Summer 1894. In autumn 1897 he contacted Professor Hermann von Tappeiner, the head of the department of Experimental Pharmacology, in order to receive a topic for preparation of a doctoral thesis (Fig. 1). Von Tappeiner was at that time in search for new antimalarials [7]. Raab therefore investigated the influence of acridine and it's derivatives on infusoria and other protozoan. He incubated the protozoan in vitro with those dyes at different concentrations determining the threshold dose of toxicity. However, although he performed more than 800 single experiments, his results with lower drug concentrations performed between Nov. 24th and 26th 1897 were inconsistent and not reproducible. Von Tappeiner and Raab then discovered that the only parameter changed during the experiments was the time of performance throughout the day, notably the influence of daylight [8]. Further experiments, also with other dyes like eosin, chinine, or phosphine led to the same results: higher toxicity on protozoan at the same level of drug concentration in the presence of light vs. absence of light.

Since it was already known that the dyes used were capable of absorption of light and the emission of fluorescence, Raab speculated that this toxic effect was mediated by fluorescence. After confirmatory experiments in order to exclude direct influences of light, mainly the infrared spectrum, and the detection of the optimal spectrum for excitation, in 1904 von Tappeiner coined the term "photodynamic reaction" for this – in his eyes – fluorescence-based effect [9].

1.3 First attempts of interpretation of the mechanism of action

After the first reports by O. Raab in January 1900, several other groups also studied those effects. However, possible interpretations of the mechanism of action were different. While in 1895 Richardson thought that hydrogen peroxide was responsible for bactericidal effects, Dieudonneé already believed that the presence of oxygen together with the dye is necessary for the toxic effects [10]. However, so far no one thought of possible catalytic effects of the sensitizers.

Von Tappeiner was convinced that the fluorescence is responsible for the effect. Together with Albert Jodlbauer he speculated that the photodynamic reaction is based on reactions by ions in close vicinity to the fluorescing dyes. Different effects of the various dyes used were therefore due to the different extent of permeation of the dyes

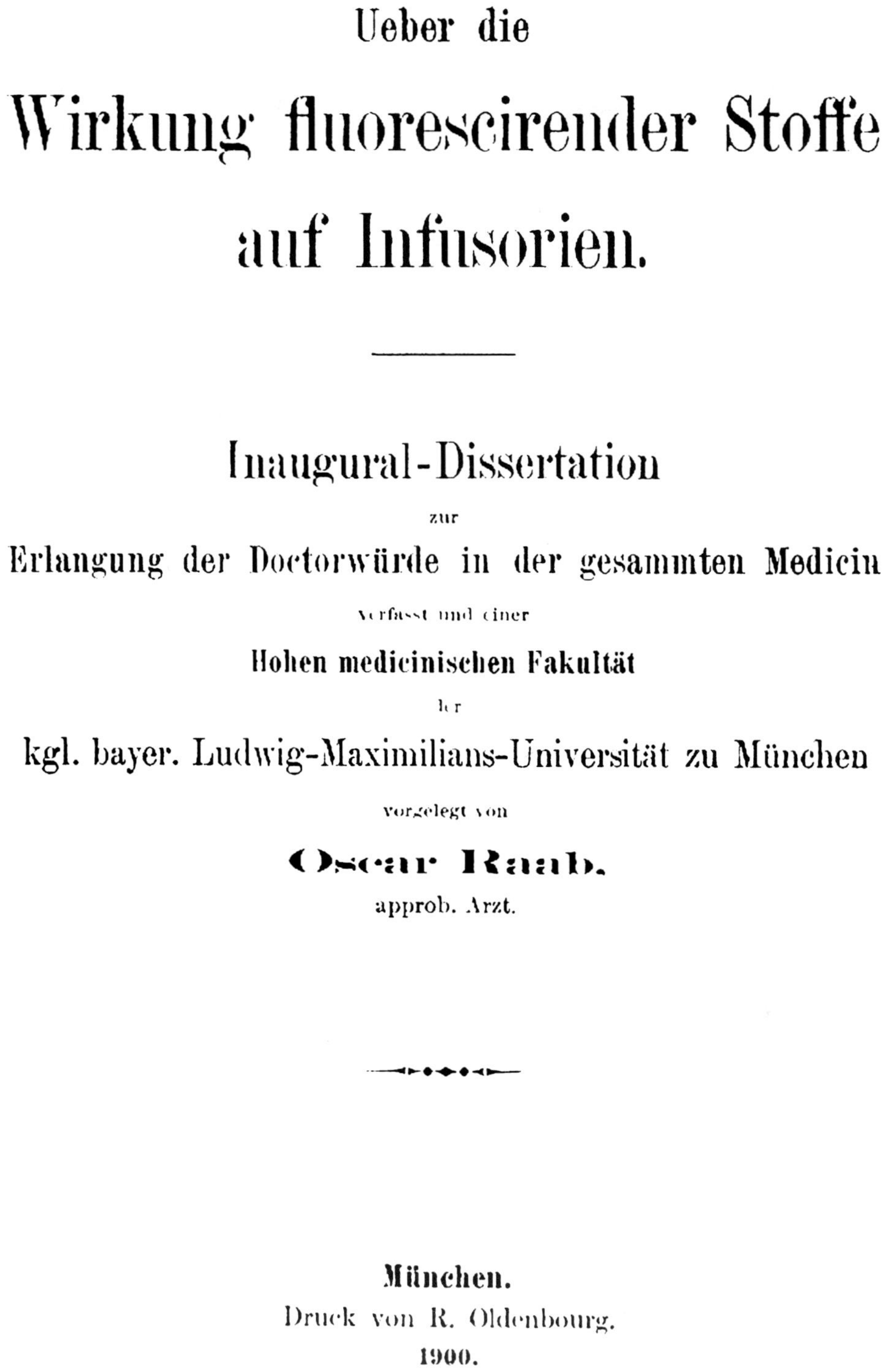

Ueber die

Wirkung fluorescirender Stoffe

auf Infusorien.

Inaugural-Dissertation

zur

Erlangung der Doctorwürde in der gesammten Medicin

verfasst und einer

Hohen medicinischen Fakultät

der

kgl. bayer. Ludwig-Maximilians-Universität zu München

vorgelegt von

Oscar Raab,

approb. Arzt.

———◄•◆•◄►———

München.

Druck von R. Oldenbourg.

1900.

Figure 1. Original title page of the doctoral thesis by Oscar Raab.

through the cellular membranes. Moreover, they assumed that the reaction is induced by absorption of different wavelengths at which each photodynamically active substance is capable of absorbing light at specific wavelengths [11]. Absorbing and fluorescing substances are both able to sensitize and to induce the photodynamic effect [10].

In contrast, Neisser in Breslau and Dreyer at the Finsen Institute already thought that the reaction is purely based on the sensitization. In parallel to the optical photosensitization of photographical plates the substrate is sensitized to light of a specific wavelength, which previously does not induce a reaction. Both believed that the fluorescence itself does not contribute to the photodynamic reaction since they observed several dyes which showed strong fluorescence but did not show sufficient photo-dynamic action [10].

Both opposite views culminated in a scientific controversy which was published in the "Deutsche Medizinische Wochenschrift" in April 1904 [12]. Von Tappeiner imputed Neisser and his coworker Halberstädter for scientific fraud. Neisser answered in the same journal in May 1904 and pointed out that Georges Dreyer in Copenhagen also believes in the concept of sensitization rather than fluorescence as a cause of the photodynamic action [13]. However, neither von Tappeiner nor Neisser refer to results by Ledoux-Lebards who proved that the presence of oxygen is necessary for PDT.

1.4 The role of oxygen for PDT

In 1902 Ledoux-Lebards observed that PDT of paramecia using eosin works better in an open flask than in a closed bottle. He therefore postulated that the presence of oxygen is a prerequisite for performing PDT [14]. Walter Straub, at that time assistant professor of Pharmacology at the University of Leipzig and later successor of von Tappeiner, also believed that oxygen is the crucial substrate for the PDT-mediated toxic effects [15]. In a later paper from 1909 von Tappeiner, in contrast to former statements, also stated that the presence of oxygen and the process of sensitization is more or less responsible for the PDT effect. However, he still argued with some doubt that the mechanism of action was not fully understood [16].

1.5 First therapeutic trials with PDT in humans

Although the mechanism of action was still unknown, it did not take long until this new therapeutic approach was tried out in patients. In a short summary of the presentation by Raab in 1900, von Tappeiner already speculated that fluorescing substances should be used together with light for therapeutic purposes. This effect should then be studied first in the skin of patients since this organ is easily accessible [17].

Together with Albert Jesionek, a young assistant professor at the Department of Dermatology, University of Munich, von Tappeiner started in February 1903 the first experiments in man. Their first paper out of three was then published already at November 24th in 1903 and dealt with the photodynamic treatment of cancerous, syphilitic and tuberculous skin conditions [18]. In their paper the authors stated that of course they would have normally waited longer until presentation but "outer circumstances", i.e. keen competition, led to early publication.

This competition was mainly Dreyer in Copenhagen and Neisser in Breslau. The Danish doctor Georges Dreyer examined since 1902 the effect of light on bacteria. His aim was to sensitize bacteria for light at longer wavelengths in order to increase the level of therapeutic depth. He believed that sensitization itself and not absorption or fluorescence is responsible for the effect. He tried several dyes and turned over to erythrosine as a dye with the lowest toxicity. Besides bacteria or skin of animals he also sensitized live human skin to demonstrate the phototoxic effect [19].

Dreyer started in March and April 1903 his first experiments in patients with lupus vulgaris by intra- and subcutaneous injection of a sterile erythrosine solution. Four to 8 hours after injection he illuminated the target sites for 15–20 minutes. Within 24 hours a severe phlegmonous reaction resulted which resolved under prominent scar formation. The patients suffered from severe pain during irradiation. Dreyer therefore terminated his experiments. These first results were then published in 1903 in the "Dermatologische Zeitschrift" which was issued bimonthly exactly two weeks before the observations by von Tappeiner and Jesionek were published [19]. Dreyer soon left the Finsen Institute and his trials were then carried on by Forchhammer who used a less concentrated dye solution and lower light doses. However, due to the ongoing severe side effects, the therapeutic trial was finally discontinued [20].

In contrast, von Tappeiner and Jesionek reported on good results using a topical application of eosin or other dyes. They started with diseases like pityriasis versicolor, psoriasis, molluscum contagiosum, skin cancer, lupus vulgaris and secondary syphilis. In a second report in 1905 both authors extended their trial on patients with superficial skin cancer and observed good results for a repetitive PDT using a topically applied 0.1 to 5% eosin dye. Sensitization and irradiation with either sunlight or light from an arc lamp took place over several weeks (Fig. 2) [21]. However, later observations discovered that this therapeutic effect was only temporary and restricted to the superficial parts of the lesions. Therefore those trials were also terminated [16].

1.6 The search for new photosensitizers

So far most of the experiments were performed with dyes like chinidine, acridine and eosin with clinically unsatisfactory results. Also at that time the influence of light on the metabolism of plants was of growing importance. Therefore fluorescing plant dyes were also studied.

Walter Hausmann, born in Meran, studied in 1908 in Vienna the photodynamic effects of chlorophyll extracts on red blood cells. Due to the structural similarity he also used hematoporphyrin (Hp), a ferric-ion free derivative of heme [22]. In 1911 he published his results on the use of hematoporphyrin on paramecia, erythrocytes and mice. Mice, which received 10 mg Hp and were kept in the dark, did not show any symptoms whereas animals exposed to sunlight after administration of 2 mg received erythema, edema, and skin necrosis [23].

1.7 Hematoporphyrin and it's first diagnostic and therapeutic use

The observations by Hausmann reactivated photodynamic research in Munich which was abandoned due to the bad clinical results in patients using eosin or other dyes [16].

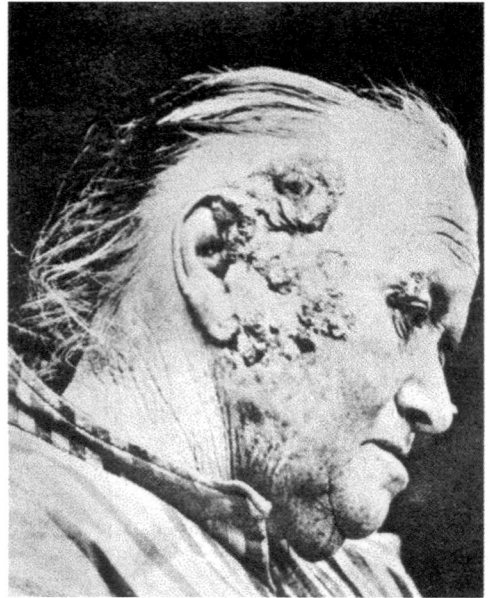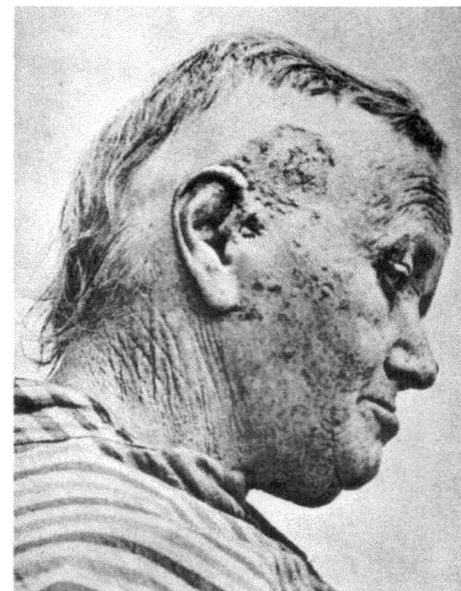

Photograph dated Sept. 10th 1903   Photograph dated Nov. 14th 1903

Figure 2a and 2b. 70-years old countrywoman with multiple skin cancers. The lesions were painted consecutively with eosin dye plus intratumural injection of eosin and were then exposed to sunlight or light from an arc lamp for 6–8 hours a day.

Friedrich Meyer-Betz, a resident at the Department of Medicine, University of Munich performed at October 12th 1912 a heroic self experiment. He injected himself with 200 mg Hp and irradiated a small area at his forearm with light from a Finsen lamp. An ulceration occurred at the irradiation site. Even after days, exposure to sunlight during a ride on a train led to a massive phototoxic reaction with swelling and burning sensation (Fig. 3) [24]. Meyer-Betz died in 1914 during the 1st World War.

Due to observations that patients with psychiatric disorders like depression profit from application of hematoporphyrin, Hp was approved in Germany in 1931 under the name "Photodyn" [25]. In cases of overdosage or posttherapeutic exposure to excessive sunlight, phototoxic reactions were reported. Henry Silver, a dermatologist from Chicago, who was familiar with phototherapeutic approaches in the treatment of psoriasis therefore tried Hp injections and subsequent application of UV-light in plaque-type psoriasis. In 1937 he reported on "about half a dozen" patients who received intramuscular Hp injections followed by oral applications. Over two courses with consecutive UV light, many small lesions and psoriatic plaques cleared [26].

Another important step was made by Albert Policard, a French physician. He discovered in the necrotic center of a freshly excised rat sarcoma, a red fluorescence related to accumulation and retention of Hp in the tumor [27]. For the first time in 1942, Auler and Banzer also found a red fluorescence in patients with tumors, metastases and in lymphatic vessels after subcutaneous or intramuscular injection of Hp [28]. Due to the 2nd World War they did not follow up on their experiments. In 1948 Figge and

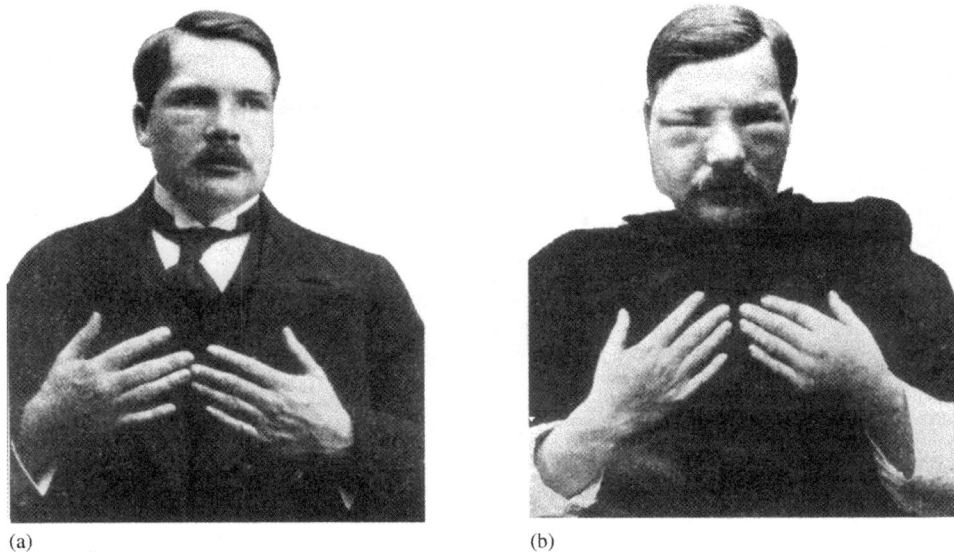

(a)   (b)

Figure 3a and 3b. (a) Friedrich Meyer-Betz before his self experiment at Oct. 14th 1912; (b) After injection of 0.2 g hematoporphyrin swelling of the sun-exposed right side of the face and left arm.

coworkers confirmed their results and proposed a use as photosensitizers for PDT due to good tumor-localizing effects [29]. However, due to impurities and the mixture of different porphyrins a potential use was accompanied by severe phototoxic reactions and therefore was judged as impractical [30]. Samuel Schwartz then in 1955 found that acetylation and reduction of hematoporphyrin led to hematoporphyrin derivative (HpD), a mixture which is enriched with hydrophobic oligomeric porphyrins [31]. Lipson then used HpD first in animals, later in human trials for the detection of various cancers [32,33,34,35].

In 1973, Thomas Dougherty reported on an impaired growth rate of transplanted mammary tumors in mice after sensitization with fluoresceine and light at 488 nm [36]. Dougherty then postulated the criteria for a suitable photosensitizer for PDT:

1. No toxicity at therapeutic doses
2. Defined uptake and accumulation in malignant tissue
3. Light activation above 600 nm
4. Photochemical activity

Using dihematoporphyrin-ether (DHE) these criteria were met in a better way. In a pioneering work in 1978 Dougherty and coworkers then reported on 25 patients with cutaneous or subcutaneous tumors treated with PDT. The patients received 2.5–5 mg/kg b.w. HpD or DHE intravenously and were irradiated with filtered light (600–700 nm) from a xenon arc lamp. One hundred eleven of the 113 tumors treated showed complete or partial remission. The application intervals of drug and light with the best ratio between skin and tumor response were 3–4 days [37].

Table 1. Timeline of reports on photodynamic reactions

1899	First report on phototoxic reaction	Raab
1900	Doctoral thesis "Über die Wirkung fluorescierender Stoffe auf Infusorien"	Raab/v. Tappeiner
1902	Presence of oxygen necessary for phototoxic reaction	Ledoux-Lebards
1900–1903	Investigations of phototoxic reactions on cells of higher organisms: frog epithelium, enzymes and toxins	v. Tappeiner/Jodlbauer
1903	First therapeutic experiments using phototoxic reaction in cancer, tuberculosis, and syphilis of the skin by painting the skin with eosin dye and subsequent irradiation with light (publication in 1905)	Jesionek/v. Tappeiner/ Posselt
1904	Confirmation and expansion of the findings by Ledoux-Lebards	Straub
1904	v. Tappeiner coins the term "photodynamic"	v. Tappeiner/Jodlbauer
1907	Report on the photodynamic phenomenon: Search for new photodynamically active substances and study of chinine, acridine and eosin	v. Tappeiner/Jodlbauer
1908	Investigation of plant dyes regarding their photodynamic activities, especially hematoporphyrin	Hausmann
1911	Publication of results on hematoporphyrin (first characterization by Scherer in 1841)	Hausmann
1910–1911	Demonstration of photodynamic effects of fluorescing dyes in mammals	Pfeiffer
1912	Heroic trial by Meyer-Betz (self-administration of hematoporphyrin)	Meyer-Betz
1918–1920	Utilization of the fluorescing reaction for desinfection of wounds	
1924	Detection of a red fluorescence inside rat sarcomas	Policard
1925	Examination of porphyrins (Fischer received Nobel prize in 1929)	Fischer
1930	Therapeutic use of hematoporphyrin in melancholy and endogenous depression	Hühnerfeld
1937	Therapeutic trial with hematoporphyrin and UV light in patients with psoriasis vulgaris	Silver
1942	Demonstration of tumor-localizing properties of porphyrins	Auler/Banzer
1948	Accumulation of porphyrins in tumors -> usefulness for tumor diagnostics	Figge
1955	Investigations on tumor localizing effects of hematoporphyrin in different tumors	Figge/Rassmussen/ Taxdal
1955	Purification of hematoporphyrin	Schwartz

Table 1. Continued

1960	Localization of hematoporphyrin derivative (HpD) in neoplastic tissue	Lipson/Baldes
1966	First report on photodynamic therapy of breast cancer	Lipson
1970	First reports on long-time cure rates in rats and mice with experimentally induced tumors after systemic application of hematoporphyrin derivative	Dougherty
1978	First report on HpD-PDT in 25 patients with cutaneous tumors/metastases	Dougherty
1990	First report on ALA-PDT in dermatologic conditions	Kennedy

1.8 Further evolutions

Meanwhile, the photosensitizer porfimer-sodium treatment has been approved for systemic PDT in several countries worldwide for different oncologic indications, however, not for a dermatologic indication. Due to the long lasting photosensitization the application in dermatology of this porphyrin derivative is very limited. In order to identify dyes which exert a high tumor to surrounding tissue ratio, researchers were attracted by natural precursors of porphyrins. The use of the topical application of 5-aminolevulinic acid (ALA) for PDT of dermatologic diseases (mainly epithelial skin cancers and premalignancies) was introduced by James Kennedy and coworkers in 1990 [38]. ALA, a metabolite of the heme biosynthesis inducing protoporphyrin IX was the first PDT agent to receive regulatory approval for the treatment of actinic keratoses in conjunction with blue light in dermatology.

References

1. Hippokrates. In: R. Fuchs (Ed.), *Sämtliche Werke*, Munich, 1895–1900, Vol. 2, (pp. 354–355).
2. J. Sontheimer (1840). *Große Zusammenstellung über die Kräfte der bekannten einfachen Heil- und Nahrungsmittel von Abn Mohammad Abdullah Ben Ahmed aus Malaga bekannt unter dem Namen Ebn Baithar*, Stuttgart.
3. A. Aggebro (1947). *Niels Finsen*, Rascher. Zürich.
4. A. Lentner (1990). *Von der Heliotherapie der Antike zur ultravioletten Phototherapie*. doctoral thesis, Düsseldorf.
5. H. Kuske (1940). Perkutane Photosensibilisierung durch pflanzliche Wirkstoffe. *Dermatologica*, **82**, 274–338.
6. J. Prime (1900). *Les accidentes toxiques par i'eosinate de sodium*. Jouve & Boyer, Paris.
7. H. v. Tappeiner (1895). Über die Wirkung von Phenylchinolinen auf niedere Organismen. *Arch. Klin. Med.*, **56**, 369–389.
8. O. Raab (1900). Über die Wirkung fluorescierender Stoffe auf Infusoria. *Z. Biol.*, **39**, 524.
9. H. v. Tappeiner, A. Jodlbauer (1904). Über die Wirkung der photodynamischen (fluoreszierenden) Stoffe auf Infusorien. *Dtsch. Arch. Klin. Med.*, **80**, 427–487.

10. E. Mettler (1905). Experimentelles über die bakterizide Wirkung des Lichtes auf Eosin, Erythrosin und Fluoreszein gefärbte Nährböden. *Arch. Hyg.*, **53**, 79–127.
11. A. Reitz (1908). Untersuchungen mit photodynamischen Stoffen. *Z. Bakt. Par. Infektkr.*, **45**, 270–285.
12. H v. Tappeiner (1904) Zur Kenntnis der lichtwirkenden (fluoreszierenden) Stoffe. *Dtsch. Med. Wochenschr.*, **16**, 579–580.
13. A. Neisser, L. Halberstaedter (1904). Mitteilung über Lichtbehandlung nach Dreyer. *Dtsch. Med. Wochenschr.*, **8**, 265–269.
14. C. Ledoux-Lebards (1902). *Annales de l'Institut Pasteur*, **16**, 593.
15. W. Straub (1904). Über chemische Vorgänge bei der Einwirkung von Licht auf fluoreszierende Substanzen (Eosin und Chinin) und die Bedeutung dieser Vorgänge für die Giftwirkung. *Münch. Med. Wochenschr.*, **23**, 1093–1096.
16. H. v. Tappeiner (1909). Die photodynamische Erscheinung. *Ergebn. Physiol.*, **8**, 698–741.
17. H. v. Tappeiner (1900). Über die Wirkung fluorescierender Stoffe auf Infusorien nach Versuchen von Raab. *Münch. Med. Wochenschr.*, **1**, 5–7.
18. H. v. Tappeiner, A. Jesionek (1903). Therapeutische Versuche mit fluoreszierenden Stoffen. *Münch. Med. Wochenschr.*, **50**, 2042–2044.
19. G. Dreyer (1903). Lichtbehandlung nach Sensibilisierung. *Dermatol. Z.*, **10**, 6.
20. H. Forchhammer (1904). Eine klinische Mitteilung über Lichtbehandlung nach Sensibilisation. *Dtsch. Med. Wochenschr.*, **38**, 1383–1384.
21. A. Jesionek, H. v. Tappeiner (1905). Zur Behandlung der Hautcarcinome mit fluoreszierenden Stoffen. *Dtsch. Arch. Klin. Med.*, **85**, 223–239.
22. W. Hausmann (1908). Die sensibilisierende Wirkung tierischer Farbstoffe und ihre physiologische Bedeutung. *Wien. Klin. Wochenschr.*, **21**, 1527–1528.
23. W. Hausmann (1911). Die sensibilisierende Wirkung des Hämatoporphyrins. *Biochem. Z.*, **30**, 276–316.
24. F. Meyer-Betz, (1913). Untersuchungen über die biologische (photodynamische) Wirkung des Hämatoporphyrins und andere Derivate des Blut- und Gallenfarbstoffes. *Dtsch. Arch. Klin. Med.*, **112**, 476–503.
25. J. Hühnerfeld (1941). *Die biologisch-klinische Bedeutung des Hämatoporphyrin-Nencki*, Johann Ambrosius Barth Verlag, Leipzig.
26. H. Silver (1937). Psoriasis vulgaris treated with hematoporphyrin. *Arch. Dermatol. Syph.*, **36**, 1118–1119.
27. A. Policard (1924). Etude sur les aspects offerts par des tumeurs expérimentales examinées à la lumière de Wood. *C.R. Soc. Biol.*, **91**, 1423–1424.
28. H. Auler, G. Banzer (1942). Untersuchungen über die Rolle der Porphyrine bei geschwulstkranken Menschen und Tieren. *Z. Krebsforsch.*, **53**, 65–68.
29. F.H.J. Figge, G.S. Weiland, L.J. Manganiello (1948). Cancer detection and therapy, affinity of neoplastic, embryonic, and traumatized tissues for porphyrins and metalloporphyrins. *Proc. Soc. Exp. Biol. Med.*, **68**, 640–641.
30. D.S. Rassmussen-Taxdal, G.E. Ward, F.H.J. Figge (1955). Fluorescence of human lymphatic and cancer tissues following high doses of intravenous hematoporphyrin. *Cancer*, **8**, 78–81.
31. S. Schwartz, K. Absolon, H. Vermund (1955) Some relationships of porphyrins, x-rays and tumors. *Med. Bull.*, **27**, 7–13.
32. R.L. Lipson (1960). *The photodynamic and fluorescent properties of a particular hematoporphyrin derivative and it's use in tumor detection*. Masters thesis, University of Minnesota, Minneapolis.
33. R.L. Lipson, E.J. Baldes, A.M. Olsen (1961). Hematoporphyrin derivative: a new aid of endoscopic detection of malignant disease. *J. Thorac. Cardiovasc. Surg.*, **42**, 623–629.

34. R.L. Lipson, E.J. Baldes, A.M. Olsen (1964). A further evaluation of the use of hematoporphyrin derivative as a new aid for the endoscopic detection of malignant disease. *Dis. Chest.*, **46**, 676–679.
35. R.L. Lipson, J.H. Pratt, E.J. Baldes, M.B. Dockerty (1964). Hematoporphyrin derivative for detection of cervical cancer. *Obstet. Gynecol.*, **24**, 78–84.
36. T.J. Dougherty (1973). Photoradiation therapy. In: *Abstracts of the American Chemical Society Meeting*. Chicago, Il.
37. T.J. Dougherty, J.E. Kaufman, A. Goldfarb, K.R. Weishaupt, D. Boyle, A. Mittleman (1978). Photoradiation therapy for the treatment of malignant tumors. *Cancer Res.*, **38**, 2628–2635.
38. J.C. Kennedy, R.H. Pottier, D.C. Pross, (1990) Photodynamic therapy with endogenous protoporphyrin IX: basic prinicples and present clinical experience. *J. Photochem. Photobiol. B: Biol.*, **6**, 143–148.

Photodynamic Therapy and Fluorescence Diagnosis in Dermatology
P.-G. Calzavara-Pinton, R.-M. Szeimies and B. Ortel, editors.

17

Chapter 2

Primary processes in photosensitization mechanisms

Béatrice M. Aveline

Table of contents

Abstract

Photosensitization processes are used in a variety of medical applications. In photodynamic therapy (PDT) of cancer and other diseases, the treatment relies on the combination of light and a tumor-localizing drug (called a photosensitizer). The drug-light interaction results in a sequence of chemical and biochemical processes, which cause irreversible damage to the target tissue. In this introductory chapter, we review the fundamental physical and chemical principles of molecular photochemistry, which are pertinent for the understanding of the primary processes of photosensitization. The nature of light and how it interacts with molecules, the generation of excited states by light activation of a photosensitizer and the subsequent (Type I and Type II) reactions undergone by these excited states are described and discussed along with the different diagnostic methods currently available for the investigation of photosensitization mechanisms.

2.1 Introduction

Photosensitization reactions are defined as processes in which absorption of light by a chromophore (the photosensitizer) induces chemical changes in another molecule (the substrate). When photosensitization causes chemical changes to essential molecules in biological systems, complex biochemical processes are initiated, which can lead to loss or alteration of the cellular activity and ultimately compromise the function of the biological system [1,2].

The phenomenon of photosensitization is widespread throughout photobiology and photomedicine [3]. Photosensitization occurs in humans in a variety of diseases such as porphyrias, age-related macular degeneration [4] and skin photosensitivity [5] as a side effect of various drugs and chemicals. In these conditions, absorption of light by the photosensitizer leads to deleterious damage to the surrounding tissue. Therapeutically, these biologically damaging reactions can be employed to target abnormal cells and eradicate diseased tissues. This has been achieved in the psoralen-UVA treatment (PUVA) of psoriasis [6] and in the photodynamic therapy (PDT) of cancer [2,7,8] and other diseases [9]. The PDT approach has also been applied in the purification of blood products [10], the photoinactivation of viruses [11] and in autologous bone marrow purging [12]. Recent efforts have concentrated on corneal repair and the treatment of age-related macular degeneration [13–15] as well as on skin cancers and non-oncologic cutaneous applications [9].

Photosensitizers can be of endogenous origin (e.g. the porphyrins, that abnormally accumulate in the blood and skin of patients suffering from porphyrias) or of exogenous nature as in the case of PDT, where the first step of the treatment involves the administration of the photosensitizer (or a metabolic precursor) to the patient. However, in all cases, the processes of photosensitization are initiated by the absorption of light by the sensitizer. Despite the frequency of photosensitization reactions in photobiology and their successful applications in a wide variety of medical areas, clarifying the mechanisms by which a light activated drug induces biological response remains a challenge in photobiological and photochemical research.

This chapter briefly introduces basic concepts of molecular photochemistry, such as the absorption of light by molecules and the resulting production of excited states, which play a crucial role as initiators of photosensitization reactions. The primary physical and chemical processes undergone by these excited states are discussed along with the parameters controlling the efficiency of the various reaction pathways (Type I/ Type II processes). The different biochemical and spectroscopic diagnostic methods available for the investigation of photosensitization mechanisms in PDT are then presented with emphasis on the underlying principles and the information that can be extracted from the results. Finally, we close this overview by giving examples, which illustrate how these diagnostic tests are currently used to unravel the action mechanism(s) of Type I and Type II photosensitizers. Excellent books and reviews have been published on various aspects of photosensitization reactions [3,16–23] and should be consulted for additional general information and when specific topics are covered too cursorily in the present chapter.

2.2 Basic concepts

The first step in all photosensitization reactions involves the absorption of light by a chromophore. A short introduction is given here to the basic physical and chemical concepts of molecular photochemistry, which are essential for the understanding of the light-molecule interactions and their consequences in photosensitization processes. A thorough review of these concepts is, however, not within the scope of this chapter and references to more detailed works and to textbooks are included for further reading [18,23–25].

2.2.1 The nature of light

Light corresponds to a small region of the electromagnetic spectrum. For the description of some of its properties, it is adequate to treat light as electromagnetic waves, while for others, a description as particles – photons – is more suitable. This is referred to as the wave-particle dualism. Most optical properties of light, such as reflection, refraction, interference and polarization, are well explained by the wave model. However, in order to understand phenomena, such as absorption and emission, the classical picture of waves is inadequate, and light has to be provided with particle-like properties. Equation (1), which combines both descriptions, shows that the energy (E, in Joules) of each photon is directly proportional to the radiation frequency (v) (i.e. inversely proportional to the light wavelength, λ). The proportionality constants are h, Planck's constant ($h = 6.63 \times 10^{-34}$ J s) and c, the velocity of propagation or speed of light ($c = 3 \times 10^8$ m s^{-1} in a vacuum).[1]

$$E = h\nu = hc/\lambda \qquad (1)$$

[1] In photochemistry, Eq. (1) is often written as: $E = Nh\nu = Nhc/\lambda$, where E is expressed in Joules per mole and N is Avogadro's number ($N = 6.023 \times 10^{23}$ mole^{-1}).

The visible part of the electromagnetic spectrum corresponds to wavelengths between 400 and 760 nm, the near ultraviolet (UV) portion spreads from 200 to 400 nm and the near infrared (IR) region, from 760 to 900 nm. In photobiology and photomedicine, the wavelengths used to initiate biological responses are in the UV and visible range. The electromagnetic radiations employed for PDT are mostly in the visible part of the spectrum and the 600–900 nm wavelength range is generally considered as the optimal therapeutic window [26,27]; at lower wavelengths, the decreased penetration of light into biological tissue becomes a limiting factor [28] whereas at higher wavelengths in the IR, the energy of the accessible excited states is too low to allow an efficient generation of cytotoxic species (vide infra). However, wavelengths outside the 600–900 nm window have also found therapeutic applications. In the PDT of non-malignant skin diseases, for example, the smaller penetration depth of UV and blue light has been used to selectively target the skin's upper layers [29].

2.2.2 Pathways of molecular excitation and deactivation

The absorption of light by a chromophore is the initial step in all photophysical and photochemical reactions: the energy of the absorbed light promotes molecules from their ground state to states of higher energy (i.e. excited states). This initial step of excitation and the subsequent deactivation processes are described below.

Electronic transitions and light absorption. An organic molecule can be visualized as a set of relatively slow-moving nuclei (which constitute the framework of the molecular structure) and of highly mobile electrons (which occupy specific volumes – orbitals – surrounding the nuclei). Each of these orbitals can hold a maximum of two electrons. Several electronic states are available to an organic molecule; each one of them corresponds to a certain spatial distribution of the electrons and is associated with a given energy. According to quantum mechanics, these energies can only have certain discrete values (i.e. they are quantized). At room temperature, essentially all the molecules are in the ground state – the electronic state associated with the lowest energy and a configuration where all the electrons are orbitally paired (i.e. where all the occupied orbitals hold two electrons). During an electronic transition, one of the electrons is promoted from an initially occupied orbital of low energy to a previously unoccupied orbital of higher energy. This process transforms the molecule from its ground state into an excited state.

An electronic transition can be induced by absorption of light provided that the energy of the incident photon is equal to the energy difference between two electronic states of the absorbing molecule. This condition is translated in the following relationship,

$$\Delta E = E_2 - E_1 = h\nu \qquad (2)$$

where h is Planck's constant, ν is the frequency at which the absorption occurs and, E_1 and E_2 are the energies of the molecule in its initial and final states, respectively.

Multiplicity of the electronic states. Aside from its orbital, each electron is characterized by a property called spin. According to quantum mechanics, the electron spin can only take two allowed orientations, which are represented by the discrete values: $+1/2$ and $-1/2$. Very often, these orientations are visualized as "spin-up" and "spin-down" configurations, denoted $\uparrow$ and $\downarrow$, respectively. When two electrons are orbitally paired (i.e. when they share the same orbital), their spins have opposite (or anti-parallel) directions, as demanded by Pauli's exclusion principle. When all the electrons of a molecule are paired, the total spin quantum number, S, which is defined as the sum of the individual "discrete values" of the electronic spins, is equal to zero. The magnetic quantum number, M, of the molecule, which is defined by $M = 2S + 1$, is then equal to one. The corresponding state of the molecule is then said to have a singlet multiplicity. Organic molecules in their ground state are generally singlet states.[2]

During an electronic transition, one of the electrons is promoted to an orbital of higher energy and consequently, the electrons become orbitally unpaired. Their spins may then be oriented in either an anti-parallel or a parallel manner. In the former case, where $S = 0$ and $M = 1$, an excited singlet state is produced whereas in the latter case, where $S = 1$ and $M = 3$, the excited state generated has a triplet multiplicity. This is schematically depicted in Fig. 1, where the three piled boxes correspond to orbitals of increasing energy and the upward and downward arrows represent the two allowed spin orientations. The concept of spin multiplicity is very important as the reactivity of an electronic excited state strongly depends on its multiplicity.

Molecular excitation and deactivation: the Jablonski diagram. The energy states available to an organic molecule can be represented by a simple energy level diagram, the so-called Jablonski diagram. The electronic state system described above is overly simplified. Nuclei are not immobile; they undergo vibrations and the entire molecule undergoes rotations, which implies that, in addition to electronic states, a molecule also has vibrational and rotational levels. The energies associated with all these states are

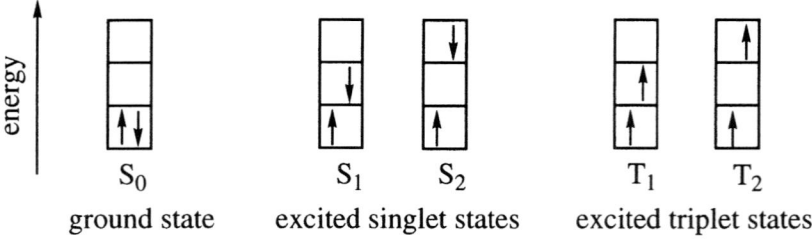

Figure 1. Schematic representation of the distribution and orientation of the spins of the electrons in different states of a molecule. In the ground state, S_0, the electrons are orbitally paired. The excited states are formed by promotion of one electron to an orbital of higher energy. In the case of the singlet states, S_1 and S_2, the orientation of the spin of the promoted electron remains unchanged whereas it is reversed in the case of the triplet states, T_1 and T_2.

---

[2] Unlike most molecules, ground state oxygen has a triplet multiplicity (vide infra).

quantized (with the energy gap between states of the same kind increasing in the order: rotation < vibration < electronic). In the Jablonski diagram, each electronic level is split into a series of vibrational levels, and each vibrational level is split into a series of rotational levels. However, very often (as for the example in Fig. 2), rotational sub-levels are not shown because they are too closely spaced. In the Jablonski diagram, the electronic states are grouped according to their spin multiplicity and numbered according to their order in the energy scale.

Absorption of a photon by a molecule results in the formation of an excited state, which has a different electronic distribution than the ground state, S_0, and is energetically less stable than S_0. Because of this relative instability, the excited state produced cannot persist indefinitely; de-excitation must take place to allow the release of the excess energy. The possible physical deactivation pathways that the excited states of a molecule can undergo are conveniently represented in the Jablonski diagram.

The excited singlet state (S_1) resulting from light absorption is extremely short-lived (a few picoseconds to several nanoseconds). Therefore, its probability of interaction with another molecule is very low and it undergoes mainly unimolecular reactions (such as intramolecular rearrangements, photoisomerization and homolytic fragmentations). However, several physical pathways leading to deactivation can also be followed. Thus, the singlet state S_1 can rapidly return to the ground state level by fluorescence emission (a radiative process) or by internal conversion (a radiationless process). During internal conversion, the excess energy of the singlet state is released as heat, which quickly dissipates into the solvent or biological tissue. In the alternative radiative process, a

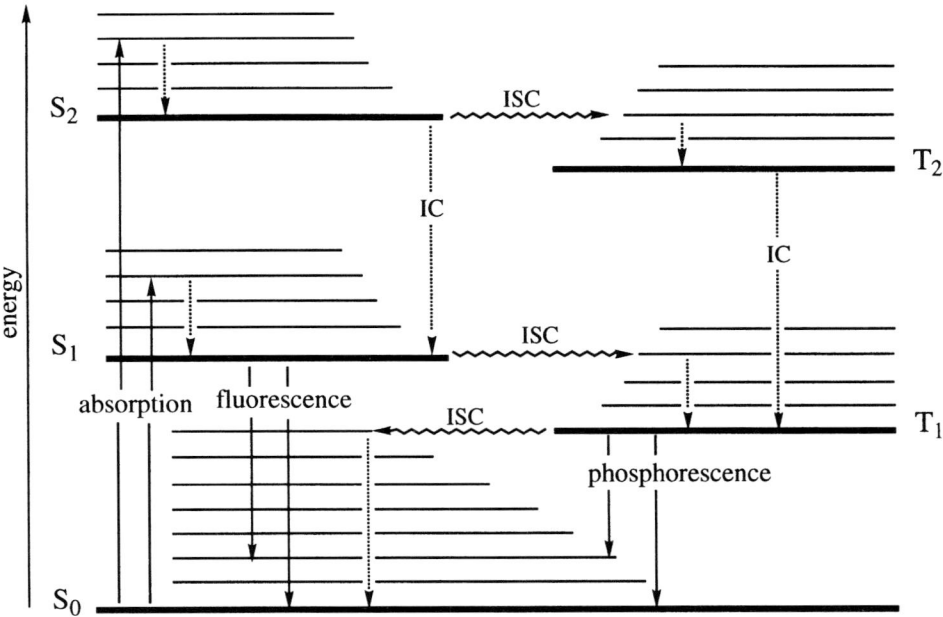

Figure 2. Jablonski diagram, where IC stands for internal conversion and ISC, for intersystem crossing. The arrows without legends correspond to vibrational relaxation processes.

photon is emitted, with an energy equal to the energy gap between the ground state (S_0) and excited singlet state (S_1) levels.[3]

In addition to fluorescence and internal conversion, S_1 can also undergo a change in spin multiplicity *via* a pathway called intersystem crossing, which gives rise to the triplet state T_1. During this process, the sign of the spin of the promoted electron is reversed (see Fig. 1). Contrary to fluorescence and internal conversion, intersystem crossing, which takes place between electronic states of different multiplicity, is spin-forbidden. As a result of the change in spin multiplicity, one observes modifications in the properties of the excited state. Thus, for example, triplet states have much longer lifetimes than singlet states. This dramatically enhances the probability of their interaction with another molecule – a property which explains the dominant role of triplet states in photosensitization processes (vide infra). The photophysical pathway, during which the excited triplet state emits a photon and returns to the ground state level, is called phosphorescence. Usually, this emission is very weak and difficult to detect at room temperature.[4] The excited triplet state T_1 can alternatively deactivate by undergoing intersystem crossing followed by a process called vibrational relaxation, which involves release of the excess vibrational energy by collisions with solvent molecules.

For most organic molecules, only the singlet state (S_1) and triplet state (T_1) of lowest energy can be considered as likely candidates for the initiation of photochemical and photophysical reactions. This is due to the fact that higher-order electronic states $(n \geqslant 2)$ undergo very rapid internal conversion from S_n to S_1 and from T_n to T_1. This generalization (which was used here in the description of the Jablonski diagram) is known as Kasha's rule.

Probability of light absorption. In photochemistry, the probability that an electronic transition will take place (i.e. the probability that a photon of wavelength λ will be absorbed) is quantitatively described by $\varepsilon(\lambda)$, the molar decadic absorption coefficient (often referred to as the molar extinction coefficient). This coefficient is defined by the Beer-Lambert law, which states that the transmission of light, $T(\lambda)$, through a non-scattering, absorbing medium varies inversely and exponentially with l, the thickness of the medium, and $[c]$, the concentration of the absorbing species. The absorbance $A(\lambda)$ (or optical density) of the sample is then defined as follows.

$$A(\lambda) = -\log[T(\lambda)] = \varepsilon(\lambda)l\,[c] \qquad (3)$$

---

[3] This implies that the fluorescence emission maximum (λ_f) of the molecule of interest does not depend on the excitation wavelength used (this is known as Vavilov's rule). It is also worth noting that, very often, due to similar spatial arrangements of the nuclei in the ground state and excited singlet state, the absorption and emission spectra of a molecule almost overlap, with, however, the emission maximum always at longer wavelengths than the absorption maximum.

[4] The energy of the photon emitted *via* phosphorescence, is equal to the energy gap between the triplet state (T_1) and the ground state (S_0) levels. Consequently, similar to fluorescence, the phosphorescence emission maximum (λ_p) is independent of the initial excitation wavelength (Vavilov's rule).

The absorption spectrum of a molecule corresponds to the plot of $\varepsilon(\lambda)$ as a function of wavelength. The spectrum and the wavelengths with the highest probability of absorption are characteristics of the molecule of interest, in a given environment.[5]

Probability of deactivation of excited states. The probability that an excited state produced by light absorption will undergo a particular chemical reaction or physical deactivation process is quantitatively represented by Φ, the quantum yield, and defined by Eq. (4).

$$\Phi = \frac{\text{number of molecular events of interest}}{\text{number of photons absorbed}} \tag{4}$$

Using this definition, the fluorescence quantum yield, for example, corresponds to the ratio of the number of photons emitted to the number of photons absorbed. A relationship also exists between the quantum yields and rate constants of the pathways followed by an excited state. The simple kinetic scheme presented below (Scheme 1) illustrates the case where upon absorption of a photon, a ground state molecule, S_0, is converted into its singlet state of lowest energy, S_1, which can deactivate by fluorescence, internal conversion and intersystem crossing.

The quantum yields of fluorescence (Φ_f) and intersystem crossing (Φ_{ISC}) of a molecule are then defined as shown below (Eq. (5)). Very often, photosensitizers have a negligible k_{IC} value compared to ($k_f + k_{ISC}$). This results from the fact that, for these photosensitizing compounds, the internal conversion from S_1 to S_0 is very slow and cannot compete with fluorescence and intersystem crossing. The sum of the fluorescence and intersystem crossing quantum yields can then be estimated to be close to unity ($\Phi_f + \Phi_{ISC} \approx 1$, which is known as Ermolev's rule).

$$\Phi_f = \frac{k_f}{k_f + k_{IC} + k_{ISC}} \quad \text{and} \quad \Phi_{ISC} = \frac{k_{ISC}}{k_f + k_{IC} + k_{ISC}} \tag{5}$$

$$S_0 \xrightarrow{h\nu} S_n \longrightarrow S_1 \qquad \text{absorption and internal conversion}$$

$$S_1 \longrightarrow S_0 + h\nu_f \qquad \text{fluorescence} \qquad k_f[S_1]$$

$$S_1 \longrightarrow S_0 + \text{heat} \qquad \text{internal conversion} \qquad k_{IC}[S_1]$$

$$S_1 \longrightarrow T_1 \qquad \text{intersystem crossing} \qquad k_{ISC}[S_1]$$

Scheme 1. Deactivation pathways available to the excited singlet state S_1, produced *via* light absorption by the ground state S_0. In this particular scheme, it is assumed that S_1 does not undergo any unimolecular reactions. The rates of the individual processes undergone by S_1 are listed on the right hand side. Each one of these rates is equal to the product of the rate constant of the process of interest and the concentration of the species involved (here, S_1).

---

[5] In biological tissue, where scattering is present, the effective pathlength of the light within the medium is increased. This generally increases the likelihood of absorption and remittance (consequently, the optical density of a scattering sample is always greater than that of a non-scattering but otherwise equivalent sample).

The physical deactivation processes that the excited states of a molecule can undergo are not all equally likely to occur. The probability of a process varies from molecule to molecule and, for a given molecule, it also strongly depends on the environment (vide infra).

2.3 Primary processes of photosensitization

2.3.1 General aspects

Photosensitized reactions. Photosensitization may be defined as a process in which light activation of a chromophore (the photosensitizer) induces chemical changes in a different molecule (the substrate). Ideally, the photosensitizer should play the role of a catalyst: it should be regenerated after its interaction with the substrate and should not interfere with the outcome of the reaction. In photobiology and photomedicine, the term "photodynamic action" is reserved to those photosensitized reactions that consume molecular oxygen [30,31].

Mechanisms of photosensitization. Photosensitization mechanisms are initiated by absorption of light by a photosensitizer, P, which generates an excited state, P*. In the presence of oxygen, P* can follow two main kinds of competing pathways called Type I and Type II reactions. According to the definition established by Foote [32] and, as shown in Scheme 2, a Type I mechanism involves the direct interaction of P* with a substrate, S, whereas in a Type II process, P* reacts first with molecular oxygen to produce highly reactive oxygen intermediates that easily initiate further reactions.

   More precisely, when a photosensitizer absorbs a photon, it is generally converted from its ground state P into $^1P^*$, its singlet state of lowest energy. Because of the short lifetime of $^1P^*$, very few photosensitized reactions take place from this electronic form of the molecule. More often, $^1P^*$ undergoes intersystem crossing to $^3P^*$, the triplet state of lowest energy. $^3P^*$, which has a much longer lifetime and therefore a greater probability of undergoing chemical reactions, generally acts as the mediator of Type I and/or Type II photosensitization processes. High intersystem crossing quantum yield (Φ_{ISC}) and long triplet state lifetime (τ_T) are therefore important pre-requisite conditions of efficient photosensitization [33].

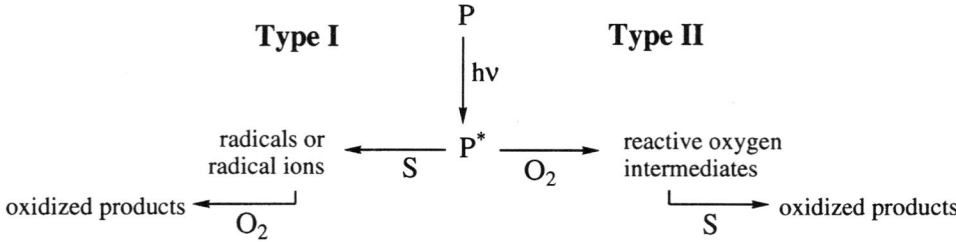

Scheme 2. Type I and Type II reaction pathways that the excited state, P*, can alternatively follow, after formation *via* light-activation of the ground state photosensitizer P.

2.3.2 Photosensitization mechanisms

Type I photosensitization processes. In a Type I mechanism, the excited triplet state of the photosensitizer, $^3P^*$, interacts directly with the substrate molecule, S. The reaction proceeds by electron or hydrogen atom transfer and leads to the formation of pairs of neutral radicals or radical ions. As shown by Eqs (6) and (7), the electron transfer can theoretically occur in either direction. However, most biological substrates undergo an oxidation (Eq. (6)).

$$^3P^* + S \rightarrow P^{\cdot-} + S^{\cdot+} \tag{6}$$

$$^3P^* + S \rightarrow P^{\cdot+} + S^{\cdot-} \tag{7}$$

Similarly, both the electronically excited photosensitizer (Eq. (8)) and the ground state substrate (Eq. (9)) can act as hydrogen donor.

$$^3PH^* + S \rightarrow P^{\cdot} + SH^{\cdot} \tag{8}$$

$$^3P^* + SH \rightarrow PH^{\cdot} + S^{\cdot} \tag{9}$$

The radical species, which result from these Type I primary processes can subsequently participate in different kinds of reactions. In the presence of oxygen, for example, oxidized forms of the sensitizer or of the substrate readily add to O_2 to give peroxyl radicals, thus initiating a radical chain auto-oxidation (as described by Eqs (10) and (11)).

$$S^{\cdot} + O_2 \rightarrow SOO^{\cdot} \tag{10}$$

$$SOO^{\cdot} + SH \rightarrow S^{\cdot} + SOOH \tag{11}$$

Semireduced forms of the photosensitizer or of the substrate also interact efficiently with oxygen and the electron transfer, which takes place between the reactants, generates the superoxide radical anion, $O_2^{\cdot-}$ (Eq. (12)).

$$P \text{ (or } S)^{\cdot-} + O_2 \rightarrow P \text{ (or } S) + O_2^{\cdot-} \tag{12}$$

Once formed, the superoxide radical anion can react directly with different substrates or act as the precursor of other reactive oxygen species (Eq. (13)), such as hydrogen peroxide (H_2O_2) and hydroxyl radical ($\bullet OH$). Under physiological conditions, H_2O_2 is produced by dismutation of $O_2^{\cdot-}$, which proceeds by protonation of $O_2^{\cdot-}$ to its conjugated acid $HO_2^{\cdot}$ (p$K_A = 4.8$ [34]), followed by reaction of $HO_2^{\cdot}$ with $O_2^{\cdot-}$. The hydroxyl radical can then be generated *via* a Fenton-like reaction between H_2O_2 and $O_2^{\cdot-}$.

$$\left. \begin{array}{l} O_2^{\cdot-} + H^+ \rightleftharpoons HO_2^{\cdot} \\ HO_2^{\cdot} + O_2^{\cdot-} + H^+ \rightarrow H_2O_2 + O_2 \\ H_2O_2 + O_2^{\cdot-} \xrightarrow[\text{catalyzed}]{\text{Fe(III)}} \bullet OH + OH^- + O_2 \end{array} \right\} \tag{13}$$

In the absence of oxygen, the radical species resulting from Type I photosensitization processes can interact with each other and yield covalent adducts of photosensitizer and substrate (Eqs (14) and (15)). These reactions are generally favored by the formation of non-covalent complexes prior to light-activation [35].

$$P^{\cdot-} + S^{\cdot+} \rightarrow P - S \tag{14}$$

$$S^{\cdot} + S^{\cdot} \rightarrow S - S \tag{15}$$

In summary, Type I photosensitization processes can produce different kinds of reactive intermediates. Although oxygen is not required in the very first step of the mechanism, its presence (or absence) in the environment strongly influences the subsequent chemical reactions and therefore determines the distribution of the photochemical reaction products. It is important to keep in mind that in the presence of oxygen, Type I photosensitization processes can induce the formation of reactive oxygen species, such as hydrogen peroxide (H_2O_2), superoxide radical anion ($O_2^{\cdot-}$) and hydroxyl radical ($\bullet OH$), which are known to efficiently oxidize a wide variety of biomolecules and ultimately cause substantial biological damage [23].

Type II photosensitization processes. Type II reaction mechanisms, on the other hand, do require the presence of molecular oxygen. In most cases, the reaction proceeds *via* energy transfer from the excited triplet state photosensitizer to the oxygen molecule. The process (Eq. (16)) regenerates the ground state sensitizer and leads to the formation of singlet oxygen (i.e. the lowest energy singlet state of O_2, which is denoted $^1O_2$ or $O_2(^1\Delta_g)$).

$$^3P^* + O_2 \rightarrow P + {}^1O_2 \tag{16}$$

As already mentioned, oxygen differs from most other organic molecules as its ground state has a triplet multiplicity. The reaction depicted by Eq. (16) is only possible if the energy of the triplet state sensitizer exceeds E_Δ, the energy of singlet oxygen [36]. Since the latter value is extremely low (E_Δ has been reported to be 94.5 kJ mole^{-1} [37]), a large number of molecules can mediate the generation of $^1O_2$ *via* triplet-triplet energy transfer.

In the simple case of a pure Type II system, the quantum yield of singlet oxygen production by a given photosensitizer can be determined by considering Scheme 3. This kinetic scheme describes the possible deactivation and reaction pathways that the triplet state photosensitizer can undergo. These include: phosphorescence, radiationless decay and interaction with oxygen by physical quenching or by energy transfer. Only the latter process leads to the formation of singlet oxygen.

The quantum yield of singlet oxygen formation of the photosensitizer is then defined as follows.

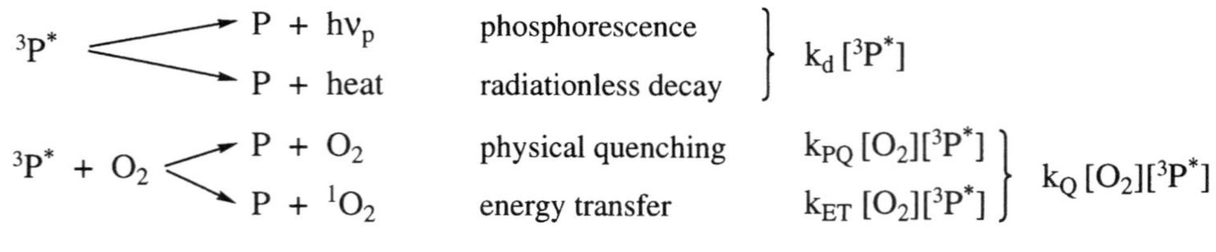

| $^3P^*$ | $\longrightarrow$ P + $h\nu_p$ | phosphorescence | $\Big\}$ $k_d\,[^3P^*]$ |
| | $\longrightarrow$ P + heat | radiationless decay | |

| $^3P^* + O_2$ | $\longrightarrow$ P + O_2 | physical quenching | $k_{PQ}\,[O_2][^3P^*]$ $\Big\}$ $k_Q\,[O_2][^3P^*]$ |
| | $\longrightarrow$ P + $^1O_2$ | energy transfer | $k_{ET}\,[O_2][^3P^*]$ |

Scheme 3. Physical and chemical processes available to the excited triplet state, $^3P^*$, and their corresponding rates. The rate constant of interaction with oxygen, k_Q, is equal to ($k_{PQ} + k_{ET}$), the sum of the rate constants for physical quenching, k_{PQ} and energy transfer, k_{ET}.

$$\Phi_\Delta = \Phi_{ISC} F_Q S_\Delta \quad \text{with} \quad F_Q = \frac{k_Q[O_2]}{k_d + k_Q[O_2]} \quad \text{and} \quad S_\Delta = \frac{k_{ET}}{k_{PQ} + k_{ET}} \tag{17}$$

F_Q corresponds to the probability that deactivation of the excited triplet state proceeds via reaction with oxygen [38] and S_Δ is the probability that the reaction between the triplet state sensitizer and oxygen results in the formation of singlet oxygen [39]. In the particular case where $k_d \ll k_Q[O_2]$, $\Phi_\Delta = \Phi_{ISC} S_\Delta$ (the form under which Eq. (17) is most commonly used).

Singlet oxygen, which is the primary product of Type II photosensitization mechanisms, is a very reactive species; it is much more electrophilic than its ground state and can oxidize biomolecules very rapidly. The reaction mechanisms and reaction products [40,41] of $^1O_2$ as well as the rate constants for its interactions with a wide variety of molecules [42] have been reviewed and published. Similarly, a compilation of the quantum yields of singlet oxygen production (Φ_Δ) by different compounds is available [43] and has recently been updated for biologically relevant molecules [44].

An alternative type of interaction between the excited triplet state sensitizer and molecular oxygen involves direct electron transfer. This reaction (Eq. (18)) yields the superoxide radical anion ($O_2^{\cdot-}$) and the radical cation of the photosensitizer.

$$^3P^* + O_2 \rightarrow P^{\cdot+} + O_2^{\cdot-} \tag{18}$$

According to the definition established by Foote [32], this reaction is a Type II process. Based on the argument that it produces radical species, some authors do not agree with this classification and prefer to consider this reaction as belonging to the Type I family [1,18]. In the present context, the resolution of this conflict is only of academic interest, since for most photosensitizers used in PDT, Eq. (18) is very unlikely.

Type I vs. Type II photosensitization processes. It follows from the previous paragraphs that the absorption of a photon by a photosensitizing molecule can induce a large variety of physical and chemical processes. For most photosensitizers used in PDT, the triplet state of lowest energy acts as the mediator of Type I and Type II reactions. The relative contribution of the different reactions (Type I/Type II mechanisms) and consequently the nature of the reactive intermediates produced (i.e. the species responsible for initiating the cascade of events that ultimately lead to damage in a biological system) are determined by the photosensitizer as well as by the reaction conditions, as can be deduced by consideration of Scheme 4.

Obviously, in the absence of a reactive substrate in the vicinity of the excited photosensitizer, $^3P^*$ will undergo a pure Type II reaction whereas under hypoxic

$^3P^*$ $\longrightarrow$	P + heat (or $h\nu_p$)	$k_d\,[^3P^*]$
$^3P^* + S$ $\longrightarrow$	Type I reaction	$k_R\,[S][^3P^*]$
$^3P^* + O_2$ $\longrightarrow$	Type II reaction	$k_Q\,[O_2][^3P^*]$

Scheme 4. Kinetic scheme of the physical and chemical reactions that $^3P^*$ can undergo after formation by intersystem crossing from $^1P^*$.

conditions, Type I processes will predominate. When the reaction conditions are such that both oxygen and substrate are present, a comparison of the efficiencies of triplet state quenching by the substrate and by oxygen (which depend on the relative values of k_R and k_O and on the concentrations [S] and [O_2]) has to be performed. Although this can easily be carried out in solutions and in simple models, where the experimental conditions are known, controlled or measurable, such a prediction is much more difficult to achieve in complex biological systems and in vivo.

2.3.3 Diagnostic methods for Type I and Type II reaction mechanisms

One of the ultimate goals in the investigation of photosensitization processes in PDT is to elucidate the mechanism of action of a given photosensitizer (i.e. to determine whether a specific reaction proceeds *via* a Type I or a Type II pathway). Over the years, different tests or diagnostic methods have been designed and developed in order to establish the relative participation of Type I and Type II processes. Different strategies have been adopted, their respective purposes being: (a) to determine the photophysical properties of the photosensitizing molecule in homogeneous solutions and to extrapolate the results obtained to biological situations, (b) to unravel the reaction mechanism by evaluating the effects of different additives on the biological or biochemical event induced by photosensitization and, (c) to identify the primary reactive species generated by light activation of the photosensitizer in a biological system by direct or indirect detection. The main experimental approaches currently in use are described below and their advantages and limitations are discussed.

Determination of Photophysical Parameters in Solution. The fundamental photophysics and photochemistry of photosensitizing molecules are now routinely studied in solution using spectroscopic techniques such as absorption, fluorescence, laser flash photolysis, pulse radiolysis, singlet oxygen luminescence detection and photothermal and photoacoustic methods [45]. These techniques provide important photophysical parameters such as the molar absorption coefficient of the photosensitizer at a suitable therapeutic excitation wavelength, its quantum yields of fluorescence (Φ_f), intersystem crossing (Φ_{ISC}) and singlet oxygen formation (Φ_Δ), and the lifetime of its excited triplet state (τ_T). The knowledge of these parameters may help predict the potential of a given photosensitizing molecule in therapeutic applications. Numerous articles, which describe the photophysical properties of photosensitizers in solution, have been published and constitute an important source of information.

However, the extrapolation of photophysical data determined in homogenous solutions to biological and biomedical situations may be misleading: a biological environment is hardly homogeneous and a photosensitizer is rarely homogeneously distributed in a cellular system or a target tissue. The microenvironment experienced by a photosensitizer changes as a function of its localization site. Factors such as oxygen concentration, pH, local concentration of sensitizer, nature and proximity of reactive substrates as well as binding to biomolecules and self-aggregation can influence the photophysical properties of the photosensitizing molecule and therefore modify its mechanisms of action. In an attempt to mimic some of these parameters, mechanistic studies have been carried out in model systems. Micelles [46] and liposomes [47] have thus been used to simulate, in an overly simplified manner, a membrane environment while considerable work has also been done to investigate the effects of protein-binding [48] on the photosensitizer photophysics. More recently, photophysical studies have also been performed on cells pre-incubated in the presence of photosensitizer [49].

Effects of additives on the photosensitization reaction. The second main strategy used to investigate photosensitization mechanisms in PDT relies upon the study of the effects of additives on biological or biochemical phenomena induced by light activation of a photosensitizer. The additives are expected to either enhance or inhibit a given photosensitization event, based on the assumption of a *specific* interaction between the additive and the reactive species suspected to be responsible for the biological response of interest. Cellular phototoxicity, membrane permeability, enzyme inactivation, lipid peroxidation, DNA damage and formation of photooxidation products are among the biological and biochemical end-points that have been studied most often [2,50–52].

Using this strategy, the probable involvement of singlet oxygen can be assessed in several ways. A biological phenomenon due to $^1O_2$ is expected to be inhibited in the presence of sodium azide, 1,4-diazabicyclo[2,2,2]octane (DABCO), 1,3-diphenyliso-benzofuran or β-carotene, which are all efficient singlet oxygen quenchers [23]. Substituting H_2O-based environments by D_2O-based media (where the lifetime of singlet oxygen is longer) is expected to lead to a significant increase in the rate of a photodynamic reaction [53]. Similarly, different tests have been developed to probe the involvement of radical species (generated by a Type I process) in photosensitization mechanisms. Commonly used quenchers include superoxide dismutase in the case of the superoxide radical anion ($O_2^{\bullet-}$), catalase for hydrogen peroxide (H_2O_2) and mannitol for the hydroxyl radical ($\bullet OH$) [54]. Unfortunately, none of these methods is completely unambiguous (e.g. singlet oxygen quenchers often react with $\bullet OH$ with greater rate constants than for interaction with $^1O_2$) and the lack of selectivity of the different additives considerably restricts their use. Furthermore, in many of these systems consideration of kinetic and localization factors often indicates that the photosensitization reaction responsible for the biological or biochemical phenomenon of interest can still occur equally efficiently in the presence of the additive. In other words, a trap or a quencher, which localizes mainly in the plasma membrane, will not affect photosensitization reactions taking place in the cytoplasm. Similarly, no effects will be observed under conditions where the concentration of the additive is not high enough (due to cytotoxicity or other reasons) to compete with endogenous substrates. Identification of the photooxidation products of cholesterol is an interesting alternative diagnostic method, which can be used as a reliable indicator of the intermediacy of $^1O_2$, the products resulting from a singlet oxygen-mediated oxidation of cholesterol have been reported to be characteristically different from those generated in radical oxidation processes [55]. However, in this case, the method suffers from a poor sensitivity, mainly because of the very low reactivity of singlet oxygen toward cholesterol.

Identification of the primary reactive species. The main goal here is to detect, in a direct or indirect manner, the primary reactive species formed upon light activation of a photosensitizer in a biological system. In the case of singlet oxygen, the only direct evidence of its presence is the observation of its weak IR luminescence. However detection of this luminescence in a biological material has proved elusive, due to the high reactivity and short lifetime of $^1O_2$ in such environments [56]. Similarly, electron spin resonance (ESR) and spin trapping are very useful techniques to detect the presence of free radicals in solution and simple in vitro systems, however, these methods are unsuitable in the case of in vivo studies [54]. More recently, intensive efforts have been directed toward the development of absorption, fluorescence and luminescence assays [57]. These tests are based on the specific interaction between a probe and a reactive oxygen species, which gives products with optical properties that can be used to prove and quantify the presence of the reactive oxygen species of interest. Lucigenin, for example, is a chemiluminescent probe reported to be sensitive to superoxide radical anion [58]. Singlet oxygen is known to bleach the fluorescence of a variety of aromatic compounds such as 1,3-diphenylisofuran [59]. Similarly, the fluorogenic probe proxyl fluorescamine used in combination with • OH-specific scavengers (such as mannitol and DMSO) provides evidence for the generation of the hydroxyl radical [60]. These probes are typical examples among a wide variety of reagents that have been designed to detect oxygen intermediates in solutions as well as in living cells and biological tissue [57]. Unfortunately, due to the lack of specificity of the trapping molecules, none of these mechanistic tests provides unequivocal information and can unambiguously discriminate between the various reactive oxygen species.

The study of the mechanisms by which a sensitizer initiates a biological response is therefore far from trivial. A thorough investigation requires the use of a judicious combination of biochemical assays and spectroscopic techniques. Since most of the available methods have inherent limitations (due to an extrapolation step, the use of overly simplified models or the lack of specificity of the probes and additives), conclusions should always be carefully drawn, even in the case of simple biological systems. The elucidation of photosensitization mechanisms under in vivo conditions can therefore be expected to be even more complex to achieve. Actually, such investigations are all the more complicated since the PDT treatment itself can progressively induce changes in the system, such as local hypoxia [61] and photobleaching [62] or local redistribution [63,64] of the photosensitizer. As a result of these changes, a switch in the major operating mechanism (from a Type II to a Type I, or the opposite) can take place.

2.3.4 Type I and Type II photosensitization processes in PDT

Ever since Weishaupt et al. [65] postulated, in 1976, that singlet oxygen is the cytotoxic agent responsible for the inactivation of tumor cells, many experimental observations supporting this hypothesis have been published. Actually, for several decades, the singlet oxygen hypothesis has worked as a powerful driving force in this research area [66] and it is now well established that $^1O_2$ plays a dominant role in the photosensitization processes induced by many of the photosensitizers used in PDT. In the case of porphyrins, phthalocyanines and methylene blue, some of the experimental results, which are considered as clear demonstrations of the involvement of singlet oxygen in photosensitized oxidations of biological systems, have been summarized by Bensasson et al. [67].

More recently, evidence concerning the important role of radical oxygen species in the damaging action of PDT has also started to accumulate. The "radical" hypothesis had already been postulated in 1968 [68]. However, the lack of assays for reactive oxygen species and for the oxidative damage that they can cause has impeded progress in the understanding of their role in photodynamic therapy. Thanks to the somewhat recent discovery of the importance of radical reactions in various human diseases as well as in normal body chemistry [23], things have quickly improved and the development of suitable mechanistic tests and techniques has stimulated a new line of investigations in this field of PDT.

Numerous photosensitizing compounds have been reported to mediate biological damage *via* a Type I mechanism, at least partially and/or under certain circumstances. Psoralens (furocoumarins), which are used in the photochemotherapy of certain skin diseases such as psoriasis, are probably one of the best known examples [69]. Thus, psoralen-induced photoinactivation of some enzymes has been shown to involve oxygen-dependent reactions and singlet oxygen has been identified as the major damaging reactive species [70]. On the other hand, photobinding of DNA to psoralens was found to be more efficient in the absence of oxygen, thereby excluding the involvement of singlet oxygen in the reaction mechanism [71]. Recently, PDT using bacteriochlorin a (BCA), a second-generation photosensitizer, was observed to induce damage to erythrocytes in vivo [72]. In order to identify the operating photosensitization mechanism(s) and assess the extent of the contribution of the different reactive oxygen species, the phenomenon was investigated in vitro using BCA-induced photohemolysis of human erythrocytes as a model system and assorted quenchers and probes of reactive oxygen species as mechanistic tools. The results obtained in these tests were supported by electron spin resonance experiments carried out using BCA in phosphate buffer. It was concluded that BCA-induced hemolysis of erythrocytes in vivo was probably caused by a mixed Type I/Type II mechanism without a predominant role of either mechanism [73]. These studies are typical of the way the elucidation of the mechanism of action of a photosensitizing molecule is currently investigated: the use of a combination of different diagnostic methods in in vitro model systems and in solutions leads to definitive, reliable conclusions, which are then cautiously extrapolated to the in vivo situation.

2.4 Summary and concluding remarks

The initial physical and chemical processes involved in photosensitization reactions include the absorption of light by the photosensitizing molecule, the formation of excited states of the photosensitizer and the interactions of the excited sensitizer with substrate molecules in a Type I mechanism or with molecular oxygen in a Type II process. In homogeneous solutions as well as in very simple biological model systems, an accurate picture of the mechanism of action of a photosensitizer can now be obtained quite easily, thanks to the development of new mechanistic biochemical assays and the use of spectroscopic techniques. However, in more complex in vitro systems and in particular under in vivo conditions, investigations of photosensitization reaction mechanisms remain extremely difficult. Consequently, the complex nature of the tissue response to PDT treatment has yet to be fully elucidated. An essential step toward this stimulating and challenging goal is to identify the primary photochemical and photophysical processes responsible for initiating the biochemical reactions under PDT conditions. In this respect, the design of specific diagnostic methods suitable for use in complex biological systems and the application of existing spectroscopic techniques under PDT conditions appear crucial. Although fundamental information about photosensitization reaction mechanisms is important for its own sake, a better understanding of these mechanisms could also be expected to allow the development of improved protocols for an effective use of PDT.

Acknowledgements

Irene Kochevar, Hans-Christian Lüdemann, Robert W. Redmond, and David Sharlin made numerous useful comments, corrections and suggestions. Their help is greatly appreciated.

References

1. J.D. Spikes (1983). Photosensitization in mammalian cells. In: J.A. Parrish, M.L. Kripke, W.L. Morison (Eds), *Photoimmunology* (pp. 23–49). Plenum Medical Book Company, New York.
2. B.W. Henderson, T.J. Dougherty (1992). How does photodynamic therapy work? *Photochem. Photobiol.*, **55**, 145–157.
3. J.D. Spikes (1977). Photosensitization. In: K.C. Smith (Ed.), *The Science of Photobiology* (pp. 87–112). Plenum, New York.
4. U.P. Andley (1987). Photodamage to the eye. *Photochem. Photobiol.*, **46**, 1057–1066.
5. G.M. Beijersbergen van Henegouwen (1997). Medicinal photochemistry: phototoxic and phototherapeutic aspects of drugs. *Adv. Drug Res.*, **29**, 79–170.
6. J.A. Parrish, T.B. Fitzpatrick, L. Tanenbaum, M.A. Pathak (1974). Phototherapy of psoriasis with oral methoxsalen and longwave ultraviolet light. *New Engl. J. Med.*, **291**, 1207–1211.
7. T. Hasan, J.A. Parrish (1996). Photodynamic therapy of cancer. In: J.F. Holland et al. (Eds) *Cancer Medicine* (4th Ed., Vol. 50(1), pp. 739–751). Williams & Wilkins, Baltimore.
8. J.G. Levy (1994). Photosensitizers in photodynamic therapy. *Seminars in Oncology*, **21**, 4–10.
9. R.M. Szeimies, P. Calzavara-Pinton, S. Karrer, B. Ortel, M. Landthaler (1996). Topical photodynamic therapy in dermatology. *J. Photochem. Photobiol. B: Biol.*, **36**, 213–219.
10. J.L. Matthews, J.T. Newman, F. Sogandares-Bernal, M.M. Judy, H. Skiles, J.E. Leveson, A.J. Marengo-Rowe, T.C. Chanh (1988). Photodynamic therapy of viral contaminants with potential for blood banking applications. *Transfusion*, **28**, 81–83.
11. F. Sieber, J.M. O'Brien, D.K. Gaffney (1992). Merocyanine-sensitized photoinactivation of enveloped viruses. *Blood Cells*, **18**, 117–127.
12. F. Sieber (1987). Merocyanine 540. *Photochem. Photobiol.*, **46**, 1035–1042.
13. E.R. Gaillard, S.J. Atherton, G. Eldred, J. Dillon (1995). Photochemical studies on human retinal lipofuscin. *Photochem. Photobiol.*, **61**, 448–453.
14. U. Schmidt-Erfurth, H. Diddens, R. Birngruber, T. Hasan (1997). Photodynamic targeting of human retinoblastoma cells using covalent low-density lipoprotein conjugates. *Br. J. Cancer*, **75**, 54–61.

15. J. Khadem, A.A. Veloso, F. Tolentino, T. Hasan, M.R. Hamblin (1999) Photodynamic tissue adhesion with chlorin e_6 protein conjugates. *Invest. Ophthalmol. Vis. Sci.*, **40**, 3132–3137.
16. J.D. Regan, J.A. Parrish (1982). *The Science of Photomedicine*. Plenum Press, New York.
17. G. Jori (1990). Photosensitized processes in vivo: proposed phototherapeutic applications. *Photochem. Photobiol.*, **52**, 439–443.
18. R.V. Bensasson, E.J. Land, T.G. Truscott (1993). *Excited States and Free Radicals in Biology and Medicine*. Oxford University Press, Oxford.
19. L.I. Grossweiner (1994). *The Science of Phototherapy*. CRC Press, Boca Raton.
20. I.E. Kochevar (1995). Primary processes in photobiology and photosensitization, In: J. Krutmann, G.A. Elmets (Eds), *Photoimmunology* (pp. 19–33). Blackwell Science Ltd, Oxford.
21. M. Ochsner (1997). Photophysical and photobiological processes in the photodynamic therapy of tumours. *J. Photochem. Photobiol. B: Biol.*, **39**, 1–18.
22. T.J. Dougherty, C.J. Gomer, B.W. Henderson, G. Jori, D. Kessel, M. Korbelik, J. Moan, Q. Peng (1998). Photodynamic therapy. Review, *J. Natl. Cancer Institute*, **90**, 889–905 and references therein.
23. B. Halliwell, J.M.C. Gutteridge (1999). *Free Radials in Biology and Medicine*. Oxford University Press, New York.
24. N.J. Turro (1978). *Modern Molecular Photochemistry*. The Benjamin/Cummings Publishing Co., Menlo Park.
25. A. Gilbert, J. Baggot (1991). *Essentials of Molecular Photochemistry*. CRC Press, Boca Raton.
26. D.C. Neckers (Ed.) (1988). New directions in photodynamic therapy. *Proc. SPIE*, Vol. 847.
27. J. Moan, V. Iani, L.W. Ma (1996). Choice of the proper wavelength for photochemotherapy. In: B. Ehrenberg, G. Jori, J. Moan (Eds) *Photochemotherapy, Photodynamic Therapy and Other Modalities. Proc. SPIE*, **2625**, 544–549.
28. J. Eichler, J. Knof, H. Henz (1977). Measurements of the depth of penetration of light in tissue, *Rad. Environ. Biophys.*, **14**, 239–245.
29. L. Emtestam, L. Berglund, B. Angelin, G.S. Drummond, A. Kappas (1989). Tin-protoporphyrin and long wavelength ultraviolet light in treatment of psoriasis. *Lancet*, **1**, 1231–1233.
30. H. von Tappeiner, A. Jodlbauer (1904). Über die Wirkung der photodynamischen (fluorescierenden) Stoffe auf Protozoen und Enzyme, *Disch. Arch. Klin. Med.*, **39**, 427–487.
31. H.F. Blum (1941). *Photodynamic Action and Diseases Caused by Light*. Reinhold, New York.
32. C.S. Foote (1991) Definition of Type I and Type II photosensitized oxidation. *Photochem. Photobiol.*, **354**, 659.
33. T. Takemura, N. Ohta, S. Nakajima, I. Sakata (1989). Critical importance of the triplet lifetime of the photosensitizer in photodynamic therapy of tumors. *Photochem. Photobiol.*, **50**, 339–344.
34. D. Behar, G. Czapski, J. Rabani, L.M. Dorfman, H.A. Schwarz (1970). The acid dissociation constant and decay kinetics of the perhydroxyl radical. *J. Phys. Chem.*, **74**, 3208–3213.
35. I.E. Kochevar (1981). Phototoxicity mechanisms: Chlorpromazine-photosensitized damage to DNA and cell membranes. *J. Invest. Derm.*, **76**, 59–64.
36. K.K. Iu, P.R. Ogilby (1987). A time-resolved study of singlet molecular oxygen ($^1\cdot_g O_2$) formation in a solution-phase photosensitized reaction: a new experimental technique to examine the dynamics of quenching by oxygen. *J. Phys. Chem.*, **91**, 1611–1617.
37. ref. [18]. (pp. 101–141).

38. K. Heihoff, R.W. Redmond, S.E. Braslavski, M. Rougée, C. Salet, A. Favre, R.V. Bensasson (1990). Quantum yield of triplet and $O_2(^1\Delta_g)$ formation of 4-thiouridine in water and acetonitrile. *Photochem. Photobiol.*, **51**, 635–641.

39. A.A. Gorman, G. Lovering, M.A.J. Rodgers (1978). A pulse radiolysis study of the triplet sensitized production of singlet oxygen: determination of energy transfer efficiency. *J. Am. Chem. Soc.*, **100**, 4527–4532.

40. H.H. Wasserman, R.W. Murray (1979). *Singlet Oxygen*. Academic Press, New York.

41. A.A. Frimer (1985). *Singlet Oxygen*. CRC Press, Boca Raton.

42. F. Wilkinson, J.G. Brummer (1981). Rate constants for the decay and reactions of the lowest electronically excited singlet state of molecular oxygen in solution. *J. Phys. Chem. Ref. Data.*, **10**, 809–1000.

43. F.W. Wilkinson, W.P. Helman, A.B. Ross (1993). Quantum yields for the photosensitized formation of the lowest electronically excited singlet state of molecular oxygen in solution. *J. Phys. Chem. Ref. Data.*, **22**, 113–262.

44. R.W. Redmond, J.N. Gamlin (1999). A compilation of singlet oxygen yields from biologically relevant molecules. *Photochem. Photobiol.*, **70**, 391–475.

45. R.W. Redmond (1993). Photophysical techniques used in photobiology and photomedicine. *NATO ASI Ser. Photobiology in Medicine*, Vol. 272, (pp. 1–28).

46. M. Craw, R.W. Redmond, T.G. Truscott (1984). Laser flash photolysis of heamatoporphyrin in some homogeneous and heterogeneous environments. *J. Chem. Soc. Faraday Trans. I*, **80**, 2293–2299. C.R. Lambert, E. Reddi, J.D. Spikes, M.A.J. Rodgers, G. Jori (1986). The effects of porphyrin structure and aggregation state on photosensitized processes in aqueous and micellar media. *Photochem. Photobiol.*, **44**, 595–601.

47. F. Richelli, G. Jori (1986). Distribution of porphyrins in the various compartments of unilamellar liposomes of dipalmitoylphophatidylcholine as probed by fluorescence spectroscopy. *Photochem. Photobiol.*, **44**, 151–157. G. Valduga, E. Reddi, G. Jori, R. Cubeddu, P. Taroni, G. Valentini (1992). Steady state and time-resolved spectroscopic studies of zinc(II) phthalocyanine in liposomes. *J. Photochem. Photobiol. B: Biol.*, **16**, 331–340.

48. E. Reddi, M.A.J. Rodgers, G. Jori (1984). Photophysical and photosensitizing properties of hematoporphyrin bound with human serum albumin. *Prog. Clin. Biol. Res.*, **170**, 373–379. B.M. Aveline, T. Hasan, R.W. Redmond (1995). The effects of aggregation, protein binding and cellular incorporation on the photophysical properties of benzoporphyrin derivative monoacid ring A (BPD-MA), *J. Photochem. Photobiol. B: Biol.*, **30**, 161–169.

49. P.A. Firey, T.W. Jones, G. Jori, M.A.J. Rodgers (1988). Photoexcitation of zinc phthalocyanine in mouse myeloma cells: the observation of triplet states but not of singlet oxygen. *Photochem. Photobiol.*, **48**, 357–360. T.C. Oldham, I.V. Eigenbrot, B. Crystall, D. Phillips (1996). Photophysics of photosensitizing dyes in living cell suspensions. *Proc. SPIE Int. Soc. Opt. Eng.*, **2625**, 266–277. B.M. Aveline, R.M. Sattler, R.W. Redmond (1998). Environmental effects on cellular photosensitization: correlation of phototoxicity mechanism with transient absorption spectroscopy measurements, *Photochem. Photobiol.*, **68**, 51–62. B.M. Aveline, R.W. Redmond (1999). Can cellular phototoxicity be accurately predicted on the basis of sensitizer photophysics? *Photochem. Photobiol.*, **69**, 306–316.

50. T.J. Dougherty (1986). Photosensitization of malignant tumors. *Semin. Surg. Oncol.*, **2**, 24–37.

51. D.P. Valenzeno (1987). Photomodification of biological membranes with emphasis on singlet oxygen mechanisms, *Photochem. Photobiol.*, **46**, 147–160.

52. D. Kessel (1997). Subcellular localization of photosensitizing agents: introduction. *Photochem. Photobiol.*, **65**, 387–388.

53. R. Nilsson, D.R. Kearns (1973). A remarkable deuterium effect on the rate of photosensitized oxidation of alcohol dehydrogenase and trypsin. *Photochem. Photobiol.*, **17**, 65–68.

54. ref. [23], (pp. 351–429).

55. W. Korytowski, G.J. Bochowski, A.W. Girotti (1992). Photoperoxidation of cholesterol in homogeneous solution, isolated membrane and cells: comparison of the 5α- and 6β-hydroperoxides as indicators of singlet oxygen intermediacy. *Photochem. Photobiol.*, **56**, 1–8.

56. A.A. Gorman, M.A.J. Rodgers (1992). Current perspectives of singlet oxygen detection in biological environments. *J. Photochem. Photobiol. B: Biol.*, **14**, 159–176.

57. R.P. Haughland (1996). Probes for reactive oxygen species, including nitric oxide, In: *Handbook of Fluorescence Probes and Research Chemicals*. Molecular Probes, Inc., Eugene.

58. J. Elsner, M. Oppermann, W. Czech, A. Kapp (1994). C3a activates the respiratory burst in human polymorphonuclear neutrophilic leukocytes via pertussis-toxin-sensitive G-proteins. *Blood*, **83**, 3324–3331.

59. M. Krieg (1993). Determination of singlet oxygen quantum yields with 1,3-diphenylisobenzofuran in model membrane systems. *J. Biochem. Biophys. Meth.*, **27**, 143–149.

60. S. Pou, A. Bhan, V.S. Bhadti, S.Y. Wu, R.S. Hosman, G.M. Rosen (1995). The use of fluorophore-containing spin traps as potential probes to localize free-radicals in cells with fluorescence imaging methods. *FASEB J.*, **9**, 1085–1090.

61. B.W. Henderson, V.H. Fingar (1989). Oxygen limitation of direct tumor cell kill during photodynamic treatment of a murine tumor model. *Photochem. Photobiol.*, **49**, 299–304.

62. T.S. Mang, T.J. Dougherty, W.R. Potter, D.G. Boyle, S. Somer, J. Moan (1987). Photobleaching of porphyrins used in photodynamic therapy and implications for therapy. *Photochem. Photobiol.*, **45**, 501–506.

63. S.R. Wood, J.A. Holroyd, S.B. Brown (1997). The subcellular localization of Zn(II) phthalocyanines and their redistribution on exposure to light. *Photochem. Photobiol.*, **65**, 397–402.

64. I. Georgakoudi, T.H. Foster (1998). Effects of the subcellular redistribution of two nile blue derivatives on photodynamic oxygen consumption. *Photochem. Photobiol.*, **68**, 115–122.

65. K.R. Weishaupt, C.J. Gomer, T.J. Dougherty (1976). Identification of singlet oxygen as the cytotoxic agent in photo-inactivation of a murine tumor. *Cancer Res.*, **36**, 2326–2329.

66. T. Ito (1978). Cellular and subcellular mechanisms of photodynamic action: the singlet oxygen hypothesis as a driving force in recent research. *Photochem. Photobiol.*, **28**, 493–508.

67. ref. [18], (p. 341).

68. K. Gollnick (1968). Type II photooxidation reactions in solution. *Adv. Photochem.*, **6**, 1–122.

69. A.Y. Potapenko (1991). Mechanisms of photodynamic effects of furocoumarins. *J. Photochem. Photobiol. B: Biol.*, **9**, 1–33.

70. H. Singh, J.A. Vadasz (1978). Singlet oxygen: A major reactive species in the furocoumarin photosensitized inactivation of E. coli ribosomes. *Photochem. Photobiol.*, **28**, 539–545.

71. G. Rodighiero, F. Dall' Acqua (1976). Biochemical and medical aspects of psoralens. *Photochem. Photobiol.*, **24**, 647–653.

72. H.L.L.M. van Leengoed, H.J. Schuitmaker, N. van der Veen, T.M.A.R. Dubbelman, W.M. Star (1993). Fluorescence and photodynamic effects of bacteriochlorin a observed in vivo in "sandwich observation" chambers. *Br. J. Cancer*, **67**, 898–903.

73. M. Hoebeke, H.J. Schuitmaker, L.E. Jannink, T.M.A.R. Dubbelman, A. Jacobs, A. van de Vorst (1997). Electron spin resonance evidence of the generation of superoxide anion, hydroxyl radical and singlet oxygen during the photohemolysis of human erythrocytes with bacteriochlorin a. *Photochem. Photobiol.*, **66**, 502–508.

Photodynamic Therapy and Fluorescence Diagnosis in Dermatology
P.-G. Calzavara-Pinton, R.-M. Szeimies and B. Ortel, editors.

Chapter 3

Mechanism of action – molecular effects

Anne C.E. Moor and Bernhard Ortel

Table of contents

Abstract

Cells exposed to photodynamic therapy (PDT) respond with activation of molecular pathways that are involved in the regulation of gene expression, cell cycle and cell death. This chapter gives a review of the molecular responses that have been documented after PDT, including effects on cell fate decision, stress responses, and intra- and intercellular signaling. The subcellular localization of the photosensitizer is an important determinant for specific molecular response patterns and for the ultimate PDT effect. Divergent and sometimes, contradictory results of consequences of PDT at the molecular level are likely due to the variability of PDT regimens with respect to photosensitizers, PDT dose, and other variables of specific experimental models. We are beginning to understand a few of the molecular effects of PDT, and this learning process will form the basis for research that aims at developing improved PDT regimens. PDT may also be usful as a tool to analyze certain cellular-molecular pathways.

3.1 Introduction

The mechanisms of PDT are subdivided into four chapters in this book, namely: (1) photophysical and photochemical reactions, (2) molecular effects, (3) cellular and vascular phototoxicity, and (4) immunological consequences. Obviously, some overlap between these chapters will be noted. On the other hand, the unavoidable repetition of some aspects offers the reader different approaches to the understanding of PDT effects. For all aspects of PDT it has to be remembered, that the interplay of the described mechanisms at the clinical target will vary depending on tissue, photosensitizer (PS) properties and other parameters of specific PDT regimens, such as photosensitizer incubation period, wavelength, total light dose (fluence) and fluence rate.

3.2 Molecular responses

For the purpose of this chapter molecular mechanisms have been defined as cellular responses at the molecular level that occur as a consequence of PDT. These molecular alterations involve those that eventually lead to (apoptotic and necrotic) cell death and those responses that are involved in cell rescue from sublethal photosensitized damage. Most studies of molecular effects have been performed in vitro, and more specifically only in one or a few model systems. Consequently, the described molecular effects may not be universal, sometimes even contradictory. A limited amount of data has been obtained from in vivo PDT, but at this time, the impact of most PDT-induced molecular responses on specific outcomes of PDT in vivo or in clinical settings remains to be documented. In this chapter we will not describe experimental details but instead try to give an overview of the main pathways that are modified by photosensitization.

As outlined in the photochemical section of this book, PDT cytotoxic effects are largely mediated by the action of singlet oxygen. The need for singlet oxygen underscores the importance of oxygen availability at the target site. Since this excited state of oxygen can only diffuse about 10 to 20 nm before returning to the ground state, the proximity of the singlet oxygen-inducing PS to the target molecule becomes important [1]. This is why mechanisms of cell death depend critically on subcellular localization of the PS. For example, predominantly mitochondrial PS accumulation will likely result in high rates of apoptosis (reviewed in [2,3]). Subcellular distribution depends primarily on physicochemical properties of the PS molecule, such as lipophilicity and electrical charge, but specific modes of PS delivery may modify uptake pathways and intracellular localization [4,5].

The significance of PS distribution within the organism and the tumor is described in the chapter on vascular and cellular mechanisms. However, we need to review some aspects of cellular localization as well, which can be cell type-dependent. This is important, because some tumors consist of large numbers of macrophages and overall uptake and intracellular PS distribution may differ considerably between macrophages and malignant tumor cells as may their response to PDT [4]. Considerations of cell type dependence are also important when non-malignant lesions and inflammatory skin disorders are treated. This aspect of cellular responses to PDT intersects with the subsequent chapter on immunological consequences of photosensitization.

3.3 Cell death pathways

There are two pathways of cellular death, necrosis and apoptosis. Some of the effector mechanisms of cell death are activated in both routes. The main difference is, that apoptosis requires several active processes within the cell, including signaling, protein modification and cleavage of genomic DNA. In short, apoptosis is an ordered process by which a cell that is damaged beyond repair disposes of itself. Necrosis is thought to be a less controlled way of cell death resulting in cell debris, while apoptotic cells (bodies) can be taken up as a whole by e.g. macrophages.

The objective in the majority of PDT applications is tumor eradication, and both mechanisms of cell death are desirable as long as they contribute to the objective of cancer treatment. In benign skin tumors and inflammatory dermatoses, lethal cytotoxicity may not be the primary objective, but rather a more or less subtle cellular modification. However, even in inflammatory disorders apoptosis might play an important role in mediating the clinical effect of PDT. Psoriasis is known to improve with several treatments that induce lymphocyte apoptosis, such as psoralen photo-chemotherapy [6] and 6-thioguanine [7]. Therefore, PDT-induced apoptosis of lymphocytes may be an important mechanism of PDT effects in e.g. inflammatory disorders, including psoriasis.

3.3.1 Apoptosis

For many years apoptosis has been a major focus of radiation biology. The demonstration of rapid induction of apoptosis by PDT has enhanced the interest and research in death mechanisms after photosensitization. Just as with many other important biological mechanisms, there is redundancy of the ways of activation and execution of apoptosis. There are several pathways of apoptosis that intersect and may substitute each other. The mitochondrial localization of many lipophilic PS seems an important factor for the induction of rapid apoptosis that does not occur with PS that localize preferentially in lysosomes or the plasma membrane. The release of cytochrome c from the mitochondria is an early event in the execution of photosensitized apoptosis [8–10]. The loss of cytochrome c is accompanied by a loss of mitochondrial membrane potential [9] and possibly by the opening of a transition pore in the mitochondrial membrane. Additionally, cytochrome c loss caused by PDT can lead to inhibition of respiration [10]. Cytochrome c release has been shown to lead to the activation of caspase 3, by the formation of a complex with apoptosis activating factor-1 (APAF-1), dATP and procaspase 9 [11,12]. Caspases are a group of proteolytic enzymes that are named after their substrate specificity (*cy*steinyl *asp*artate-specific protein*ase*). Caspase 3 and other caspases play a key role in initiating and executing apoptotic mechanisms and are themselves activated by proteolytic cleavage of their pro-forms [13]. Caspases degrade many proteins that normally support cellular structure, function, and survival. Cleavage products of caspase substrates indicate activation of apoptotic pathways. Several of these have been demonstrated after PDT, including poly(ADP-ribose)poly-merase (PARP) [14–16] DNA fragmentation factor and Bap31 [17]. Caspase 3 plays a very important role in PDT-induced apoptosis, but overall PDT efficacy as determined by clonogenic survival need not be affected by caspase 3 deficiency (Oleinick, N.L. unpublished results).* In this case, cell death is not accompanied by all biochemical markers of apoptosis, but it is associated with caspase 9 activation (Fig. 1). This specific pathway may or may not be termed apoptosis, but its existence clearly demonstrates redundancy of effector mechanisms that remove those cells that are damaged by PDT beyond repair.

In addition to the cytochrome c release-induced caspase 3 activation, an alternative route of caspase activation has been demonstrated after PDT. This pathway acts through caspase 8, the downstream effector of FAS/CD95 receptor. The activation of caspase 8 by PDT can lead to activation of caspase 3 and the subsequent effector mechansisms leading to apoptosis [17,18]. Important modulators of apoptosis are the members of the Bcl-2 family of proteins. Bcl-2 is a protein that is found in the outer membrane of the mitochondria, as well as in the endoplasmic reticulum and the nuclear envelope [19]. Bcl-2 is a member of a large family of pro- and anti-apoptotic proteins, which can have profound influence on the sensitivity of many cell types to apoptosis induction by a host of agents [20]. Bcl-2 has been shown in various studies to protect against apoptosis induction and the loss of clonogenic potential by photodynamic treatment with several photosensitizers [21–23]. The mechanism of the protective role of Bcl-2 in PDT induced apoptosis is not clear. Initially, it was suggested [22] that this could be due to the known anti-oxidant effect of Bcl-2 [24], but also to its ability to interfere with calcium homeostasis, which in itself has been shown to play a role in PDT-induced cell death [25]. Alternatively, Bcl-2 might be involved in the inhibition of cytochrome c release after PDT [26,27]. However, it has also been shown that Bcl-2 has an inhibitory effect on apoptosis induced by PDT, which is not at the level of cytochrome c release, but more

---

* Added in proof: L.Y. Xue, S.M. Chiu, N.L. Oleinick (2001). Photodynamic therapy-induced death of MCF-7 human breast cancer cells: a role for caspase-3 in the late steps of apoptosis but not for the critical lethal event. *Exp. Cell. Res.*, **263**, 145–155.

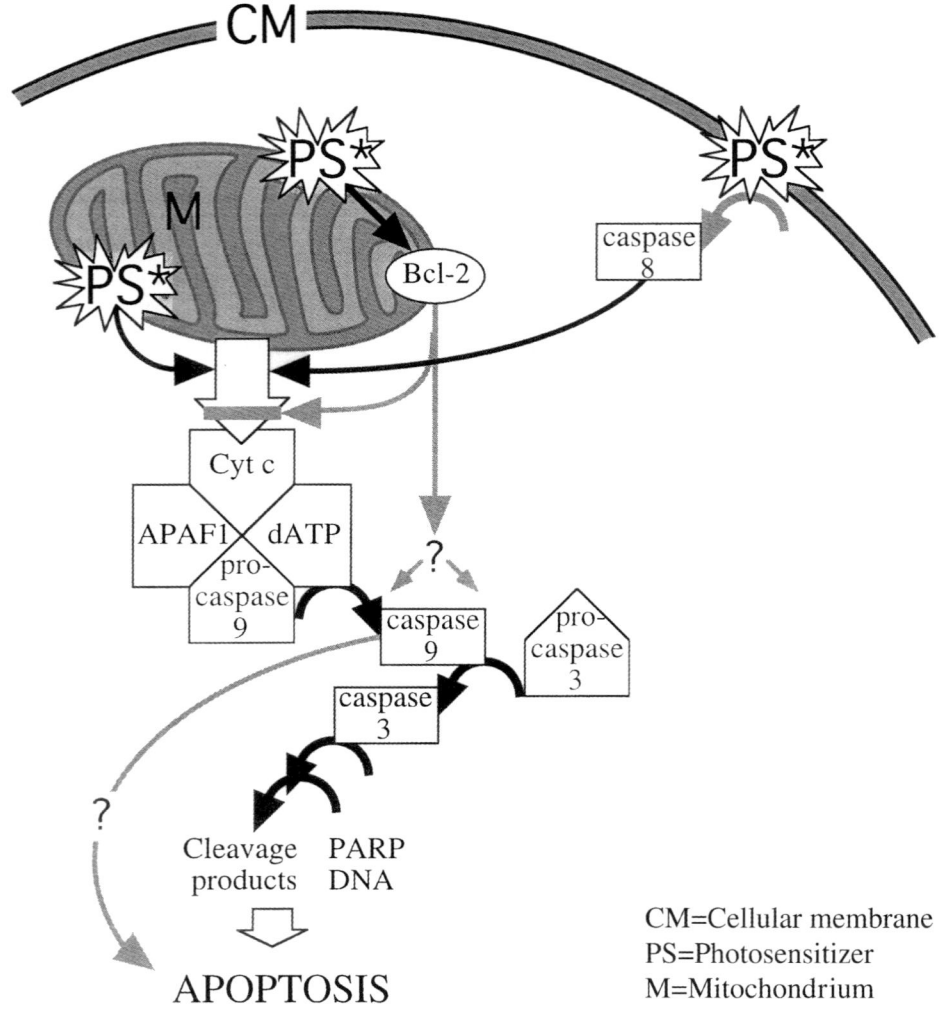

CM=Cellular membrane
PS=Photosensitizer
M=Mitochondrium

Figure 1.

downstream [28]. In apparent contrast to these studies, Kim et al. [29] showed enhanced sensitivity to PDT in cells, which had been transfected with Bcl-2. The authors explained that this contradictory result might be due to a simultaneous increase in Bax, a pro-apoptotic member of the Bcl-2 family. They postulated that Bcl-2 might be preferentially damaged by PDT thereby increasing the Bax:Bcl-2 ratio which subsequently would lead to enhanced apoptosis.

Mitochondria are not the only subcellular sites, where PDT induced apoptosis might be initiated. Although the sequence of events and the relative contribution of the different targets have not been completely elucidated, it is clear that plasma membrane events, as well as some lysosomal processes might be involved as well. Early studies of the induction of apoptosis by PDT showed involvement of the membrane-associated phospholipases (PL), such as PLA2 and PLC, which take part in a number of signaling events [30]. Increased calcium levels in the cells can activate PLA2 after PDT [31]. It has been shown that PLA2 may be involved both in mediating apoptosis and in providing a protective function *via* the activation of the second messengers, prostaglandin E2 and cyclic AMP [31–33]. PLC plays a role in protein kinase C (PKC) activation through the formation of diacylglycerol. PKC, has been shown to be a modulator of apoptosis induced by PDT using photosensitizers with different subcellular distribution patterns [9,34–36]. Using inhibitors and inducers of PKC in these studies, it has been concluded that PKC activation protects most cells from undergoing apoptosis.

Lysosomally localized photosensitizers have been thought to be mainly involved in necrosis after PDT (reviewed in [37]), possibly by direct release of the lysosomal contents in the cytosol. It is not clear if this is in fact an important mechanism in PDT, because the disintegration of lysosomes results in a loss of those conditions such as low pH, which optimize enzyme function inside the lysosome. In addition, lysosomally localized photosensitizers can relocalize soon after the start of illumination, thereby being enabled to redistribute within the cell and to exert their effects at different subcellular sites [37]. There is an acid isoform of sphingomyelinase (SMase) located in lysosomes. SMases generate ceramide from sphingomyelin, and ceramide has received considerable attention in recent years as a potential mediator of apoptosis [38]. In a series of studies, ceramide production has been implicated in the induction of apoptosis after PDT [39–41]. These reports were supported by the finding that cells inherently defective in lysosomal SMase where protected from PDT-induced apoptosis, while supplementation of exogenous SMase restored the apoptotic pathway [41]. The relative contribution to the induction of apoptosis by ceramide generation in vivo has not been analyzed at this time.

3.3.2 Necrosis

Necrosis is a less controlled way for cells to undergo death. The mechanisms involved in this process are also much less studied. However, several studies have clearly demonstrated that PDT can induce necrosis and that this pathway will take over, if the apoptotic pathway is blocked for whatever reason [28,42] (Oleinick, N.L. unpublished results).* In addition, photosensitizers localized in the lysosomes are thought to induce cell death predominantly *via* a necrotic pathway [37]. In living animals and in clinical situations early vascular damage may be an important mechanism of PDT efficacy. In this setting, hypoxia-induced necrosis may play an important role for the overall tumor control by PDT and should therefore not be neglected. Vascular mechanisms in PDT are addressed in another chapter.

---

* Added in proof: L.Y. Xue, S.M. Chiu, N.L. Oleinick (2001). Photodynamic therapy-induced death of MCF-7 human breast cancer cells: a role for caspase-3 in the late steps of apoptosis but not for the critical lethal event. *Exp. Cell. Res.*, **263**, 145–155.

3.4 Cell cycle and growth arrest

Cellular damage by PDT can result in growth arrest. It has to be mentioned that a series of these studies were performed at PDT doses that result in very substantial cell killing. The relevance of cell cycle control by PDT has to be evaluated at doses that allow cells to survive. PDT has been shown to increase p21 (CIP1/WAF1) protein levels [43]. This small protein works as an inhibitor of the cdk (cyclin dependent kinase) thus halting progression of the cell cycle in the G0/G1 phase. An involvement of nitrous oxide in p21 induction by PDT has been suggested but the exact mechanisms have not been worked out [44]. Retinoblastoma protein (Rb) is a cell cycle-controlling protein that is regulated by phosphorylation. PDT-induced hypophosphorylation of Rb inhibits free E2F that in turn is not available for supporting G1-S transition [45]. This also results in G0/G1 phase growth arrest. In one study it was shown that growth arrest and longterm survival were independent of p53 function, although the apoptosis rate was reduced in cells deficient in p53 function [42]. A better understanding of PDT effects on cell cycle control may be important for developing applications of PDT where cell death is not the primary objective, such as benign neoplasms or inflammatory disorders.

3.5 Signaling pathways

Cellular proliferation is a central process of life and cell cycle is therefore regulated at many levels and by a variety of exogenous and endogenous signals. Regulation of cell growth by receptor-mediated signals involves molecular cascades that amplify and balance the exogenous signal. The epidermal growth factor receptor (EGFR) has attracted considerable attention in this context. The EGFR is a member of the receptor tyrosine kinase family. Upon ligand binding EGFR dimerizes and is autophosphory-lated, which initiates activation of the extracellular signal-regulated kinase (ERK) proteins of the mitogen-activated protein kinase (MAPK) family. Subsequent steps are also mediated by kinases and eventually result in a mitogenic or differentiating response. PDT at high (lethal) doses has been shown to inhibit phosphorylation of EGFR and the downstream molecule *src* [46]. Downstream phosphorylation and activation of ERK does not seem to be effected in most cases [47–50] and thus the significance of EGFR modulation by PDT is not clear at this point.

Other pathways inducing MAPK activation are the stress-activated pathways involving stress-activated kinase/c-jun NH_2 terminal kinase (SAPK/JNK) and p38/ HOG1. Singlet oxygen is an important activator of these stress pathways [48] and, as discussed in detail in Chapter 2, the photodynamic process involves singlet oxygen formation. In photosensitization experiments using inhibitors of SAPK and p38, phototoxicity was increased in those cells in which the stress-activated pathways were blocked [50,51]. These findings indicate that these signaling cascades may provide a protective function in cells exposed to PDT. On the other hand it has been suggested that SAPK activation may result in phosphorylation and inactivation of the anti-apoptotic protein Bcl-2, which would support apoptosis [52,53]. In contrast, p38 inhibition was shown to reduce apoptotic response in two different PDT models. This demonstrates that activation of stress pathways may support or counteract apoptosis depending on cell line and other specific conditions.

The transcription factor NF-κB is involved in pro-and anti-apoptotic activity and is activated by PDT and reactive oxygen species. NF-κB activation by PDT has been demonstrated in several cellular systems. NF-κB expression is important in PDT-induced cytokine expression. NF-κB activity is regulated by its inhibitors IκBα and IκBβ. Lack of IκBα function may result in increased caspase 3 activation, DNA fragmentation, and apoptosis rates after PDT, but a mechanistic understanding of this finding has not been reached. Differential effects of PDT on NF-κB and its inhibitors may determine the overall outcome with respect to gene expression and apoptosis in different PDT regimens and, as most of the discussed PDT effects, is likely highly cell type-dependent.

3.6 Induction of stress proteins

A number of proteins involved in cellular stress responses have been shown to be regulated by PDT, notably members of the heat shock protein family, the glucose-regulated proteins and heme oxygenase. The heat shock proteins, which are known chaperones for damaged proteins, might be involved in the rescue response of cells after PDT damage. Heat shock protein 1 (HS1), which may be involved in cellular rescue responses, was found to be phosphorylated after PDT [54]. Previously it was shown that PDT using different photosensitizers induced a number of heat shock proteins in RIF-1 cells [55,56]. HSP70, induction was most efficient with photosensitizers that localize preferentially in lysosomes [56], although another study showed a downregulation of HSP70 instead [57]. The PDT-induced activation of the HSP promoter may have a potential for future use in gene activation. The possibility to control and spatially confine HSP promoter-dependent gene induction by PDT makes this approach very promising [58].

Glucose regulated proteins (GRP) have a chaperon function in the endoplasmatic reticulum (ER). In addition they can serve as intracellular calcium stores and are involved in resistance against chemotherapeutic agents [59]. Under certain PDT conditions, which lead to a preferential localization of the photosensitizer in the mitochondria, lysosomes and ER, both GRP-78 and -94 were found to be upregulated [60]. However, the exact role of the GRPs remains unclear since studies in which GRP is upregulated show either a protective effect against PDT induced cell death [60] or a potentiation of the PDT effect [61].

A group of stress inducible proteins that are less studied for their regulation by PDT, are the heme oxygenase (HO) proteins. PDT has been shown to induce HO-34 protein [62], an effect that was also observed at the mRNA level in a different study [63].

3.7 Induction of cytokines and regulation of cell surface molecules

Although in the chapter on immunological effects of PDT, induction of cytokines is described more extensively, it is important to discuss the regulation of cytokines in the context of molecular events induced by PDT. Specifically, sub-lethal doses of PDT may not lead to cell death, but still influence cell physiology.

(1) Cytokine induction

Both in vitro and in vivo data indicate an altered expression of both interleukin (IL)-10 and IL-6 after PDT. Interestingly, these effects were organ-dependent in tumor-bearing mice. In subcutaneous tumors IL-6 expression was enhanced after PDT, while IL-10 was decreased [64]. In normal PDT-treated skin both cytokines were upregulated. Among others, upregulation of IL-6 and IL-10 has been documented after ultraviolet B exposure [65]. Recently it has been suggested that UVB-induced membrane photo-sensitization may play an important role in NFκB-mediated induction of IL-6 [66]. IL-10 induction in the skin results in suppression of the CHS response and this has been demonstrated for UVB and PDT [67]. The opposite effects of PDT on interleukin expression in different tissues [64] indicate, that overall immunological consequences of PDT in vivo depend on multiple factors in addition to the usual variables of PDT. At the upstream molecular level, Kick et al. [68], studied the role of the transcriptional activator AP-1 in PDT-induced IL-6 production. In this specific model system, IL-6 was upregulated after PDT. It was shown that PKC inhibition by calphostin C had no effect on PDT induction of c-fos and the binding of AP-1 to the promoter region of the IL-6 gene. In addition, prolonged c-fos and c-jun induction was observed in these cells after PDT with PF, mainly by mRNA stabilization. Only a slightly increased DNA binding by AP-1 was observed indicating the tight control of this transcription factor [69]. Another transcription factor associated with cytokine regulation is NFκB, which can be activated by reactive oxygen species and is counteracted by inhibitory ligands (IκBs). PDT with different photosensitizers has been shown to induce NF-κB in various cellular systems [70–72]. The activation of NF-κB has important implications for the regulation of interleukins by ultraviolet radiation, reactive oxygen species, and PDT [66,73–75].

(2) Surface molecules

Soluble excreted molecules such as cytokines or growth factors are important for cell-to-cell communication and their modulation confers effects of many therapies, e.g. ultraviolet phototherapy (IL-6). Adhesion molecules mediate physical interactions between different cells and between cells and extracellular matrix and are involved in cellular signaling [76]. Adhesion molecules play an important role in cell trafficking [77], and in cancers certain integrin expression patterns have been associated with altered tumor growth pattern and metastatic behavior [78].

PDT has been studied to some extent as to its effect on cell surface molecules. At sublethal PDT doses adhesion molecules such as ICAM-1 and integrin expression were downregulated in vitro [79,80] and in vivo [80]. In one study PDT resulted in a transient decrease in adhesion molecule expression, and injection of in vitro treated cells resulted in reduction of metastases when compared to injection of untreated cells [81]. PDT can inhibit cellular adhesion properties without an alteration of integrin expression, possibly due to inhibition of the signaling function associated with focal adhesion plaques [82]. To understand implications for metastatic behavior after PDT, studies will be needed that investigate how PDT effects on adhesion molecules in primary tumors in situ affect their interactions with other cells and extracellular matrix.

The expression of cell surface antigens (e.g. MHC class I) was found downregulated after PDT in several studies [79,81,83,84]. It seems that this mostly transitory effect depends on cell type, activation status and PDT regimen [83]. The implications of altered surface antigen expression on immunomodulatory effects of PDT are discussed in Chapter 5.

3.8 Summary

PDT has developed from its early beginnings in the treatment of skin cancers to a highly versatile approach to managing a variety of diseases of the skin and many other organs. Current research targets the consequences of sublethal PDT at all levels. This approach is important for the understanding of longterm sequelae of PDT of malignancies, as well as for the development of new therapeutic strategies for cancer and non-malignant disease. This chapter has focused on PDT as modulator of a variety of molecular responses. Cells that are exposed to PDT exhibit modulation of pathways that are involved in the regulation of gene expression, cell cycle and cell death. The ultimate outcome i.e. survival or death depends on the balance of these different effects. For many PDT-induced alterations, seemingly contradictory results have been reported. In this context, several issues need to be considered. Firstly, different photosensitizers and PDT regimens may result in different cellular response patterns even in the same experimental model. Secondly, many specific responses may be effected only in a narrow window of time, PDT dose, or both. In addition, cell type and state (e.g. proliferation or differentiation) may predispose cells to specific response patterns. The understanding of PDT effects at the molecular level is the basis for developing improved treatment regimens. Conversely, specific PDT regimens may be useful tools to study and manipulate molecular responses. Ultimately these findings will help to design rationally-based PDT and combination treatments for benign disorders as well as malignant diseases.

Acknowledgements

Support of the authors was provided by Office of Naval Research Contract N00014-94-1-0927 (B.O.) and National Institutes of Health Grant R01 AR40352-03 (A.M.) during the preparation of this chapter.

References

1. Q. Peng, J. Moan, J.M. Nesland (1996). Correlation of subcellular and intratumoral photosensitizer localization with ultrastructural features after photodynamic therapy. *Ultrastruct. Pathol.*, **20**, 109–129.
2. A.C.E. Moor (2000). Signaling pathways in cell death and survival after photodynamic therapy. *J. Photochem. Photobiol. B.*, **57**, 1–13.
3. N.L. Oleinick, H.H. Evans (1998). The photobiology of photodynamic therapy: cellular targets and mechanisms. *Radiat. Res.*, **150**, S146–156.
4. M.R. Hamblin, J.L. Miller, B. Ortel (2000). Scavenger-receptor targeted photodynamic therapy. *Photochem. Photobiol.*, **72**, 533–540.
5. T. Hasan (1992). Photosensitizer delivery mediated by macromolecular carrier systems. In: T.J. Dougherty, B.W. Henderson (Eds), *Photodynamic Therapy: Basic Principals and Clinical Applications* (pp. 187–200). Marcel Dekker New York, NY.

6. T.R. Coven, I.B. Walters, I. Cardinale, J.G. Krueger (1999). PUVA-induced lymphocyte apoptosis: mechanism of action in psoriasis. *Photodermatol Photoimmunol Photomed.*, **15**, 22–27.

7. F.P. Murphy, T.R. Coven, L.H. Burack, P. Gilleaudeau, I. Cardinale, R. Auerbach, J.G. Krueger. (1999). Clinical clearing of psoriasis by 6-thioguanine correlates with cutaneous T-cell depletion via apoptosis: evidence for selective effects on activated T lymphocytes. *Arch. Dermatol.*, **135**, 1495–1502.

8. D.J. Granville, C.M. Carthy, H. Jiang, G.C. Shore, B.M. McManus, D.W. Hunt (1998). Rapid cytochrome c release, activation of caspases 3, 6, 7 and 8 followed by Bap31 cleavage in HeLa cells treated with photodynamic therapy. *FEBS Lett.*, **43**, 7 5–10.

9. D. Kessel, Y. Luo (1999). Photodynamic therapy: a mitochondrial inducer of apoptosis. *Cell Death Differ.*, **6**, 28–35.

10. M.E. Varnes, S.M. Chiu, L.Y. Xue, N.L. Oleinick (1999). Photodynamic therapy-induced apoptosis in lymphoma cells: translocation of cytochrome c causes inhibition of respiration as well as caspase activation. *Biochem. Biophys. Res. Commun.*, **255**, 673–679.

11. N.A. Thornberry, Y. Lazebnik (1998). Caspases: enemies within. *Science*, **281**, 1312–1316.

12. G. Nunez, M.A. Benedict, Y. Hu, N. Inohara (1998). Caspases: the proteases of the apoptotic pathway. *Oncogene*, **17**, 3237–3245.

13. A.G. Porter, R.U. Janicke (1999). Emerging roles of caspase-3 in apoptosis. *Cell Death Differ.*, **6**, 99–104.

14. Y. Luo, D. Kessel (1997). Initiation of apoptosis versus necrosis by photodynamic therapy with chloroaluminum phthalocyanine. *Photochem. Photobiol.*, **66**, 479–483.

15. J. He, C.M. Whitacre, L.Y. Xue, N.A. Berger, N.L. Oleinick (1998). Protease activation and cleavage of poly(ADP-ribose) polymerase: an integral part of apoptosis in response to photodynamic treatment. *Cancer Res.*, **58**, 940–946.

16. D.J. Granville, J.G. Levy, D.W. Hunt (1997). Photodynamic therapy induces caspase-3 activation in HL-60 cells. *Cell Death Diff.*, **4**, 623–629.

17. D.J. Granville, J.R. Shaw, S. Leong, C.M. Carthy, P. Margaron, D.W. Hunt, B. McManus (1999). Release of cytochrome c, Bax migration, Bid cleavage, and activation of caspases 2, 3, 6, 7, 8, and 9 during endothelial cell apoptosis. *Am. J. Pathol.*, **155**, 1021–1025.

18. S. Zhuang, M.C. Lynch, I.E. Kochevar (1999). Caspase-8 mediates caspase-3 activation and cytochrome c release during singlet oxygen-induced apoptosis of HL-60 cells. *Exp. Cell Res.*, **250**, 203–212.

19. S. Krajewski, S. Tanaka, S. Takayama, M.J. Schibler, W. Fenton, J.C. Reed (1993). Investigation of the subcellular distribution of the bcl-2 oncoprotein: residence in the nuclear envelope, endoplasmic reticulum, and outer mitochondrial membranes. *Cancer Res.*, **53**, 4701–4714.

20. A. Gross, J.M. McDonnell, S.J. Korsmeyer (1999). BCL-2 family members and the mitochondria in apoptosis. *Genes Dev.*, **13**, 1899–1911.

21. D.J. Granville, H. Jiang, M.T. An, J.G. Levy, B.M. McManus, D.W. Hunt (1999). Bcl-2 overexpression blocks caspase activation and downstream apoptotic events instigated by photodynamic therapy. *Br. J. Cancer*, **79**, 95–100.

22. J. He, M.L. Agarwal, H.E. Larkin, L.R. Friedman, L.Y. Xue, N.L. Oleinick (1996). The induction of partial resistance to photodynamic therapy by the protooncogene BCL-2. *Photochem. Photobiol.*, **64**, 845–852.

23. W.G. Zhang, L.P. Ma, S.W. Wang, Z.Y. Zhang, G.D. Cao (1999). Antisense bcl-2 retrovirus vector increases the sensitivity of a human gastric adenocarcinoma cell line to photodynamic therapy. *Photochem. Photobiol.*, **69**, 582–586.

24. C.M. Payne, C. Bernstein, H. Bernstein (1995). Apoptosis overview emphasizing the role of oxidative stress, DNA damage and signal-transduction pathways. *Leuk. Lymphoma.*, **19**,

43–93.

25. E. Ben-Hur, T.M. Dubbelman (1993). Cytoplasmic free calcium changes as a trigger mechanism in the response of cells to photosensitization. *Photochem. Photobiol.*, **58**, 890–894.

26. J. Yang, X. Liu, K. Bhalla, C.N. Kim, A.M. Ibrado, J. Cai, T.I. Peng, D.P. Jones, X. Wang (1997). Prevention of apoptosis by Bcl-2: release of cytochrome c from mitochondria blocked. *Science*, **275**, 1129–1132.

27. R.M. Kluck, E. Bossy-Wetzel, D.R. Green, D.D. Newmeyer (1997). The release of cytochrome c from mitochondria: a primary site for Bcl-2 regulation of apoptosis. *Science*, **275**, 1132–1136.

28. C.M. Carthy, D.J. Granville, H. Jiang, J.G. Levy, C.M. Rudin, C.B. Thompson, B.M. McManus, D.W. Hunt (1999). Early release of mitochondrial cytochrome c and expression of mitochondrial epitope 7A6 with a porphyrin-derived photosensitizer: Bcl-2 and Bcl-xL overexpression do not prevent early mitochondrial events but still depress caspase activity. *Lab. Invest.*, **79**, 953–965.

29. H.R. Kim, Y. Luo, G. Li, D. Kessel (1999). Enhanced apoptotic response to photodynamic therapy after bcl-2 transfection. *Cancer Res.*, **59**, 3429–3432.

30. M.L. Agarwal, H.E. Larkin, S.I. Zaidi, H. Mukhtar, N.L. Oleinick (1993). Phospholipase activation triggers apoptosis in photosensitized mouse lymphoma cells. *Cancer Res.*, **535**, 897–5902.

31. L.C. Penning, M.H. Rasch, E. Ben-Hur, T.M. Dubbelman, A.C. Havelaar, J. Van der Zee, J. Van Steveninck (1992). A role for the transient increase of cytoplasmic free calcium in cell rescue after photodynamic treatment. *Biochim. Biophys. Acta.*, **1107**, 255–260.

32. L.C. Penning, M.J. Keirse, J. VanSteveninck, T.M. Dubbelman (1993). Ca(2 +)-mediated prostaglandin E2 induction reduces haematoporphyrin-derivative-induced cytotoxicity of T24 human bladder transitional carcinoma cells in vitro. *Biochem. J.*, **292**, 237–240.

33. L.C. Penning, J. VanSteveninck, T.M. Dubbelman (1993). HPD-induced photodynamic changes in intracellular cyclic AMP levels in human bladder transitional carcinoma cells, clone T24. *Biochem. Biophys. Res. Commun.*, **194**, 1084–1089.

34. M.H. Rasch, K. Tijssen, J.W. Lagerberg, W.E. Corver, J. VanSteveninck, T.M. Dubbelman (1997). The role of protein kinase C activity in the killing of Chinese hamster ovary cells by ionizing radiation and photodynamic treatment. *Photochem. Photobiol.*, **66**, 209–213.

35. S. Zhuang, M.C. Lynch, I.E. Kochevar (1998). Activation of protein kinase C is required for protection of cells against apoptosis induced by singlet oxygen. *FEBS Lett.*, **437**, 158–162.

36. Y. Luo, D. Kessel (1996). The phosphatase inhibitor calyculin antagonizes the rapid initiation of apoptosis by photodynamic therapy. *Biochem. Biophys. Res. Commun.*, **221**, 72–76.

37. K. Berg, J. Moan (1997). Lysosomes and microtubules as targets for photochemotherapy of cancer. *Photochem. Photobiol.*, **65**, 403–409.

38. K. Hofmann, V.M. Dixit (1998). Ceramide in apoptosis – does it really matter? *Trends Biochem. Sci.*, **23**, 374–377.

39. D. Separovic, J. He, N.L. Oleinick (1997). Ceramide generation in response to photodynamic treatment of L5178Y mouse lymphoma cells. *Cancer Res.*, **57**, 1717–1721.

40. D. Separovic, K.J. Mann, N.L. Oleinick (1998). Association of ceramide accumulation with photodynamic treatment-induced cell death. *Photochem. Photobiol.*, **68**, 101–109.

41. D. Separovic, J.J. Pink, N.A. Oleinick, M. Kester, D.A. Boothman, M. McLoughlin, L.A. Pena, A. Haimovitz-Friedman (1999). Niemann-Pick human lymphoblasts are resistant to phthalocyanine 4-photodynamic therapy-induced apoptosis. *Biochem. Biophys. Res. Commun.*, **258**, 506–512.

42. A.M. Fisher, A. Ferrario, N. Rucker, S. Zhang, C.J. Gomer (1999). Photodynamic therapy

sensitivity is not altered in human tumor cells after abrogation of p53 function. *Cancer Res.*, **59**, 331–335.

43. N. Ahmad, D.K. Feyes, R. Agarwal, H. Mukhtar (1998). Photodynamic therapy results in induction of WAF1/CIP1/P21 leading to cell cycle arrest and apoptosis. *Proc. Natl. Acad. Sci. USA*, **95**, 6977–6982.

44. S. Gupta, N. Ahmad, H. Mukhtar (1998). Involvement of nitric oxide during phthalocyanine (Pc4) photodynamic therapy-mediated apoptosis. *Cancer Res.*, **58**, 1785–1788.

45. N. Ahmad, S. Gupta, H. Mukhtar (1999). Involvement of retinoblastoma (Rb) and E2F transcription factors during photodynamic therapy of human epidermoid carcinoma cells A431. *Oncogene*, **18**, 1891–1896.

46. K. Kalka, N. Ahmad, T. Kermode, H (1999). Mukhtar. Involvement of epidermal growth factor receptor (EGF-R)-tyrosine kinase pathway in photodynamic therapy-mediated cell death. *J. Invest. Dermatol.*, **112**, 659.

47. L.O. Klotz, C. Fritsch, K. Briviba, N. Tsacmacidis, F. Schliess, H. Sies (1998). Activation of JNK and p38 but not ERK MAP kinases in human skin cells by 5-aminolevulinate-photodynamic therapy. *Cancer Res.*, **58**, 4297–4300.

48. L.O. Klotz, C. Pellieux, K. Briviba, C. Pierlot, J.M. Aubry, H. Sies (1999). Mitogen-activated protein kinase (p38-, JNK-, ERK-) activation pattern induced by extracellular and intracellular singlet oxygen and UVA. *Eur. J. Biochem.*, **260**, 917–922.

49. J. Tao, J.S. Sanghera, S.L. Pelech, G. Wong, J.G. Levy (1996). Stimulation of stress-activated protein kinase and p38 HOG1 kinase in murine keratinocytes following photodynamic therapy with benzoporphyrin derivative. *J. Biol. Chem.*, **271**, 27107–27115.

50. L. Xue, J. He, N.L. Oleinick (1999). Promotion of photodynamic therapy-induced apoptosis by stress kinases. *Cell Death Diff.*, **6**, 855–864.

51. Z. Assefa, A. Vantieghem, W. Declercq, P. Vandenabeele, J.R. Vandenheede, W. Merlevede, P. de Witte, P. Agostinis (1999). The activation of the c-Jun N-terminal kinase and p38 mitogen-activated protein kinase signaling pathways protects HeLa cells from apoptosis following photodynamic therapy with hypericin. *J. Biol. Chem.*, **274**, 8788–8796.

52. M. Verheij, G.A. Ruiter, S.F. Zerp, W.J. van Blitterswijk, Z. Fuks, A. Haimovitz-Friedman, H. Bartelink (1998). The role of the stress-activated protein kinase (SAPK/JNK) signaling pathway in radiation-induced apoptosis. *Radiother. Oncol.*, **47**, 225–232.

53. R.K. Srivastava, Q.S. Mi, J.M. Hardwick, D.L. Longo (1999). Deletion of the loop region of Bcl-2 completely blocks paclitaxel-induced apoptosis. *Proc. Natl. Acad. Sci. USA*, **96**, 3775–3780.

54. L.Y. Xue, J. He, N.L. Oleinick (1997). Rapid tyrosine phosphorylation of HS1 in the response of mouse lymphoma L5178Y-R cells to photodynamic treatment sensitized by the phthalocyanine Pc 4. *Photochem. Photobiol.*, **66**, 105–113.

55. P.M. Curry, J.G. Levy (1993). Stress protein expression in murine tumor cells following photodynamic therapy with benzoporphyrin derivative. *Photochem. Photobiol.*, **58**, 374–379.

56. C.J. Gomer, S.W. Ryter, A. Ferrario, N. Rucker, S. Wong, A.M. Fisher (1996). Photodynamic therapy-mediated oxidative stress can induce expression of heat shock proteins. *Cancer Res.*, **56**, 2355–2360.

57. L.Y. Xue, M.L. Agarwal, M.E. Varnes (1995). Elevation of GRP-78 and loss of HSP-70 following photodynamic treatment of V79 cells: sensitization by nigericin. *Photochem. Photobiol.*, **62**, 135–143.

58. M.C. Luna, A. Ferrario, S. Wong, A.M. Fisher, C.J. Gomer (2000). Photodynamic therapy-mediated oxidative stress as a molecular switch for the temporal expression of genes ligated to the human heat shock promoter. *Cancer Res.*, **60**, 1637–1644.

59. H. Liu, R.C. Bowes, III, B. van de Water, C. Sillence, J.F. Nagelkerke, J.L. Stevens (1997).

Endoplasmic reticulum chaperones GRP78 and calreticulin prevent oxidative stress, Ca^{2+} disturbances, and cell death in renal epithelial cells. *J. Biol. Chem.*, **272**, 21751–21759.

60. C.J. Gomer, A. Ferrario, N. Rucker, S. Wong, A.S. Lee (1991). Glucose regulated protein induction and cellular resistance to oxidative stress mediated by porphyrin photosensitization. *Cancer Res.*, **51**, 6574–6579.

61. J. Morgan, J.E. Whitaker, A.R. Oseroff (1998). GRP78 induction by calcium ionophore potentiates photodynamic therapy using the mitochondrial targeting dye victoria blue BO. *Photochem. Photobiol.*, **67**, 155–164.

62. C.J. Gomer, M. Luna, A. Ferrario, N. Rucker (1991). Increased transcription and translation of heme oxygenase in Chinese hamster fibroblasts following photodynamic stress or Photofrin II incubation. *Photochem. Photobiol.*, **53**, 275–279.

63. D. Bressoud, V. Jomini, R.M. Tyrrell (1992). Dark induction of haem oxygenase messenger RNA by haematoporphyrin derivative and zinc phthalocyanine; agents for photodynamic therapy. *J. Photochem. Photobiol. B.*, **14**, 311–318.

64. S.O. Gollnick, X. Liu, B. Owczarczak, D.A. Musser, B.W. Henderson (1997). Altered expression of interleukin 6 and interleukin 10 as a result of photodynamic therapy in vivo. *Cancer Res.*, **57**, 3904–3909.

65. T. Scholzen, M. Hartmeyer, M. Fastrich, T. Brzoska, E. Becher, T. Schwarz, T.A. Luger (1998). Ultraviolet light and interleukin-10 modulate expression of cytokines by transformed human dermal microvascular endothelial cells (HMEC-1). *J. Invest. Dermatol.*, **111**, 50–56.

66. D. Kulms, B. Poppelmann, T. Schwarz (2000). Ultraviolet radiation-induced interleukin 6 release in HeLa cells is mediated via membrane events in a DNA damage-independent way. *J. Biol. Chem.*, **275**, 15060–15066.

67. G.O. Simkin, J.S. Tao, J.G. Levy, D.W. Hunt (2000). IL-10 contributes to the inhibition of contact hypersensitivity in mice treated with photodynamic therapy. *J. Immunol.*, **164**, 2457–2462.

68. G. Kick, G. Messer, A. Goetz, G. Plewig, P. Kind (1995). Photodynamic therapy induces expression of interleukin 6 by activation of AP-1 but not NF-kappa B DNA binding. *Cancer Res.*, **55**, 2373–2379.

69. G. Kick, G. Messer, G. Plewig, P. Kind, A.E. Goetz (1996). Strong and prolonged induction of c-jun and c-fos proto-oncogenes by photodynamic therapy. *Br. J. Cancer*, **74**, 30–36.

70. S. Legrand-Poels, S. Schoonbroodt, J.Y. Matroule, J. Piette (1998). Nf-kappa B: an important transcription factor in photobiology. *J. Photochem. Photobiol. B.*, **45**, 1–8.

71. B. Piret, S. Legrand-Poels, C. Sappey, J. Piette (1995). NF-kappa B transcription factor and human immunodeficiency virus type 1 (HIV-1) activation by methylene blue photosensitization. *Eur. J. Biochem.*, **228**, 447–455.

72. S.W. Ryter, C.J. Gomer (1993). Nuclear factor kappa B binding activity in mouse L1210 cells following photofrin II-mediated photosensitization. *Photochem. Photobiol.*, **58**, 753–756.

73. S. Legrand-Poels, L. Zecchinon, B. Piret, S. Schoonbroodt, J. Piette (1997). Involvement of different transduction pathways in NF-kappa B activation by several inducers. *Free Radic. Res.*, **27**, 301–309.

74. J.Y. Matroule, G. Bonizzi, P. Morliere, N. Paillous, R. Santus, V. Bours, J. Piette (1999). Pyropheophorbide-a methyl ester-mediated photosensitization activates transcription factor NF-kappaB through the interleukin-1 receptor-dependent signaling pathway. *J. Biol. Chem.*, **274**, 2988–3000.

75. K. Abeyama, W. Eng, J.V. Jester, A.A. Vink, D. Edelbaum, C.J. Cockerell, P.R. Bergstresser, A. Takashima (2000). A role for NF-kappaB-dependent gene transactivation in sunburn. *J. Clin. Invest.*, **105**, 1751–1759.

76. J. Bajorath (2000). Molecular organization, structural features, and ligand binding

characteristics of CD44, a highly variable cell surface glycoprotein with multiple functions. *Proteins*, **39**, 103–111.

77. P. Drillenburg, S.T. Pals (2000). Cell adhesion receptors in lymphoma dissemination. *Blood*, **95**, 1900–1910.
78. J.P. Johnson (1999). Cell adhesion molecules in the development and progression of malignant melanoma. *Cancer Metastasis. Rev.*, **18**, 345–357.
79. D.E. King, H. Jiang, G.O. Simkin, M.O. Obochi, J.G. Levy, D.W. Hunt (1999). Photodynamic alteration of the surface receptor expression pattern of murine splenic dendritic cells. *Scand. J. Immunol.*, **49**, 184–192.
80. J.M. Runnels, N. Chen, B. Ortel, D. Kato, T. Hasan (1999). BPD-MA-mediated photosensitization in vitro and in vivo: cellular adhesion and beta1 integrin expression in ovarian cancer cells. *Br. J. Cancer.*, **80**, 946–953.
81. N. Rousset, V. Vonarx, S. Eléouet, J. Carré, E. Kerninon, Y. Layat, T. Patrice (1999). Effects of photodynamic therapy on adhesion molecules and metastasis. *J. Photochem. Photobiol. B:Biol.*, **52**, 65–73.
82. P. Margaron, R. Sorrenti, J.G. Levy (1997). Photodynamic therapy inhibits cell adhesion without altering integrin expression. *Biochim. Biophys. Acta*, **1359**, 200–210.
83. D.W. Hunt, H. Jiang, D.J. Granville, A.H. Chan, S. Leong, J.G. Levy (1999). Consequences of the photodynamic treatment of resting and activated peripheral T lymphocytes. *Immunopharmacology*, **41**, 31–44.
84. D.J. Blom, H.J. Schuitmaker, I. de Waard-Siebinga, T.M. Dubbelman, M.J. Jager (1997). Decreased expression of HLA class I on ocular melanoma cells following in vitro photodynamic therapy. *Cancer Lett.*, **112**, 239–243.

Photodynamic Therapy and Fluorescence Diagnosis in Dermatology
P.-G. Calzavara-Pinton, R.-M. Szeimies and B. Ortel, editors.

Chapter 4

Correlation of intracellular and intratumoral photosensitizer distribution with photodynamic effect

Qian Peng

Table of contents

4.1 Introduction

Photodynamic therapy (PDT) is the selective destruction of tumor with a combination of a tumor-localizing photosensitizer and selective delivery of visible light. This results primarily in a singlet oxygen-induced photodamage to the tumor. The basis of PDT of cancer depends thus on the distribution of the photosensitizer, the propagation of the photoactivating light and the amount of oxygen in cancerous tissue. PDT has recently become an established cancer modality, but it can be further improved through a better understanding of the determinants affecting its efficiency. Among the determinants are: (1) photochemical and photophysical properties of a given photosensitizer; (2) the distribution of the photosensitizer in cells and tissues; (3) the relationships between the chemical properties of the photosensitizer and the characteristics of its cellular/tissular distribution; and (4) the relationships between the distribution patterns of the photosensitizer within cells and tissues and the actual target sites of its PDT effect. By the term 'distribution' is meant the concentration and/or the localization pattern of a photosensitizer within a given cell/tissue at a given time after its administration. In this communication we will discuss correlation of intracellular and intratumoral photosensitizer distribution with photodynamic effect.

4.2 Intracellular localization of AlPcS$_n$s and their subcellular targets of PDT in cells in vitro

Several components of cells (such as mitochondria, lysosomes, biomembranous systems, etc.) are affected by PDT with HpD and Photofrin [1]. Since HpD and Photofrin contain several porphyrin components with various hydrophobicity and probably different subcellular localization patterns as well, pure dyes with a well-defined lipophilicity are better suited for such investigation. For example, sulfonated aluminium phthalocyanines (AlPcS$_n$s) are second-generation photosensitizers with several advantages over HpD/Photofrin. The number of sulfonate groups determines the hydrophobicity of AlPcS$_n$s. The subcellular localization patterns of AlPcS$_n$s have been studied by use of confocal laser scanning microscopy in LOX human melanoma cells in vitro [2]. Hydrophilic AlPcS$_4$ (and AlPcS$_3$) localized in lysosomes, while more hydrophobic AlPcS$_2$ (and AlPcS$_1$) distributed mainly in mitochondria and other biomembraneous structure (Figs. 1, 2). Ultrastructural studies have demonstrated that the primary target of AlPcS$_4$-mediated PDT is the lysosomes, whereas AlPcS$_2$-based PDT destroys mitochondria and membranes. Moreover, Fig. 3 shows that AlPcS$_2$ and AlPcS$_1$ are more efficient in photosensitizing the cells than AlPcS$_3$ and AlPcS$_4$. This indicates that the efficiency of PDT-induced cell inactivation is much larger for the membrane-localizing lipophilic dyes than for the lysosomally localized hydrophilic dyes. Since they have similar yields of $^1O_2$ the results clearly suggest the importance of the subcellular localization pattern of the dyes in PDT.

Like most dyes phthalocyanine-induced singlet oxygen can mediate oxidation of lipids, amino acid and crosslinkage of protein components of cell/organellae membranes. The modifications of these biomolecules in cellular membranes can cause alterations in membrane permeability, loss of fluidity and inhibition/inactivation of

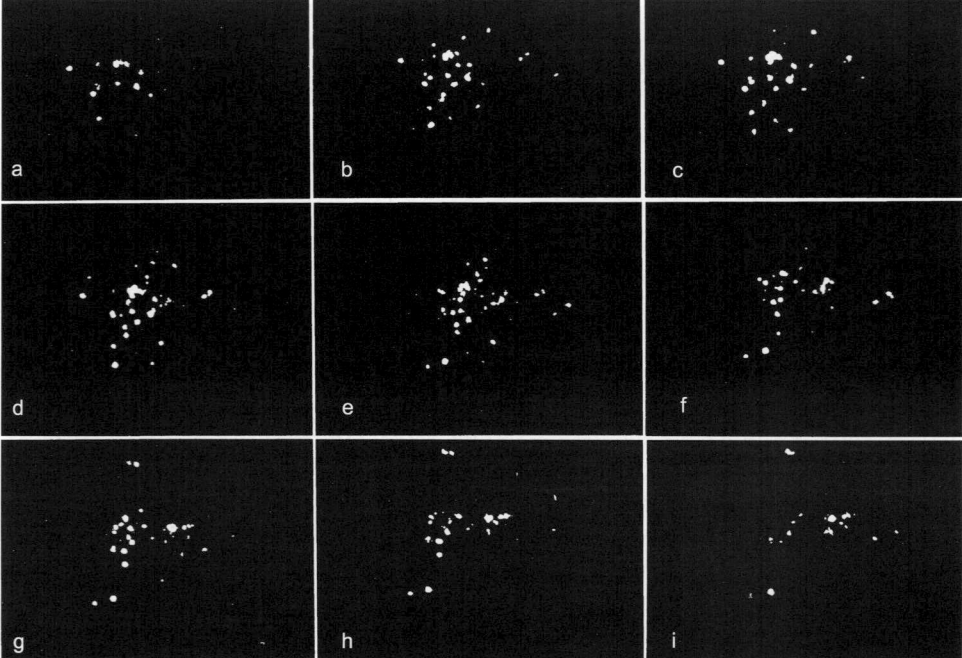

Figure 1. Confocal fluorescence microphotographs of a LOX cell optically serially sectioned into 9 slices (a-i) from top to bottom (0.5 μm apart) after incubation with 2.5 μg/ml of AlPcS$_4$ for 18 h in the dark, showing that AlPcS$_4$ had only dot fluorescence in the cytoplasm of the cell, an identical localization pattern of acridine orange which is known to be localized in the lysosomes.

membrane-associated enzyme systems and receptors, and finally result in cellular necrosis and apoptosis [3–5].

4.3 Distribution of a photosensitizer in tumor

The initial findings of Policard and Leulier [6] and Figge [7] stimulated early work on accumulation of porphyrins in tumors, but the mechanism of the selective distribution of a photosensitizer in tumor tissue is still not fully understood. In the blood various chemical properties of dyes bind different serum proteins and thus affect their transport [8,9]. It appears that the distribution pattern of a dye in tumor depends, to a great extent, upon its hydrophobicity. Generally, relative hydrophilic dyes are largely transported by albumin and globulins and mainly distributed in the vascular stroma of tumor tissue; they peak rapidly in tumor, but also disappear quickly. More hydrophobic drugs are preferentially incorporated into lipoproteins, particularly, low density lipoproteins (LDL) and localized in neoplastic cells of tumor tissue [10]. They usually reach a maximal level in tumor later with a long retention. The amphiphilic dyes may differ in their hydrophobicity as well as pharmacokinetic patterns. Moreover, a drug delivery

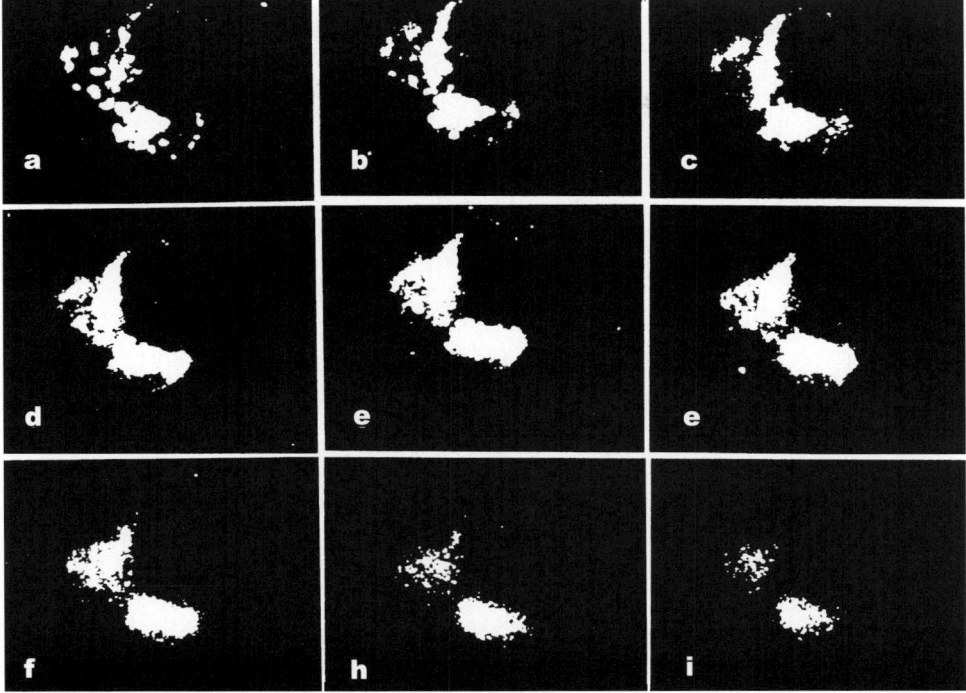

Figure 2. Confocal fluorescence microphotographs of a LOX cell optically serially sectioned into 9 slices (a-i) from top to bottom (0.5 μm apart) after incubation with 2.5 μg/ml of AlPcS$_2$ for 18 h in the dark, showing that AlPcS$_2$ had a combined diffuse and granular localization pattern in the cytoplasm of the cell.

system, such as LDL, liposomes, oil emulsions and inclusion complexes can significantly change the distribution pattern of a drug in the components of tumor [8].

Since HpD was taken up in vitro to similar extents by normal and cancer cells with varying oncogenic potential [11,12], some special properties of stromal compartment rather than neoplastic cells may account for preferential uptake of a photosensitizer by tumor. They include a high number of LDL receptors [8,13,14], low pH values [15–18] and the presence of many macrophages [19].The abnormal pathohistological structure of tumor stroma characterized by a large interstitial space, a high amount of newly synthesized collagen [20–22], a high interstitial fluid pressure, a poor lymphatic network and an increased vascular permeation [23,24] also favors a slow rate of disappearance of the photosensitizer.

4.4 PDT effect on tumor

Two main types of mechanisms are involved in photodynamic effect on tumor destruction [1,25–28].

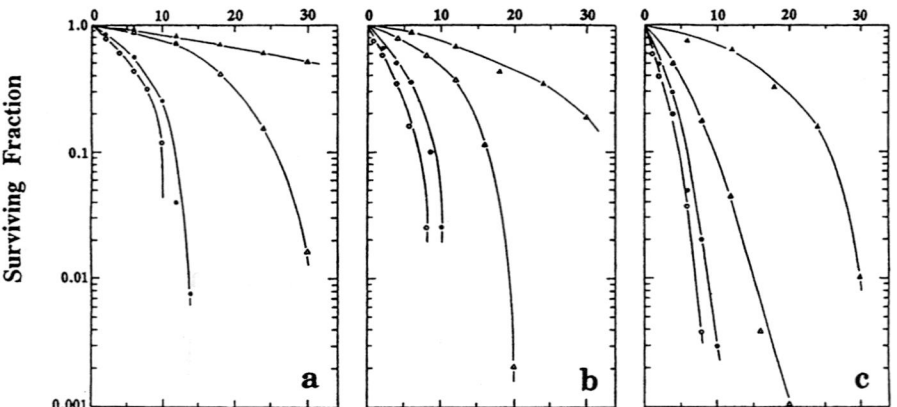

Figure 3. Dose-response curves for inactivation of LOX cells incubated for 18 hr with 10 μg/ml (a), 20 μg/ml (b) and 40 μg/ml (c) analogues of $AlPcS_n$ followed by irradiation as indicated. The light source was a bank of 4 fluorescent tubes (Philips TLD/83 filtered through a Cinemoid 35 filter) emitting light around 612 nm. The fluence rate of the light reaching the cells was 36 W/m^2. Symbols: - ● -, $AlPcS_1$; - ○ -, $AlPcS_2$; - △ -, $AlPcS_3$ and - ▲ -, $AlPcS_4$. The error limits were less than 20% of the mean values.

(1) Photodamage to stroma of a tumor. The stroma of a tumor is composed of vasculature (endothelium and vascular wall), cellular components (e.g. tumor associated macrophage) and intercellular matrix. All the stromal components can be targeted by PDT. Initially, functional disturbance (such as vasoconstriction, vasodilatation and aggregation of blood cells) of arterioles, capillaries and post-capillary venules induced by PDT can slow down the blood flow and cause complete stasis [29–32]. As a result, hypoxia/anoxia and subsequent death of neoplastic cells occur [1,33].

A number of endogenous vasoactive mediators can be produced during PDT to induce a strong inflammatory response. Injured endothelial cells can release von Willebrand Factor (vWF), a glycoprotein known to mediate thrombosis and ischemia [34,35]. Arachidonic acid is released from membrane phospholipids through the activation of phospholipase A_2 by membrane lipid damage [36], prostaglandin endoperoxides is then formed via the action of cyclooxygenase on arachidonic acid and finally thromboxane (a potent vasoconstrictor) is made from the prostaglandin endoperoxides [32,36]. PDT can also cause a release of histamine and other mediators (e.g. prostaglandin D_2, platelet active factor) from degranulation of mast cells [37]. Histamin is a powerful vasodilator, which increases the capillary porosity, allowing leakage of fluid and proteins into the tissues.

It should be noted that the subendothelial zone of the capillary wall, which consists of collagen fibers and other connective tissue elements, appears to be a sensitive target for PDT [38]. This is consistent with the findings that collagen, elastin and fibrin have a high affinity for some porphyrins and phthalocyanines [21,39]. Generally, PDT can significantly slow the rate of tumor regrowth [33,40,41], a similar effect to so-called tumor bed effect (TBE) induced by ionizing radiation. TBE is well known to be due to

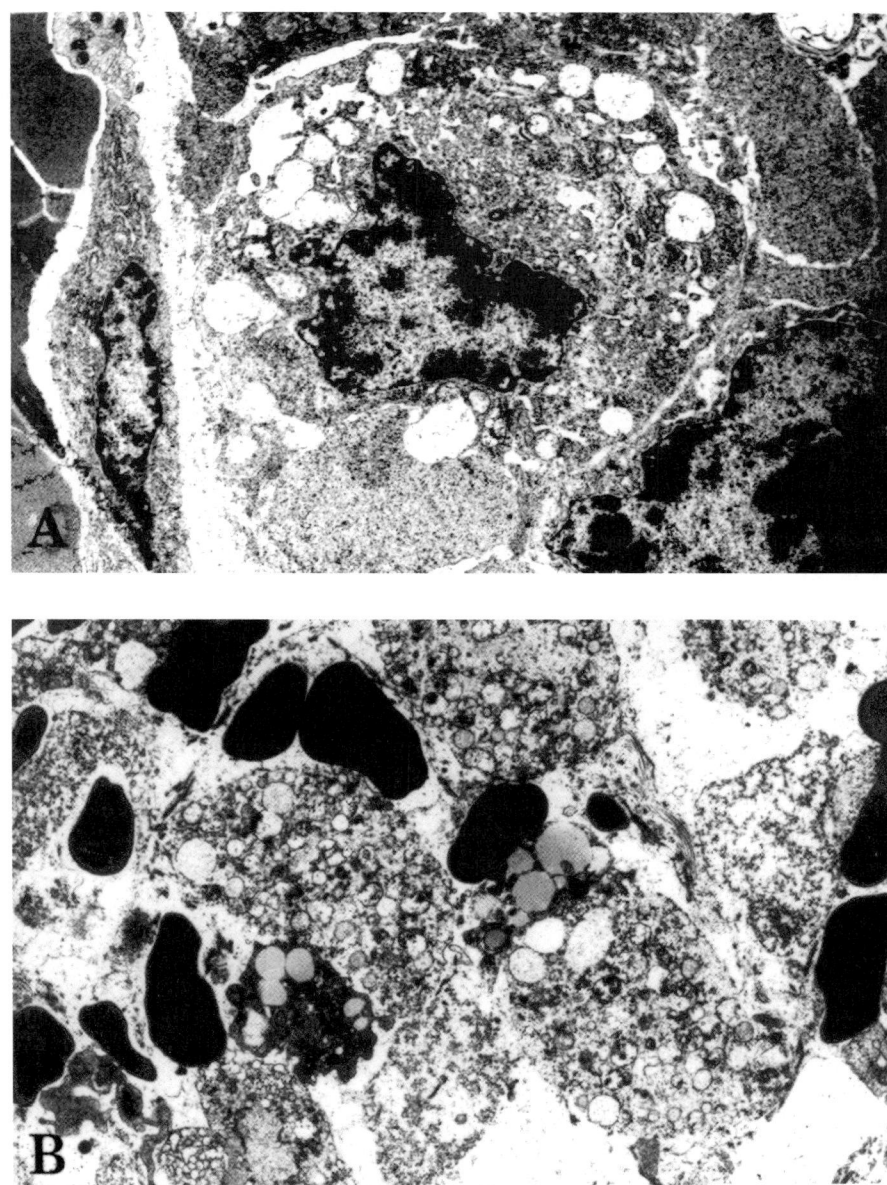

Figure 4. Electron microphotographs of CaD2 mouse mammary carcinoma were taken 1 (A) and 24 (B) hrs after treatment with 1 mg/kg b.w. i.p. administration of tetra(m-hydroxyphenyl)chlorin (m-THPC) and 50 J/cm^2 of light (652 nm). (A) shows a damaged tumor cell next to an undamaged endothelial cell although congestion exists (magnification $\times$ 8500). (B) shows hemorrhage and severe damage to neoplastic cells (magnification $\times$ 4850).

a destruction of endothelium of tumor bed, so that the endothelium is unable to proliferate to support the growth of the tumor.

PDT-induced specific immune response is also involved in eliminating neoplastic cells. This may be particularly effective for small foci of tumor cells that have survived initial PDT effects. Furthermore, this reaction is not limited to the original PDT-treated site but can include disseminated and metastatic lesions. Importantly, PDT induced immunity can be against weak immunogenic tumors [28]. It has been reported that tumor-associated macrophages can release tumor necrosis factor (TNF) after Photofrin-PDT in vitro [42]. This factor may contribute to tumor regression by direct inactivation of tumor cells and/or by mediating hemorrhagic necrosis [42]. In other cases, however, PDT has been shown to inhibit the activity of natural killer cells [43] and also to result in immune suppression [44,45].

(2) Direct photodamage to neoplastic cells of a tumor. Many early reports showed a direct PDT effect upon cancer cells [46–51] including inactivation of some cellular enzymes [52–57]. Recently, it has become apparently that such direct effects are induced by photosensitizers that localize in neoplastic cells of tumor [41,58–60]. However, the direct PDT effect may be very limited and less than 2 logs of neoplastic cells be killed, far short of 6–8 logs needed for tumor cure [28]. This limitation appears to be due to insufficient availability of oxygen within tumor tissues [28].

In addition, the PDT efficacy appears to be related to the intratumoral localization pattern of a given sensitizer. Damage at certain sites is more efficient in mediating cell death than that at other sites. For example, $AlPcS_2$ was localized in the vascular wall and in the tumor cells at early stages after systemic administration. Such a dual localization pattern seems to be a favourable factor for high PDT efficiency [41,60].

Finally, the mechanisms involved in the killing of tumors by PDT appear to be a complex interplay between direct and indirect (via stroma) effects on neoplastic cells. Figure 4 shows such complexity in the CaD2 mouse mammary carcinoma treated with m-THPC-mediated PDT. Basically, the intratumoral localization pattern of the applied dye is a crucial determinant for the effects since $^1O_2$ can only diffuse less than 0.1 μm during its lifetime in biological systems [61] and the targets of PDT are thus the sites where the dyes are localized.

References

1. J. Moan, K. Berg (1992). Photochemotherapy of cancer: Experimental research. *Photochem. Photobiol.*, **55**, 931–948.
2. Q. Peng, G. Farrants, K. Madslien, J. Moan, H.E. Danielsen, J.M. Nesland (1991a). Subcellular localization, redistribution and photobleaching of derivatives of sulfonated aluminum phthalocyanines in a human melanoma cell line. *Int. J. Cancer*, **49**, 290–295.
3. J.D. Spikes, J.C. Bommer (1986). Zinc tetrasulphophphthalocyanine as a photodynamic sensitizer for biomolecules. *Int. J. Radiat. Biol.*, **50**, 41–45.
4. A.W. Girotti (1990). Photodynamic lipid peroxidation in biological systems. *Photochem. Photobiol.*, **51**, 497–509.

5. M.L. Agarwal, M.E. Clay, E.J. Harvey, H.H. Evans, A.R Antunez, N.L. Oleinick (1991). Photodynamic therapy induces rapid cell death by apoptosis in L5178Y mouse lymphoma cells. *Cancer Res.*, **51**, 5993–5996.

6. A. Policard, A. Leulier (1924). Caracterisation de l'haemato-porphyrine et de l'urobiline urinaire par la lumiere de wood and etude sur les aspects offerts par des tumour experimentales examinees a la lumiere de wood. *C.R. Soc. Biol.*, **91**, 1422.

7. F.H.J., Figge (1942). Near ultraviolet rays and fluorescence phenomena as aids to discovery and diagnosis in medicine. *Univ. Maryland Med. bull.*, **26**, 165–176.

8. M. Kongshaug (1992). Minireview: Distribution of tetrapyrrole photosensitizers among human plasma proteins. *Int. J. Biochem.*, **24**, 1239–1265.

9. D. Kessel, K. Woodburn (1993). Biodistribution of photosensitizing agents. *Int. J. Biochem.*, **25**, 1377–1383.

10. D. Kessel, P. Thompson, K. Saatio, K.D. Nantwi (1987). Tumor localization and photosensitization by sulfonated derivatives of tetraphenylporphine. *Photochem. Photobiol.*, **45**, 787–790.

11. C.T. Chang, T.J. Dougherty (1978). Photoradiation therapy: kinetics and thermodynamics of porphyrin uptake and loss in normal and malignant cells in culture. *Radiat. Res.*, **74**, 498 (Abstract).

12. J. Moan, H.B. Steen, K. Feren, T. Christensen (1981). Uptake of hematoporphyrin derivative and sensitized photoinactivation of C3H cells with different oncogenic potential. *Cancer Lett.*, **14**, 291–296.

13. D. Gal, P.C. McDonald, J.C. Porter, E.R. Simpson (1981). Cholesterol metabolism in cancer cells in monolayer culture-III. Low density lipoprotein metabolism. *Int. J. Cancer*, **29**, 315–319.

14. G., Norata, G. Canti, L. Ricci, A. Nicolin, E. Trezzi, A.L. Catapona (1984). In vivo assimilation of low density lipoproteins by a fibrosarcoma tumor line in mice. *Cancer Lett.*, **25**, 203–208.

15. L. Wike-Hooley, J. Haveman, H.S. Reinhold, (1984). The relevance of tumor pH to the tretament of malignant disease. *Radiother. Oncol.*, **2**, 343–366.

16. J.P. Thomas, A.W. Girotti (1989). Glucose administration augments in vivo uptake and phototoxicity of the tumor-localizing fraction of hematoporphyrin derivative. *Photochem. Photobiol.*, **49**, 241–247.

17. D. Brault (1990). Physical chemistry of porphyrins and their interaction with membranes: the importance of pH. *J. Photochem. Photobiol.*, **6**, 79–86.

18. Q. Peng, J. Moan, L-S. Cheng (1991b). The effect of glucose administration on the uptake of Photofrin II in a human tumor xenograft. *Cancer Lett.*, **58**, 29–35.

19. S. Eccles, P. Alexander (1974). Macrophages content of tumors in relation to metastatic spread and host immune reaction. *Nature*, **250**, 667–669.

20. P.M. Gullino (1975). Extracellular compartments of solid tumors. In: F.F. Becker (Ed.), *Cancer, A Comprehensive Treatise* (Vol. 3, Biology of Tumors, pp. 327–354). Plenum Press, New York.

21. D.A., Musser, J.M. Wagner, N. Datta-Gupta (1982). The interaction of tumor localizing porphyrins with collagen and elastin. *Res. Commun. Chem. Pathol. Pharmacol.*, **2**, 251–259.

22. M.A. El-Far, N.R. Pimstone (1985). The interaction of tumor-localizing porphyrins with collagen, elastin, gelatin, fibrin and fibrinogen. *Cell Biochem. Function*, **3**, 115–119.

23. R.K. Jain (1987a). Transport of molecules in the tumor interstitium: a review. *Cancer Res.*, **47**, 3039–3051.

24. R.K. Jain (1987b).Transport of molecules across tumor vasculature. *Cancer and Metastasis Reviews*, **6**, 559–594.

25. T.J. Dougherty (1987). Photosensitization therapy and detection of malignant tumors. *Photochem. Photobiol.*, **45**, 879–890.
26. C. Zhou (1989). Mechanisms of tumor necrosis induced by photodynamic therapy. *J. Photochem. Photobiol, B:Biol.*, **3**, 299–318.
27. Q. Peng, J. Moan, J.M. Nesland (1996). Correlation of subcellular and intratumoral photosensitizer localization with ultrastructural features after photodynamic therapy. *Ultrastruct. Pathol.*, **20**(3–4), 109–129.
28. J. Dougherty, C.J. Gomer, B.W. Henderson, G. Jori, D. Kessel, M. Korbelik, J. Moan, Q. Peng (1998). Review: Photodynamic therapy. *J. Natl. Cancer Inst.*, **90**, 889–905.
29. S.H. Selman, R.W. Keck, J.E. Klauning, M. Kreimer-Birnbaum, P.J. Goldblatt, S.L. Britton (1983). Acute blood flow changes in transplantable FANFT-induced urothelial tumors treated with hematoporphyrin derivative and light. Surg. *Forum*, **34**, 676–678.
30. W.M. Star, H.P.A. Marijnissen, A.E. van den Berg-Blok, J.A.C. Versteeg, K.A.P. Franken, H.S. Reinhold (1986). Destruction of rat mammary tumor and normal tissue microcirculation by hematoporphyrin derivative photoradiation observed in vivo in sandwich observation chambers. *Cancer Res.*, **46**, 2532–2540.
31. K. Chandhuri, R.W. Keck, S. Selman (1987). Morphological changes of tumor micro-vasculature following hematoporphyrin derivative sensitizer photodynamic therapy. *Photochem. Photobiol.*, **46**, 823–827.
32. V.H. Fingar, T.J. Wieman, S.A. Wiehle, P.B. Cerrito (1992). The role of microvascular damage in photodynamic therapy: the effect of treatment on vessel constriction, permeability, and leukocyte adhesion. *Cancer Res.*, **52**, 4914–4921.
33. B.W. Henderson, T.J. Dougherty (1992). How does photodynamic therapy work? *Photochem. Photobiol.*, **55**, 145–157.
34. T.H. Foster, M.C. Primavera, V.J. Marder, R. Hilf, L.A. Sporn (1991). Photosensitized release of von Willebrand Factor from cultured human endothelial cells. *Cancer Res.*, **51**, 3261–3266.
35. H.-B. Ris, H.J. Altermatt, R. Inderbitzi, R. Hess, B. Nachbur, J.C.M. Stewart, Q. Wang, C.K. Lim, R. Bonnett, M.C. Berenbaum, U. Althaus (1991). Photodynamic therapy with chlorins for diffuse malignant mesothelioma: initial clinical results. *Br. J. Cancer*, **64**, 1116–1120.
36. M.W.R. Reed, T.J. Wieman, K.W. Doak, K. Pietsch, D.A. Schuschke (1989). The microvascular effects of photodynamic therapy: evidence for a possible role of cyclooxigenase products. *Photochem. Photobiol.*, **50**, 419–423.
37. F.A. Kerdel, N.A. Soter, H.W. Lim (1987). In vivo mediator release and degranulation of mast cells in hematoporphyrin derivative-induced phototoxicity in mice. *J. Invest. Dermatol.*, **88**, 277–280.
38. J.S., Nelson, L.H. Liaw, A. Orenstein, W.G. Roberts, M.W. Berns (1988). Mechanism of tumor destruction following photodynamic therapy with hematorphyrin derivative, chlorin and phthalocyanine. *J. Natl. Cancer Inst.*, **80**, 1599–1605.
39. Q. Peng, J.M. Nesland, J. Moan, J.F. Evensen, M. Kongshaug, C. Rimington (1990a). Localization of fluorescent Photofrin II and aluminum phthalocyanine tetrasulfonate in transplanted human malignant tumor LOX and normal tissues of nude mice using highly light-sensitive video intensification microscopy. *Int. J. Cancer*, **45**, 972–979.
40. B.W. Henderson (1990). Probing the effects of photodynamic therapy through in vivo-in vitro methods. In: D. Kessel (Ed.), *Photodynamic Therapy of Neoplastic Disease* (Vol. I, pp. 169–188), CRC Press, Boca Raton, Florida.
41. Q. Peng, J. Moan (1995a). Correlation of distribution of sulfonated alumin Phthalocyanine with their photodynamic effect in tumor and skin of mice bearing CaD2 Mammary carcinoma. *Br. J. Cancer*, **72**, 565–574.

42. S. Evans, W. Matthews, R. Perry, D. Frankel, J. Norton, H.I., Pass (1990). Effects of photodynamic therapy on tumor necrosis factor production by murine macrophages. *J. Natl. Cancer Inst.*, **82**, 34–39.

43. J.F. Marshall, W.S. Chan, I.R. Hart (1989). Effect of photodynamic therapy on anti-tumor immune defenses: comparison of the photosensitizers hematoporphyrin derivative and chloro-aluminum sulfonated phthalocyanine. *Photochem. Photobiol.*, **49**, 627–632.

44. C. Jolles, M.J. Ott, R.C. Straight, D.H. Lynch (1988). Systemic immunosuppression induced by peritoneal photodynamic therapy. *Am. J. Obstet. Gynecol.*, **158**, 1446–1453.

45. D.A. Musser, R.J. Fiel (1991). Cutaneous photosensitizing and immunosuppressive effects of a series of tumor localizing porphyrins. *Photochem. Photobiol.*, **53**, 119–123.

46. D.S. Rasmussen-Taxdal, G.E. Ward, F.H. Figge (1955). Fluorescence of human lymphatic and cancer tissues following high doses of intravenous hematoporphyrin. *Cancer*, **8**, 78–81.

47. R.L. Lipson, E.J. Baldes, A.M. Olsen (1961). The use of a derivative of haematoporphyrin in tumor detection. *J. Natl. Cancer Inst.*, **26**, 1–11.

48. J. Winkelman (1961). Intracellular localization of 'hematoporphyrin' in a transplanted tumor. *J. Natl. Cancer Inst.*, **27**, 1369–1377.

49. T.J. Dougherty, G.B. Grindey, K.R. Fiel, K.R. Weishaupt, D.G. Boyle (1975). Photoradiation therapy. II. Cure of animal tumors with hematoporphyrin and light. *J. Natl. Cancer Inst.*, **55**, 115–121.

50. M. Tsutsui, C. Carrano, E.A. Tsutsui (1975). Tumor localizers: porphyrins and related compounds (unusual metalloporphyrins XIII). *Ann. N.Y. Acad. Sci.*, **244**, 674–684.

51. M.W. Berns, A. Dahlman, F.M. Johnson, R. Burns, D. Sperling, M. Guiltinan, A. Siemens, R. Walter, W. Wright, M. Hammer-Wilson, A. Wile (1982). In vitro cellular effects of hematoporphyrin derivative. *Cancer Res.*, **42**, 2325–2329.

52. S.L. Gibson, R. Hilf (1983). Photosensitization of mitochondrial cytochrome c oxidase by hematoporphyrin derivative and related porphyrins in vitro and in vivo. *Cancer Res.*, **43**, 4191–4197.

53. T.M.A.R. Dubbelman, J. van Steveninck (1984). Photodynamic effect of hematoporphyrin derivative on transmembrane transport systems of murine L929 fibroblasts. *Biochim. Biophys. Acta*, **771**, 201–207.

54. R. Hilf, D.B. Smail, R.S. Murant, P.B. Leake, S.L. Gibson (1984). Hematoporphyrin derivative-induced photosensitivity of mitochondrial succinate dehydrogenase and selected cytosolic enzymes of R3230 AC mammary adenocarcinomas of rats. *Cancer Res.*, **44**, 1483–1488.

55. D.S. Perlin, R.S. Murant, S.L. Gibson, R. Hilf (1985). Effects of photosensitization by hematoporphyrin derivative on mitochondrial triphosphate-mediated proton transport and membrane integrity of R3230 AC mammary adenocarcinoma. *Cancer Res.*, **45**, 653–658.

56. T.L. Ceckler, R.C. Bryant, D.P. Penney, S.L. Gibson, R. Hilf (1986). 31P-NMR spectroscopy demonstrates decreased ATP levels in vivo as an early response to photodynamic therapy. *Biochem. Biophys. Res. Commun.*, **140**, 273–279.

57. R. Hilf, S.L. Gibson, D.P. Penney, T.L. Ceckler, R.G. Bryant (1987). Early biochemical response to photodynamic therapy monitored by NMR spectroscopy. *Photochem. Photobiol.*, **46**, 809–817.

58. Q. Peng, J. Moan, J.M. Nesland, C. Rimington (1990b). Aluminum phthalocyanines with asymmetrical lower sulfonation and with symmetrical higher sulfonation: A comparison of localizing and photosensitizing mechanism in human tumor LOX xenografts. *Int. J. Cancer*, **46**, 719–726.

59. Q. Peng, J. Moan, T. Warloe, J.M. Nesland, C. Rimington (1992). Distribution and photosensitizing efficiency of porphyrins induced by application of exogenous 5-aminolevulinic acid in mice bearing mammary carcinoma. *Int. J. Cancer*, **52**, 433–443.

60. Q. Peng, J. Moan, Ma Li-wei, J.M. Nesland (1995b). Uptake, localization and photodynamic effect of meso-tetra(hydroxyphenyl)porphine and its corresponding chlorin in normal and tumor tissues of mice bearing CaD2 mammary carcinoma. *Cancer Res.*, **55**, 2620–2626.

61. J. Moan, K. Berg (1991). The photodegradation of porphyrins in cells can be used to estimate the lifetime of singlet oxygen. *Photochem. Photobiol.*, **53**, 549–553.

Photodynamic Therapy and Fluorescence Diagnosis in Dermatology
P.-G. Calzavara-Pinton, R.-M. Szeimies and B. Ortel, editors.

Chapter 5

Immunologic actions of PDT

David W.C. Hunt, P. Mark Curry and John R. North

Table of contents

5.1 Introduction

Photodynamic therapy (PDT) is a clinically effective technique with which to treat tumors and abnormal vasculature accessible to activating light [1]. PDT uses light-absorbing compounds that catalyze the generation of highly reactive oxygen species to produce localized toxic effects. Drawbacks of chemotherapy and radiation based anti-cancer therapies include marrow depletion and reduced anti-microbial immunity. In contrast, there is little evidence that PDT has general immunosuppressive or myelosuppressive activity at clinically effective doses, although it has become apparent that PDT does have immuno-modulatory attributes. Experimental animals have been shown to develop specific anti-tumor immunity after PDT [2]. Furthermore, PDT can modify immune reactions and lessen the severity of experimentally induced immune diseases at doses below those that cause skin inflammation or erythema [3,4], perhaps by altering the function of immune and non-immune cell types.

Two areas are of interest to clinicians and researchers regarding the influence of PDT on the immune system. Firstly, what effect does PDT have on the immunological state of the cancer patient and on the potential for bolstering anti-tumor immunity? Secondly, can PDT be used to treat immune mediated diseases?

5.2 Photosensitizer action at the cellular level

The site of photosensitizer localization influences subsequent photodynamic effects on cells. Many photosensitizers localize to the mitochondria and mitochondrial damage has been widely described for PDT-treated cells [1,5–8]. Higher concentrations generally are required to elicit phototoxicity when using photosensitizers that localize in the plasma membrane, Golgi apparatus, lysosomes and/or endoplasmic reticulum than when using photosensitizers that localize to mitochondria [1,7,8]. Most photosensitizers appear to be excluded from the nucleus [1,7], an important feature that may minimize the mutagenic potential of PDT.

A continuum of responses is evident in PDT-treated cells, related to the dose used. In addition, there is an inverse relationship between photosensitizer and light doses. Thus, comparable biological effects can be produced by combining a relatively high photosensitizer dose with a low light dose or, conversely, a lower drug dose with a higher light dose. At the cellular level, with low intensity PDT, viability may be maintained while other traits (signaling activity, cytokine formation, receptor expression) may be altered. At more intense PDT levels, cells may undergo apoptosis rapidly. At high levels, PDT may kill cells by disrupting the membrane of organelles such as lysosomes. This less controlled form of cell death, designated necrosis, may contribute to the formation of an inflammatory state within PDT-treated tissue [9].

Most work examining the cellular effects of PDT has involved in vitro studies with transformed cell types, where the sub-lethal effects and apoptosis-inducing actions are typically evident over a relatively narrow range of photosensitizer concentrations, or conversely, light doses. The following sub-sections deal with the capacity of PDT to trigger cell-signaling events and induce cell death, particularly with regard to the effects on immune cells and their role in various disease processes.

5.2.1 Cell signaling and stress responses

PDT has been shown to activate specific cell signaling pathways. Although the full significance of this effect is unclear, effects on specific gene transcription and, ultimately, cell viability are possible. For example, nuclear factor-kappa B (NF-κB) regulates the transcription of immuno-regulatory cytokine and receptor genes, as well as those that encode proteins that influence cell viability [10]. Intracellular oxidative stress appears to be a unifying characteristic of exogenous NF-κB-activating agents [10]. NF-κB activity is normally regulated by members of a group of inhibitory (IκB) proteins that sequester NF-κB in the cytoplasm; IκB degradation liberates NF-κB for nuclear translocation [10]. NF-κB activation has been demonstrated in vitro for transformed immune cell types treated with PDT [11,12]. NF-κB was shown to translocate to the nucleus of Photofrin®-PDT treated murine L1210 lymphoma cells [11]. Similarly, PDT with verteporfin caused IκBα degradation and the nuclear translocation of the p50 and RelA NF-κB species in HL-60 myeloid leukemia cells [12]. Identification of the genes affected by PDT-mediated NF-κB activation will assist our understanding of the effects of PDT on immunologic reactions.

PDT with several types of photosensitizers stimulated mitogen-activated protein kinase (MAPK) signaling within various cell types. Activation of stress activated protein kinase (SAPK) and p38 MAPKs was described initially for mouse Pam212 keratinocytes treated with verteporfin and red light [13]. Subsequently, similar responses were delineated for various cell types treated with aminolevulinic acid (ALA) [14], phthalocyanine Pc4 [15], or hypericin [16] and light. The significance of these signaling events triggered by PDT is under debate. Blockade of p38 activity with pyridinyl imidazole compounds inhibited apoptosis induced by PDT with phthalocyanine Pc4 in mouse lymphoma cells [15]. In contrast, inhibition of SAPK and p38 activity exaggerated the extent of apoptosis produced in HeLa cells treated with hypericin and red light [16]. Some of these kinase-mediated cell signaling events associated with PDT could also affect gene transcription.

An effect of PDT on cell signaling activity in normal cell types has not been investigated to date. It may be misleading to extrapolate observations on PDT-induced signaling obtained in transformed cell lines to normal immune cells. Nevertheless, it is likely that there are several cell-type specific signaling responses to PDT and that these signaling pathways may interact to influence the functional behavior and/or survival of cells. Elucidation of the consequences of PDT-mediated cell signaling events, especially those that lead to gene transcription, will be important for determining their relationship to immune reactions. Of particular interest for dermatologic indications is the relationship between PDT-induced signaling and cytokine production.

Cells suffering damage from a number of diverse agents may protect themselves by increasing the expression of several families of stress molecules, including the heat shock proteins (HSP) and glucose-regulated proteins (GRP) [17]. The increased expression of these stress proteins may impart a level of resistance to subsequent insult. In response to elevated temperature, oxidative stress, and chemical agents, certain cell proteins lose their native conformation. This change in conformation mobilizes the heat shock factor (HSF) family of transcription factors. HSF bind the heat shock element (HSE) consensus sequence within the promoter region of genes encoding stress proteins. These genes are transcribed rapidly in response to the presence of improperly folded intracellular proteins [17]. Some stress proteins are normally expressed and they help maintain protein conformation and translocate polypeptides across membranes, thereby serving as molecular chaperones [18]. PDT has been shown to alter expression levels of stress proteins in tumor cells in vivo and in vitro [19–24], which has led to heightened heat tolerance by PDT-treated cells [22]. In non-PDT studies, HSP70, HSP90 and GRP94 stress proteins contributed to the development of protective immunity against tumors in mice [25]. HSP expression in tumor cells treated with PDT may, therefore, supply an activating signal for the immune system.

5.2.2 Apoptosis

Remarkably rapid induction of apoptosis has been a consistently reported effect of PDT. This marked capacity of PDT is believed to be related to the localization of many porphyrin photosensitizers to the mitochondrion [1,5–8], a primary site for the regulation of apoptotic stimuli [9,26]. Mitochondrial cytochrome c appeared immediately in the cytosolic fraction of HeLa cells [27,28] and human umbilical vein endothelial cells [29] treated with cytotoxic levels of verteporfin and light. This pivotal event has been associated with the onset of apoptosis induced by many non-PDT agents [9]. Rapid release of mitochondrial cytochrome c has also been shown for various cell types treated with light and the photosensitizers hypericin [30], phthalocyanine Pc4 [31] or porphycene [7].

Treatment with verteporfin and red light rapidly triggered a cascade of intracellular caspase (*cy*steinyl *asp*artate-specific protein*ase*) activity, leading to the irreversible cleavage of specific cellular proteins with structural, signaling or transcriptional function for various cell types [16,26–29,32–34]. Cytochrome c release preceded the processing and activation of caspases and the degradation of key cellular proteins after PDT [27–29]. The involvement of caspases in PDT-mediated apoptosis was also shown for light-irradiated mouse lymphoma cells treated with phthalocyanine Pc4 [35].

The capacity of PDT to induce apoptosis in non-transformed cells, particularly cells of the immune system, has been studied to a much lesser extent than in tumor cell lines. The occurrence of apoptosis has been documented for mouse dendritic cells (DC) [36] as well as immature [37], resting, and activated [38] T lymphocytes treated with verteporfin-based PDT. Hypericin induced apoptosis in normal, virally transformed and malignant human T cells treated with either UVA or white light [39]. The potent apoptosis-inducing action of PDT indicates that it may be effective for the treatment of immune-mediated skin diseases in which the removal of unwanted cells is desired.

5.3 Direct action of PDT on immune cell types

5.3.1 T cells

T lymphocytes have a central role in orchestrating many aspects of normal immune and autoimmune responses. PDT has been shown to selectivity decrease the viability of activated T cells in blood samples obtained from patients infected with the human immunodeficiency virus (HIV) [40]. Treatment with verteporfin and light reduced the percentage of mononuclear cells (MNC) bearing the IL-2 receptor (CD25) and HLA-DR, proteins that are highly expressed by activated human lymphoid cells [40]. In contrast, CD25 and HLA-DR levels on blood MNC of normal donors were relatively unchanged after PDT [40].

Mouse spleen cells activated with a T cell mitogen were more sensitive to photodynamic killing with verteporfin [41] or hematoporphyrin derivative [42] than were non-activated spleen cells, an effect that was correlated with greater verteporfin uptake by the activated cell population [41]. Highly purified mouse T cells activated with anti-CD3 antibody exhibited greater verteporfin uptake and underwent apoptosis at lower photosensitizer concentrations than non-activated T cells [38]. Mitogen-activated human T cells synthesized larger amounts of protoporphyrin IX (PPIX) than their resting equivalents following incubation with ALA and were killed selectively following light exposure [43,44]. Activated T cells may be valuable targets for the treatment of T cell-mediated diseases, such as psoriasis, with PDT.

Apoptosis may be triggered in many cell types, including immune cells that express the Fas (APO-1, CD95) receptor. The triggering agent can be the natural ligand (FasL) or agonistic anti-Fas antibodies [9,37]. Fas-FasL interactions limit the proliferation of activated T cells, promote the lysis of virally infected cells by cytotoxic T cells, and contribute to the state of immune privilege of certain tissues by killing activated inflammatory cells [9]. An anti-Fas antibody augmented the cytotoxic effect of verteporfin-based PDT against CD4 + CD8 + thymocytes in vitro [37]. In combination, PDT and anti-Fas antibody further increased thymocyte caspase activity and the percentage of cells undergoing apoptosis [37]. The initial biochemical events instigated by Fas ligation and PDT are distinct, but likely converge into a common biochemical pathway that leads to apoptosis. Cells receiving sub-lethal photodynamic stress could be directed towards apoptosis if ligands for death receptors such as FasL or tumor necrosis factor-α (TNF-α) are present and/or become more highly expressed in response to PDT. For example, if PDT promoted keratinocyte expression of FasL, it might lead to increased depletion of activated, pathogenic T cells from psoriatic plaques.

5.3.2 Antigen presenting cells (APC)

In some cases, treatment at sub-lethal levels of photosensitizer and light specifically modifies the surface receptor expression pattern of immune as well as non-immune cells. Human MNC treated with HpD and light in vitro had reduced levels of MHC Class II, but not MHC Class I antigens; these MNC poorly stimulated allo-reactive T cells in the mixed leukocyte reaction (MLR) [45]. Similarly, exposure of normal human monocytes or U937 monocytic cells to HpD and light lowered the binding of IgG$_{2a}$ antibodies to the IgG Fc receptor (FcγRI, CD64) but did not affect the status of several other surface molecules [46]. In this case, disturbance of the FcγRI by PDT was caused by an oxidation of specific amino acid residues produced by superoxide anions [46]. It appears that alteration of APC receptor expression levels is one way by which PDT can modify immune reactions.

DC are bone marrow-derived APC that perform a sentinel role for the immune system [47,48]. Abundant expression of MHC, adhesion and co-stimulatory molecules endows DC lineage cells with the capacity to activate naïve autologous and allogeneic T lymphocytes *via* the engagement of specific counter-receptors on T cells [47,48]. DC normally are present in low numbers within most tissues and are important sources of IL-12, a pro-inflammatory cytokine that promotes the formation of cellular immunity [47]. Langerhans cells (LC) from the DC lineage are the major resident APC type within skin and are responsible for the stimulation of T cell-mediated skin graft rejection reactions. Photodynamic treatment with HpD depleted LC from mouse skin and led to a prolonged acceptance of skin grafts on allogeneic recipients [49]. Pre-treatment of murine skin sections with verteporfin and red light prolonged their survival on MHC-mismatched mice [50]. These results suggested that donor LC may be compromised by PDT. Indeed, LC isolated from PDT-treated skin sections had reduced levels of MHC and co-stimulatory molecules but undiminished expression of several other receptors [50].

Mouse spleen DC treated with sub-lethal levels of verteporfin and light in vitro expressed markedly lowered levels of MHC, co-stimulatory, and adhesion molecules; they poorly stimulated allogeneic T cells in the MLR [36]. Similarly, human blood DC incubated with ALA accumulated PPIX and had a reduced capacity to stimulate the proliferation of allogeneic T cells following light exposure [51]. Interestingly, in contrast to T cells, DC did not require a deliberate activation step to synthesize PPIX from ALA [51]. Although mature DC are not in a proliferative state, they express significant levels of the receptor (CD71) for the iron transport protein transferrin [52], indicating an active iron-heme metabolic pathway.

The mechanism by which PDT modifies LC and DC surface receptor levels is not yet defined, but presumably results from the oxidative stress placed upon the cell by the treatment. The biochemical changes triggered by PDT may produce an oxidative stress response that lowers the immuno-stimulatory properties of these cells.

It was suggested that the marked abundance of DC within the synovial tissue and fluid of patients with rheumatoid arthritis (RA) [53,54] indicates that DC have a role in the pathogenesis of this autoimmune disease. Hence, it is plausible that PDT might have a beneficial effect against immune-mediated conditions such as RA [55–57], as well as skin diseases, by lowering DC-T cell interactions.

5.4 PDT in experimental immune models

5.4.1 Contact hypersensitivity (CHS) response

Treatment of mice with various photosensitizers and light has been shown to inhibit the immunologically-mediated CHS response to topically applied haptens such as dinitrofluorobenzene (DNFB). The CHS reaction has been a useful model to elucidate immunological effects of PDT in vivo because the ear swelling response generated by hapten sensitization provides a readily measured indicator of the immunologic impact of a treatment. Initial studies showed that HpD [58] or Photofrin® [59] combined with light inhibited the CHS response. It was suggested that splenic adherent cells (probably macrophages) played a role in this inhibition [59]. These findings engendered a view that PDT was an immunosuppressive therapy, although no other immune parameters were evaluated. Subsequently, this perception was dismissed when it was shown that verteporfin and light doses that reduced the CHS response still preserved central immune responses and hematopoietic activity [60].

Treatment of mice with Photofrin® and whole body blue light irradiation induced an extended expression of interleukin-10 (IL-10) within the skin [61]. This cytokine inhibits several T cell functions, lowers the T cell-activating function of APC, including their production of IL-12, and impairs the CHS response [62,63]. Localized high intensity PDT with phthalocyanine Pc4 inhibited the CHS response when hapten was applied to a non-irradiated site, thereby suggesting a generalized suppression [64]. However, this effect was unchanged by the administration of neutralizing antibodies to IL-10 or TNF-α [64], indicating that these cytokines were not responsible for the inhibitory effect of PDT on the CHS response. In contrast, a separate study showed full-fledged CHS responses in IL-10 gene knockout B6 mice treated with doses of verteporfin and red light that significantly impaired the anti-hapten response of normal animals [65]. In the wild-type mice, skin IL-10 levels were elevated 72 to 96 hours after PDT [65]. In this study, the influence of PDT on the CHS response of the wild type mice was neutralized by the administration of an anti-IL-10 antibody [65].

IL-12 contributes significantly to the development of murine CHS responses [66]. The administration of recombinant IL-12 prevented the PDT-related inhibition of the CHS response [65]. IL-12 influences T cell cytokine production by promoting the formation and release of pro-inflammatory factors, including interferon-γ (IFN-γ) [67]. For PDT-treated mice, IL-12 supplementation may counteract the inhibitory effect of IL-10 on the development of CHS responses. When PDT is delivered to the skin, the function of cutaneous APC may be modified and the availability of IL-10 heightened, thereby contributing to the deficient CHS responses of PDT-treated animals. In addition, it cannot be discounted that the CHS modifying effects of PDT in mice could be partially attributable to a penetration of activating light to internal organs, including primary lymphoid tissues. Nevertheless, an increase in skin IL-10 levels following PDT might contribute to the amelioration of disease in a variety of immune-mediated skin conditions.

5.4.2 Experimental Autoimmune Encephalomyelitis (EAE)

PDT was assessed with an adoptively transferable form of EAE in PL strain mice [68]. In this model, neurodegenerative symptoms are evident by three weeks following transfer of myelin basic protein (MBP)-sensitized spleen cells to syngeneic animals. Mice given verteporfin and whole body red light treatment 24, 48 or 120 hours after cell injection exhibited less severe disease symptoms than the controls [68]. Spinal cord tissues were analyzed by semi-quantitative polymerase chain reaction (PCR) for mRNA encoding the variable (V) α4 region of the T cell receptor (TCR). This receptor is expressed by the pathogenic MBP-specific T cells of PL strain mice. Vα4 TCR mRNA transcripts were detectable in virtually all samples from light-treated control mice, but in fewer than one-half of the samples prepared from mice treated with PDT 24 h after injection of spleen cells [68]. These observations indicated that PDT had affected the entry of MBP-specific Vα4 TCR T cells into the central nervous system, or inhibited their expansion. Mice given PDT 24 hours after cell administration also maintained greater body mass, indicating that the treatment had protected against the wasting effects of this condition [68]. This study indicated that PDT could constrain the action of autoreactive T cells in vivo.

5.4.3 Adjuvant-induced arthritis

PDT was tested in a model of adjuvant-induced autoimmune arthritis in MRL-*lpr* strain mice [55]. Animals given three whole body PDT treatments one week apart, after administration of complete Freund's adjuvant, exhibited reduced levels of disease-related symptoms, similar to the control mice [55]. The therapeutic effect was comparable to that produced by several immuno-modulatory agents [56]. Hematopoietic progenitor activity and spleen cell mitogenic responses were unaffected by the photodynamic treatments [55]. It was suggested that PDT may have targeted activated, arthritogenic T lymphocytes to lower the incidence of disease, although this possibility was not studied directly [55].

5.4.4 Antigen-induced arthritis

PDT has been evaluated in an antigen-induced arthritis model in rabbits [57]. When a protein antigen is administered into the knee joint of animals pre-sensitized to the antigen, an immune-mediated response develops that produces synovial inflammation and pannus formation, as well as bone and cartilage degradation. Antigen-challenged rabbits were given verteporfin systemically, then red light was applied to opposing sides of the joint [57]. Synovial inflammation, pannus formation, and cartilage and bone destruction were significantly less in PDT treated rabbits than in rabbits given no treatment, photosensitizer, or light alone [57]. Verteporfin was shown to associate with T cells and macrophages within the synovium of antigen challenged rabbits [57]. PDT was shown to induce apoptosis in T cells and macrophages within synovial tissue obtained six hours after treatment [57].

5.5 Anti-tumor immunity and PDT

Various attributes of cancer may limit the development of effective anti-tumor immunity. These characteristics include the presence of cytokines that impair immune cell function, a reduced expression of MHC and accessory molecules, and the loss of tumor-specific antigen expression. A growing body of evidence indicates that tumor resolution with PDT involves the immune system [2,69–71]. Furthermore, PDT may promote the formation of immunity against tumors that are poorly or non-immunogenic [2].

   PDT was less effective in the cure of tumors in severe, combined immunodeficient (scid) [2] or T cell-deficient nude [69] mice than in immune competent mice, providing evidence for the involvement of the immune system in PDT-mediated tumor clearance. Large numbers of granulocytes and macrophages infiltrated murine tumors shortly after PDT [2,61]. An adjuvant preparation derived from a *Mycobacterium* cell wall extract administered locally immediately after photo-irradiation led to improved anti-tumor activity as a result of PDT with a variety of photosensitizers [70]. Adjuvant preparations may amplify PDT-induced anti-tumor immunity by stimulating and sustaining the infiltration of granulocytes and other myeloid cell types to the treatment site [70]. Treatment of tumor-bearing mice with Photofrin®-based PDT increased myelopoiesis in the spleen and marrow, and this effect was augmented by the co-administration of the hematopoietic growth factor granulocyte colony stimulating factor (G-CSF) [72]. Thus, the anti-tumor effect of PDT can be augmented by the use of a variety of immune mobilizing factors.

IL-1β, IL-2 and TNF-α were present in the urine of bladder cancer patients treated with localized Photofrin®-based PDT suggesting that a localized immune, inflammatory response ensued after PDT [73]. Photofrin®, in the absence of light, increased TNF-α and IFN-γ release by human MNC and mouse spleen cells cultured in medium alone or with sub-optimal concentrations of mitogen [74]. Moreover, Photofrin® stimulated hematopoietic activity within the marrow and spleens of normal mice held under ambient light conditions [75,76]. The mechanisms by which Photofrin® stimulates the immuno-hematopoietic system in the absence of overt light treatment are not known.

The phagocytic activity of macrophages was stimulated by PDT with low levels of photosenstizer [77] and tumor cells were more susceptible to lysis by macrophages [78]. Photofrin®-based PDT enhanced the release of prostaglandin-E2 [79] and TNF-α [80] from murine peritoneal macrophages. Sub-lethal PDT treatment of human macrophage-differentiated U937 cells with meta-tetra(hydroxyphenyl)chlorin (mTHPC) elevated TNF-α and nitric oxide production, as well as phagocytotic activity, in a light-dose dependent fashion [81]. Hence, the function of macrophages present within tumor tissue may be augmented if they receive sub-lethal PDT, an effect that could contribute to tumor clearance.

PDT-induced expression of HSP by tumor cells may promote the formation of anti-cancer immunity. Antigens chaperoned by stress proteins are avidly taken up by APC [18,82]. It is significant that exogenous antigens taken up in association with stress proteins are capable of stimulating cytotoxic T cell immunity [2,9,10]. HSP expression induced in PDT-treated tumor cells may provide a strong activation signal for the immune system and, as such, contribute to the clearance of residual disease.

An up-regulation of IL-6 and a decrease in IL-10 mRNA levels occurred in murine tumors treated with PDT, accompanied by a dramatic increase in IL-10 expression in the skin [61]. It has been suggested that IL-6 has a role in the localized inflammatory effect produced by PDT and may modulate anti-tumor immunity [61].

In mouse models, anti-tumor immunity provoked by PDT appears to be T cell dependent [69,71]. When tumor-bearing scid mice were given tumor-sensitized spleen cells from immunologically intact Balb/c mice, the tumors were eliminated following Photofrin®-based PDT [2]. Antibody depletion experiments indicated that tumor-specific CD8+ T cells were primarily responsible for the immunologic anti-tumor effect associated with PDT, although contributions from natural killer (NK) and CD4+ T cells also were identified [2]. Antibody depletion of CD8+ cytotoxic T cells following PDT led to reduced cure rates, but removal of CD4+ T cells had a lesser effect [2]. For EMT-6 mammary fibrosarcoma-bearing Balb/c mice treated with PDT using benzophenothiazine and red light, tumor resolution was strongly dependent on the presence of CD8+ T cells and, to a lesser extent, NK cells [69]. Thus, anti-tumor immunity fostered by PDT requires cytotoxic T cells.

The presentation of tumor cell antigens by certain APC initiates anti-tumor immunity. DC take up antigens efficiently and, in the process, mature into cells that effectively activate antigen-specific T lymphocytes [47,48]. The capacity of DC to instigate de novo immune responses led to their designation as "nature's adjuvant" [48,83,84]. DC phagocytose dying cells and can subsequently stimulate the proliferation of tumor antigen-specific cytotoxic T cells [85,86]. Cellular events provoked by PDT may unveil previously hidden tumor-associated antigens in forms that are then recognizable by the immune system. A preliminary study showed that tumor cell antigens were detectable within the draining lymph nodes of tumor-bearing mice after treatment with PDT, suggesting that PDT had initiated events that could promote anti-tumor immunity [87]. The capacity of PDT to elicit rapid tumor cell killing may provide a source of antigenic material that can be presented effectively by APC to tumor-specific T cells. These cytotoxic T cells may eliminate local tumor cell foci that have escaped the direct action of PDT, and may even eliminate those at distant sites.

5.6 Summary

Historically, cancer has been the major therapeutic focus for PDT. The growing understanding of how PDT affects the behavior and viability of a variety of cell types has fostered the evaluation and use of PDT for non-oncologic indications, including ocular, cardiovascular, and immune diseases. The formulation, pharmacokinetics, and type of photosensitizer, the duration between its administration and light application, and the region or extent of body surface area exposed to activating light may influence the impact of PDT on immune reactions. Among immune-mediated diseases, psoriasis is a promising candidate indication for PDT, since the skin lesions and the immunocytes associated with active disease are readily accessible to activating light. The beneficial effect of PDT on psoriasis may arise from a reduction in the number of pathogenic T cells, a modification of APC function, and/or an alteration of keratinocyte cytokine production within afflicted skin.

Continued efforts to optimize PDT regimens will reveal the ultimate potential of this technique to effectively treat immune-mediated diseases. Likewise, there is potential to exploit the capacity of PDT to stimulate anti-tumor immunity. Both of these therapeutic areas will continue to provide opportunities to understand the underlying biological processes and develop improved treatment regimens.

Acknowledgments

The authors thank Karen Munro for her expert assistance in the preparation of this manuscript and Leslie Ratkay, Ruth Salmon, and Guillermo Simkin for their helpful review comments.

References

1. T.J. Dougherty, C.J. Gomer, B.W. Henderson, G. Jori, D. Kessel, M. Korbelik, J. Moan, Q. Peng (1998). Photodynamic therapy. *J. Natl. Cancer Inst.*, **90**, 889–905.
2. M. Korbelik, G.J. Dougherty (1999). Photodynamic therapy-mediated immune response against subcutaneous mouse tumors. *Cancer Res.*, **59**, 1941–1946.
3. D.W.C. Hunt, J.G. Levy (1998). Immunomodulatory aspects of photodynamic therapy, *Expert Opin. Invest. Drugs*, **7**, 57–64.
4. D.W. Hunt, A.H. Chan (2000). Influence of photodynamic therapy on immunological aspects of disease – an update, *Expert Opin. Invest. Drugs*, **9**, 807–817.
5. N.L. Oleinick, H.H. Evans (1998). The photobiology of photodynamic therapy: cellular targets and mechanisms. *Radiat. Res.*, **150**, S146–156.
6. D. Kessel, Y. Luo (1998). Mitochondrial photodamage and PDT-induced apoptosis. *J. Photochem. Photobiol. B Biol.*, **42**, 89–95.
7. D. Kessel, Y. Luo (1999). Photodynamic therapy: a mitochondrial inducer of apoptosis. *Cell Death Differ.*, **6**, 28–35.
8. D. Kessel, Y. Luo, P. Mathiew, J.J. Reiners (2000). Determinants of the apoptotic response to lysosomal photodamage. *Photochem. Photobiol.*, **71** 196–200.
9. D.J. Granville, C.M. Carthy, D.W. Hunt, B.M. McManus (1998). Apoptosis: molecular aspects of cell death and disease. *Lab. Invest.*, **78**, 893–913.
10. S. Legrand-Poels, S. Schoonbroodt, J.Y. Matroule, J. Piette (1999). NF-κB: an important transcription factor in photobiology. *J. Photochem. Photobiol. B Biol.*, **45**, 1–8.
11. S.W. Ryter, C.J. Gomer (1993) Nuclear factor kappa B binding activity in mouse L1210 cells following photofrin II-mediated photosensitization. *Photochem. Photobiol.*, **58**, 753–756.
12. D.J. Granville, C.M. Carthy, H. Jiang, J.G. Levy, B.M. McManus, J.-Y. Matroule, J. Piette, D.W.C. Hunt (2000) NF-κB activation by the photochemotherapeutic agent verteporfin. *Blood*, **95**, 256–262.
13. J. Tao, J.S. Sanghera, S.L. Pelech, G. Wong, J.G. Levy (1996). Stimulation of stress-activated protein kinase and p38 HOG1 kinase in murine keratinocytes following photodynamic therapy with benzoporphyrin derivative. *J. Biol. Chem.*, **271**, 27107–27115.
14. L.O. Klotz, C. Fritsch, K. Briviba, N. Tsacmacidis, F. Schliess, H. Sies (1998). Activation of JNK and p38 but not ERK MAP kinases in human skin cells by 5-aminolevulinate-photodynamic therapy. *Cancer Res.*, **58**, 4297–4300.
15. L.-Y. Xue, J. He, N.L. Oleinick (1999). Promotion of photodynamic therapy-induced apoptosis by stress kinases. *Cell Death Differ.*, **6**, 855–864.
16. Z. Assefa, A. Vantieghem, W. Declercq, P. Vandenabeele, J.R. Vandenheede, W. Merlevede, P. de Witte, P. Agostinis (1999). The activation of the c-Jun terminal kinase and p38 mitogen-activated protein kinase signaling pathways protects HeLa cells from apoptosis following photodynamic therapy with hypericin, *J. Biol. Chem.*, **274**, 8788–8796.
17. A. Stephanou, D.S. Latchman (1999) Transcriptional regulation of the heat shock protein genes by STAT family transcription factors, *Gene Expr.*, **7**, 311–319.
18. P.K. Srivastava, H. Udono (1994). Heat shock protein-peptide complexes in cancer immunotherapy *Current Opin. Immunol.*, **6**, 728–732.
19. C.J. Gomer, A. Ferrario, N. Hayachi, N. Rucker, B.C. Szirth, A.L. Murphree (1988). Molecular, cellular, and tissue responses following photodynamic therapy, *Lasers Surg. Med.*, **8**, 450–463.
20. C.J. Gomer, N. Rucker, A. Ferrario, S. Wong (1989). Properties and applications of photodynamic therapy, *Radiat. Res.*, **120**, 1–18.
21. P.M. Curry, J.G. Levy (1993). Stress protein expression in murine tumor cells following photodynamic therapy with benzoporphyrin derivative. *Photochem. Photobiol.*, **58**, 374–379.

22. C.J. Gomer, S.W. Ryter, A. Ferrario, N. Rucker, S. Wong, A.M. Fisher (1996). Photodynamic therapy-mediated oxidative stress can induce expression of heat shock proteins. *Cancer Res.*, **56**, 2355–2360.

23. L.Y. Xue, M.L. Agarwal, M.E. Varnes (1995). Elevation of GRP-78 and loss of HSP-70 following photodynamic treatment of V79 cells: sensitization by nigericin. *Photochem. Photobiol.*, **62**, 135–143.

24. M.C. Luna, A. Ferrario, S. Wong, A.M. Fisher, C.J. Gomer (2000). Photodynamic therapy-mediated oxidative stress as a molecular switch for the temporal expression of genes ligated to the human heat shock promoter. *Cancer Res.*, **60**, 728–732.

25. P.K. Srivastava, H. Udono, N.E. Blachere, Z. Li (1994) Heat shock proteins transfer peptides during antigen processing and CTL priming. *Immunogenetics*, **39**, 93–98.

26. S.A. Susin, H.K. Lorenzo, N. Zamzami, I. Marzo, C. Brenner, N. Larochette, M.C. Prevost, P.M. Alzari, G. Kroemer (1999). Mitochondrial release of caspase-2 and-9 during the apoptotic process. *J. Exp. Med.*, **189** 381–394.

27. D.J. Granville, C.M. Carthy, H. Jiang, G.C. Shore, B.M. McManus, D.W. Hunt (1998). Rapid cytochrome c release, activation of caspases 3, 6, 7 and 8 followed by Bap31 cleavage in HeLa cells treated with photodynamic therapy. *FEBS Lett.*, **437**, 5–10.

28. C.M. Carthy, D.J. Granville, H. Jiang, J.G. Levy, C.M. Rudin, C.B. Thompson, B.M. McManus, D.W. Hunt (1999). Early release of mitochondrial cytochrome c and expression of mitochondrial epitope 7A6 with a porphyrin-derived photosensitizer: Bcl-2 and Bcl-xL overexpression do not prevent early mitochondrial events but still depress caspase activity. *Lab. Invest.*, **79**, 953–965.

29. D.J. Granville, J.R. Shaw, S. Leong, C.M. Carthy, P. Margaron, D.W. Hunt, B.M. McManus (1999). Release of cytochrome c, Bax migration, Bid cleavage, and activation of caspases 2, 3, 6, 7, 8, and 9 during endothelial cell apoptosis. *Am. J. Pathol.*, **55**, 1021–1025.

30. A. Vantieghem, Z. Assefa, P. Vandenabeele, W. De Clercq, S. Courtois, J.R. Vandenheede, P. de Witte, P. Agostinis (1998). Hypericin-induced photosensitization of HeLa cells leads to apoptosis or necrosis. Involvement of cytochrome *c* and procaspase-3 activation in the mechanism of apoptosis. *FEBS Lett.*, **440** 19–24.

31. M.E. Varnes, S.M. Chiu, L.Y. Xue, N.L. Oleinick (1999). Photodynamic therapy-induced apoptosis in lymphoma cells: translocation of cytochrome c causes inhibition of respiration as well as caspase activation. *Biochem. Biophys. Res. Commun.*, **255**, 673–679.

32. D.J. Granville, J.G. Levy, D.W.C. Hunt (1997). Photodynamic therapy induces caspase-3 activation in HL-60 cells, *Cell Death Differ.*, **4**, 623–628.

33. D.J. Granville, H. Jiang, M.T. An, J.G. Levy, B.M. McManus, D.W. Hunt (1998). Overexpression of Bcl-X(L) prevents caspase-3-mediated activation of DNA fragmentation factor (DFF) produced by treatment with the photochemotherapeutic agent BPD-MA. *FEBS Lett.*, **422**, 151–154.

34. D.J. Granville, H. Jiang, M.T. An, J.G. Levy, B.M. McManus, D.W.C. Hunt (1999). Bcl-2 overexpression blocks caspase activation and downstream apoptotic events instigated by photodynamic therapy. *Brit. J. Cancer*, **79**, 95–100.

35. J. He, C.M. Whitacre, L. Xue, N.A. Berger, N.L. Oleinick (1998). Protease activation and cleavage of poly(ADP-ribose) polymerase: an integral part of apoptosis in response to photodynamic treatment. *Cancer Res.*, **58**, 940–946.

36. D.E. King, H. Jiang, G.O. Simkin, M.O. Obochi, J.G. Levy, D.W. Hunt (1999) Photodynamic alteration of the surface receptor expression pattern of murine splenic dendritic cells. *Scand. J. Immunol.*, **49**, 184–192.

37. H. Jiang, D.J. Granville, B.M. McManus, J.G. Levy, D.W. Hunt (1999). Selective depletion of a thymocyte subset in vitro with an immunomodulatory photosensitizer. *Clin. Immunol.*, **91**, 178–187.

38. D.W. Hunt, H. Jiang, D.J. Granville, A.H. Chan, S. Leong, J.G. Levy (1999). Consequences of the photodynamic treatment of resting and activated peripheral T lymphocytes. *Immunopharmacology*, **41**, 31–44.
39. F.E. Fox, Z. Niu, A. Tobia, A.H. Rook (1998). Photoactivated hypericin is an anti-proliferative agent that induces a high rate of apoptotic death of normal, transformed and malignant T lymphocytes: implications for the treatment of cutaneous lymphoproliferative and inflammatory disorders. *J. Invest. Dermatol.*, **111**, 327–332.
40. J. North, H. Neyndorff, J.G. Levy (1993). Photosensitizers as virucidal agents, *J. Photochem. Photobiol. B Biol.*, **17**, 99–108.
41. M.O. Obochi, A.J. Canaan, A.K. Jain, A.M. Richter, J.G. Levy (1995). Targeting activated lymphocytes with photodynamic therapy: susceptibility of mitogen-stimulated splenic lymphocytes to benzoporphyrin derivative (BPD) photosensitization. *Photochem. Photobiol.*, **62**, 169–175.
42. G. Canti, O. Marelli, L. Ricci, A. Nicolin (1981). Haematoporphyrin-treated murine lymphocytes: in vitro inhibition of DNA synthesis and light-mediated inactivation of cells responsible for GVHR. *Photochem. Photobiol.*, **34**, 589–594.
43. E.A. Hryhorenko, K. Rittenhouse-Diakun, N.S. Harvey, J. Morgan, C.C. Stewart, A.R. Oseroff (1998). Characterization of endogenous protoporphyrin IX induced by delta-aminolevulinic acid in resting and activated peripheral blood lymphocytes by four-color flow cytometry. *Photochem. Photobiol.*, **67**, 565–572.
44. E.A. Hryhorenko, A.R. Oseroff, J. Morgan, K. Rittenhouse-Diakun (1999). Deletion of alloantigen-activated cells by aminolevulinic acid-based photodynamic therapy. *Photochem. Photobiol.*, **69**, 560–565.
45. S. Gruner, H. Volk, F. Noack, H. Meffert, R. von Baehr (1986). Inhibition of HLA-DR antigen expression and of the mixed leukocyte reaction by photochemical treatment. *Tissue Antigens*, **27**, 147–154.
46. J. Krutmann, M. Athar, D.B. Mendel, I.U. Khan, P.M. Guyre, H. Mukhtar, C.A. Elmets (1989). Inhibition of the high affinity Fc receptor (Fc gamma RI) on human monocytes by porphyrin photosensitization is highly specific and mediated by the generation of superoxide radicals, *J. Biol. Chem.*, **264**, 11407–11413.
47. J. Banchereau, R.M. Steinman (1998). Dendritic cells and the control of immunity. *Nature*, **392**, 245–252.
48. R.M. Steinman (1991). The dendritic cell system and its role in immunogenicity. *Annu. Rev. Immunol.*, **9**, 271–296.
49. S. Gruner, H. Meffert, D. Volk, R. Grunow, S. Jahn (1985). The influence of haematoporphyrin derivative and visible light on murine skin graft survival, epidermal Langerhans cells and stimulation of the allogeneic mixed leukocyte reaction, *Scand. J. Immunol.*, **21**, 267–273.
50. M.O. Obochi, L.G. Ratkay, J.G. Levy (1997). Prolonged skin allograft survival after photodynamic therapy associated with modification of donor skin antigenicity, *Transplantation*, **63**, 810–817.
51. E.A. Hryhorenko, A.R. Oseroff, J. Morgan, K. Rittenhouse-Diakun (1998). Antigen specific and nonspecific modulation of the immune response by aminolevulinic acid based photodynamic therapy. *Immunopharmacology*, **40**, 231–240.
52. R. Gieseler, D. Heise, A. Soruri, P. Schwartz, J.H. Peters (1998). In vitro differentiation of mature dendritic cells from human blood monocytes. *Dev. Immunol.*, **6**, 25–39.
53. R. Thomas, K.P. MacDonald, A.R. Pettit, L.L. Cavanagh, J. Padmanabha, S. Zehntner (1999). Dendritic cells and the pathogenesis of rheumatoid arthritis. *J. Leukocyte Biol.*, **66**, 286–292.

54. A.R. Pettit, R. Thomas (1999). Dendritic cells: the driving force behind autoimmunity in rheumatoid arthritis? *Immunol. Cell Biol.*, **77**, 420–427.
55. R.K. Chowdhary, L.G. Ratkay, H.C. Neyndorff, A. Richter, M. Obochi, D.J. Waterfield, J.G. Levy (1994). The use of transcutaneous photodynamic therapy in the prevention of adjuvant-enhanced arthritis in MRL/lpr·mice. *Clin. Immunol. Immunopathol.*, **72** 255–263.
56. L.G. Ratkay, R.K. Chowdhary, H.C. Neyndorff, J. Tonzetich, J.D. Waterfield, J.G. Levy (1994). Photodynamic therapy; a comparison with other immunomodulatory treatments of adjuvant-enhanced arthritis in MRL/lpr mice. *Clin. Exp. Immunol.*, **95**, 373–377.
57. L.G. Ratkay, R.K. Chowdhary, A. Iamaroon, A.M. Richter, H.C. Neyndorff, E.C. Keystone, J.D. Waterfield, J.G. Levy (1998). Amelioration of antigen-induced arthritis in rabbits by induction of apoptosis of inflammatory cells with local application of transdermal photodynamic therapy. *Arthritis Rheum.*, **41**, 525–534.
58. C.A. Elmets, K.D. Bowen (1986). Immunological suppression in mice treated with hematoporphyrin derivative photoradiation. *Cancer Res.*, **46**, 1608–1611.
59. D.H. Lynch, S. Haddad, V.J. King, M.J. Ott, R.C. Straight, C.J. Jolles (1989). Systemic immunosuppression induced by photodynamic therapy (PDT) is adoptively transferred by macrophages. *Photochem. Photobiol.*, **49**, 453–458.
60. G.O. Simkin, D.E. King, J.G. Levy, A.H. Chan, D.W. Hunt (1997). Inhibition of contact hypersensitivity with different analogs of benzoporphyrin derivative. *Immunopharmacology*, **37** 221–230.
61. S.O. Gollnick, X. Liu, B. Owczarczak, D.A. Musser, B.W. Henderson (1997). Altered expression of interleukin 6 and interleukin 10 as a result of photodynamic therapy in vivo. *Cancer Res.*, **57**, 3904–3909.
62. A.H. Enk, V.L. Angeloni, M.C. Udey, S.I. Katz (1993). Inhibition of Langerhans cell antigen-presenting function by IL-10. A role for IL-10 in induction of tolerance. *J. Immunol.*, **151**, 2390–2398.
63. F. Koch, U. Stanzl, P. Jennewein, K. Janke, C. Heufler, E. Kampgen, N. Romani, G. Schuler (1996). High level IL-12 production by murine dendritic cells: up-regulation via MHC Class II and CD40 molecules and downregulation by IL-4 and IL-10, *J. Exp. Med.*, **184**, 741–746.
64. J.C. Reddan, C.Y. Anderson, X. Hui, S. Hrabovsky, K. Freye, R. Fairchild, K.A. Tubesing, C.A. Elmets (1999). Immunosuppressive effects of silicon phthalocyanine photodynamic therapy. *Photochem. Photobiol.*, **70**, 72–77.
65. G.O. Simkin, J.-S. Tao, J.G. Levy, D.W.C. Hunt (2000). IL-10 contributes to the inhibition of contact hypersensitivity in mice treated with photodynamic therapy. *J. Immunol.*, **164**, 2457–2462.
66. H. Riemann, A. Schwarz, S. Grabbe, Y. Aragane, T.A. Luger, M. Wysocka, M. Kubin, G. Trinchieri, T. Schwarz (2000). Neutralization of IL-12 in vivo prevents induction of contact hypersensitivity and induces hapten-specific tolerance. *J. Immunol.*, **156**, 1799–1803.
67. R.R. Caspi (1998). IL-12 in autoimmunity. *Clin. Immunol. Immunopathol.*, **88**, 4–13.
68. S. Leong, A.H. Chan, J.G. Levy, D.W.C. Hunt (1996). Transcutaneous photodynamic therapy alters the development of an adoptively transferred form of murine experimental autoimmune encephalomyelitis. *Photochem. Photobiol.*, **64**, 751–757.
69. J.A. Hendrak-Henion, T.L. Knisely, L. Cincotta, E. Cincotta, A.H. Cincotta (1999). Role of the immune system in mediating the antitumor effect of benzophenothiazine photodynamic therapy. *Photochem. Photobiol.*, **69**, 575–581.
70. M. Korbelik, I. Cecic (1998). Enhancement of tumour response to photodynamic therapy by adjuvant mycobacterium cell-wall treatment. *J. Photochem. Photobiol. B Biol.*, **44**, 151–158.

71. M. Korbelik, I. Cecic (1999). Contribution of myeloid and lymphoid host cells to the curative outcome of mouse sarcoma treatment by photodynamic therapy. *Cancer Lett.*, **137**, 91–98.

72. J. Golab, G. Wilczynski, R. Zagozdzon, T. Stoklosa, A. Dabrowska, J. Rybczynska, M. Wasik, E. Machaj, T. Olda, K. Kozar, R. Kaminski, A. Giermasz, A. Czajka, W. Lasek, W. Feleszko, M. Jakobisiak (2000). Potentiation of the anti-tumour effects of Photofrin-based photodynamic therapy by localized treatment with G-CSF. *Br. J. Cancer*, **8**, 1485–1491.

73. U.O. Nseyo, R.K. Whalen, M.R. Duncan, B. Berman, S.L. Lundal (1990). Urinary cytokines following photodynamic therapy for bladder cancer. A preliminary report. *Urology*, **36**, 167–171.

74. S. Herman, Y. Kalechman, U. Gafter, B. Sredni, Z. Malik (1996). Photofrin II induces cytokine secretion by mouse spleen cells and human peripheral mononuclear cells. *Immunopharmacology*, **31**, 195–204.

75. D.W. Hunt, R.A. Sorrenti, C.B. Smits, J.G. Levy (1993). Photofrin, but not benzoporphyrin derivative, stimulates hematopoiesis in the mouse. *Immunopharmacology*, **26**, 203–212.

76. D.W. Hunt, H. Jiang, J.G. Levy (1998). Photofrin increases murine spleen cell transferrin receptor expression and responsiveness to recombinant myeloid and erythroid growth factors. *Immunopharmacology*, **38**, 267–278.

77. N. Yamamoto, T.W. Sery, J.K. Hoober, N.P. Willett, D.D. Lindsay (1994). Effectiveness of Photofrin II in activation of macrophages and in vitro killing of retinoblastoma cells. *Photochem. Photobiol.*, **60**, 160–164.

78. M. Korbelik, G. Krosl (1994). Enhanced macrophage cytotoxicity against tumor cells treated with photodynamic therapy. *Photochem. Photobiol.*, **60**, 497–502.

79. B.W. Henderson, J.M. Donovan (1989). Release of prostaglandin E2 from cells by photodynamic treatment in vitro. *Cancer Res.*, **49**, 6896–6900.

80. S. Evans, W. Matthews, R. Perry, D. Fraker, J. Norton, H.I. Pass (1990). Effect of photodynamic therapy on tumor necrosis factor production by murine macrophages. *J. Natl. Cancer Inst.*, **82**, 34–39.

81. S. Coutier, L. Bezdetnaya, S. Marchal, V. Melnikova, I. Belitchenko, J.L. Merlin, F. Guillemin (1999). Foscan (mTHPC) photosensitized macrophage activation: enhancement of phagocytosis, nitric oxide release and tumour necrosis factor-alpha-mediated cytolytic activity. *Br. J. Cancer*, **81**, 37–42.

82. D. Przepiorka, P.K. Srivastava (1998). Heat shock protein-peptide complexes as immuno-therapy for human cancer. *Mol. Med. Today*, **4**, 478–484.

83. G. Schuler, R.M. Steinman (1997). Dendritic cells as adjuvants for immune-mediated resistance to tumors, *J. Exp. Med.*, **186**, 1183–1187.

84. J.W. Young, K. Inaba (1998). Dendritic cells as adjuvants for Class I major histocompatibility complex-restricted antitumour immunity. *J. Exp. Med.*, **183**, 7–11.

85. M.L. Albert, B. Sauter, N. Bhardwaj (1998). Dendritic cells acquire antigen from apoptotic cells and induce Class I- restricted CTLs. *Nature*, **392**, 86–89.

86. M.L. Albert, S.F. Pearce, L.M. Francisco, B. Sauter, P. Roy, R.L. Silverstein, N. Bhardwaj (1998). Immature dendritic cells phagocytose apoptotic cells via alphavbeta5 and CD36, and cross-present antigens to cytotoxic T lymphocytes. *J. Exp. Med.*, **188**, 1359–1368.

87. B.Y. Lee, S.O. Gollnick, B. Owczarczak, B.W. Henderson (1999). Detection of tumor antigens in the lymph node following Photofrin photodynamic therapy (PDT). *Photochem. Photobiol.*, **69**, 10S.

Photodynamic Therapy and Fluorescence Diagnosis in Dermatology
P.-G. Calzavara-Pinton, R.-M. Szeimies and B. Ortel, editors.

Chapter 6

Light sources for photodynamic therapy and fluorescence diagnosis in dermatology

Wolfgang Bäumler

Table of contents

Abstract

Photoactivation of sensitizers is a basic requirement in photodynamic therapy (PDT) and fluorescence diagnosis (FD). After absorption of light by a sensitizer the energy of the light is stored in the singlet or triplet state of the sensitizer molecule. In the singlet state the energy is converted to heat or is emitted as light used for fluorescence diagnosis, whereas the triplet state can provide energy to generate reactive oxygen species (ROS) necessary for a successful PDT.

The light used for PDT or FD can be provided by incoherent light sources (ILS) or laser systems. The efficacy of PDT or FD is not affected by the coherence of light. The emission spectrum $(S(\lambda))$ of a ILS or the emission wavelength (λ_{em}) of a laser has to be adjusted to the absorption spectrum of the sensitizer used.

For PDT the light source has to provide sufficient intensity of light, up to 50 mW/cm^2 (up to 40 J/cm^2) for non-oncologic indications and up to 200 mW/cm^2 (up to 150 J/cm^2) for oncologic indications. The optical power of the light source must be adequate for the irradiation of areas up to 500 cm^2 and the intensity applied to the skin should be as uniform as possible on the entire area irradiated. By using light in the red or infrared range of the spectrum the interactions with chromophors of the skin (melanin, hemoglobin) are minimized leading to a sufficient penetration of light into skin. Since most of the sensitizers absorb in the visible part of the electromagnetic spectrum a compromise is found by using the long-wave part of the sensitizer absorption.

In contrast to PDT blue light is mainly applied for fluorescence diagnosis of diseased skin due to its separability from the red fluorescence (e.g. Protoporphyrin IX).

6.1 Introduction

One hundred years ago the first light source used for PDT was natural sunlight. Later on arc lamps (carbon, xenon) or slide projectors were applied [1]. The spectral output of these light sources were either white light or narrowed to a range of wavelength which coincide with the absorption spectrum of the sensitizer used. In the last years incoherent light sources came into the market specially designed for PDT in dermatology [2].

The efficacy of PDT and fluorescence diagnosis (FD) is correlated to the amount of light energy absorbed in the sensitizer. In a first step the light applied is absorbed in the singlet state of the sensitizer molecule. In a second step the excited molecules turn back to the groundstate by emitting fluorescence (fluorescence quantum yield, FD), performing intersystem crossing to the triplet state (triplet quantum yield, PDT) or generating heat. The latter leads to an undesired photothermal effect which can be avoided by using light intensities of less than 200 mW/cm^2. In a third step a fraction of molecules transfer its energy present in the triplet state to generate oxygen radicals or singlet oxygen (ROS quantum yield, PDT) leading to a modulation of cellular processes or even cell death.

Regarding PDT the amount of ROS generated by a photoactivated sensitizer is correlated to the absorbed light energy and therefore the application of the appropriate light parameters (fluence, intensity) either for oncologic or non-oncologic indications is important for the efficacy of PDT. It is obvious that only a fraction of the light energy applied is used for the generation of ROS. In order to achieve a high efficacy in PDT it is an important goal to provide sufficient absorption of light in the sensitizer by adjusting the emission of the light source to the absorption spectrum of the sensitizer used. The light applied to the skin surface penetrates the tissue to ensure the photoactivation of preferably all sensitizer molecules present in the diseased tissue. However, the penetration decreases with increasing wavelength and is maximal at about 1000 nm. Since most of the sensitizers absorb in the visible part of the electromagnetic spectrum a compromise is found by using the long-wave part of the sensitizer absorption. Regarding porphyrins this is the red part of the visible spectrum.

Regarding FD usually blue light is applied due to the high absorption of porphyrins at 400 nm (Soret band) and the good distinction of the red fluorescence generated [3]. However, the penetration of blue light into tissue is only a few tenth of millimeter and it must be noticed that the respective fluorescence is generated only in the superficial part of the skin. The light intensity used for FD is at least 20 fold lower as compared to PDT.

Consequently, a light source used for PDT or FD in dermatology has to be designed according to the basic requirements as follows. The emission spectrum of the light must be adjusted to the absorption spectrum of the sensitizer used. When using porphyrins this is usually blue light for FD and red light for PDT. In order to ensure uniform conditions of treatment the intensity applied to the skin must be as homogeneous as possible on the entire area irradiated. Regarding PDT the intensity and the fluence of the light is splitted in two ranges, a high dose range for oncologic indications and a low dose range for non-oncologic indications, whereas the intensity used for FD is remarkably lower than for PDT.

6.2 Light

Light can be described as electromagnetic waves with the field strength $\vec{E}$

$$\vec{E} = \vec{E}_0 \cdot \sin(2\pi \cdot v \cdot t) = \vec{E}_0 \cdot \sin\left(\frac{2\pi \cdot c \cdot t}{\lambda}\right)$$

where v is the frequency (s^{-1}), λ the wavelength (m) and c the speed of the light $(m\ s^{-1})$ connected by $c = v\lambda$. The intensity I_0 (W/m^2) is defined by (ε dielectric constant)

$$I_0 = \varepsilon \cdot c \cdot \vec{E}?$$

The intensity of light is defined by the optical power (Watt) applied to a area with a well-defined size (m^2). Likewise the light can be described as packets of energy (photons). The energy of light E (Joule) is inversely proportional to its wavelength and given by

$$E = \frac{h \cdot c}{\lambda}$$

where h is Planck's constant $(6.63 \times 10^{-34}\ Js)$. The energy of one photon is $5 \times 10^{-19}\ J$ (400 nm) and $3.2 \times 10^{-19}\ J$ (630 nm). The application of an intensity of $100\ mW/cm^2$ (630 nm) results in a flux of 3×10^{17} photons per cm^2 and second.

The emission spectrum of a light source is determined by the spectral width ($\Delta\lambda$) and the spectral intensity $S(\lambda)$. The spectral width of a laser is usually confined to a value less than 1 nm yielding a high spectral intensity, whereas the light emission of lamp systems shows a broad spectral distribution of hundreds of nanometers. Therefore, the spectral intensity of the light $S(\lambda)$ [W cm^{-2} nm^{-1}] is remarkable different for lasers and ILS when using an equal intensity.

6.3 Absorption of light

The absorption of light is determined by the transition of the sensitizer molecule from the electronic ground state (S_0) to excited states (S_n). Due to the complex organic structure of the sensitizer molecules each electronic state shows a variety of vibrational states leading to a variety of optical transitions (Fig. 1). The energy (ΔE) required for the transition is

$$\Delta E = E(S_n) - E(S_0) = E_n - E_0 = \frac{h \cdot c}{\lambda_n}$$

where λ_n is the wavelength corresponding to the energy difference of the respective transition. The absorption spectra of the sensitizer is the arrangement of all transition wavelengths. In order to adjust the emission spectrum of the light source to the absorption spectrum $\alpha(\lambda)$ of the sensitizer, its absorption spectrum is measured by using Lambert-Beer's law

$$\alpha(\lambda) = \frac{\ln T}{d}$$

where T is the transmission of light through a sample of thickness d. The measurement of the absorption cross section $\sigma(\lambda)$ is performed using a solution of a sensitizer at a concentration c

$$\sigma(\lambda) = \frac{\ln T}{N \cdot c \cdot d}$$

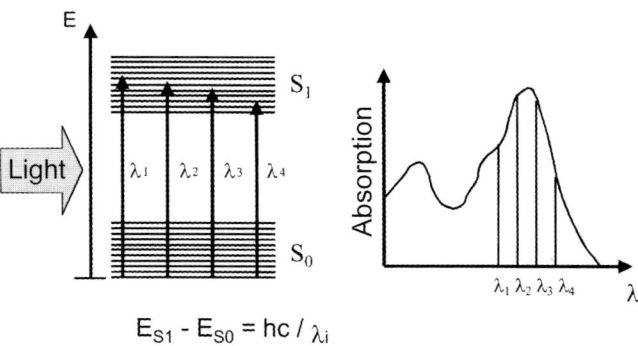

$$E_{S1} - E_{S0} = hc / \lambda_i$$

Figure 1. Due to the complex organic structure of the sensitizer molecules each electronic state shows a variety of vibrational states leading to a variety of optical transitions.

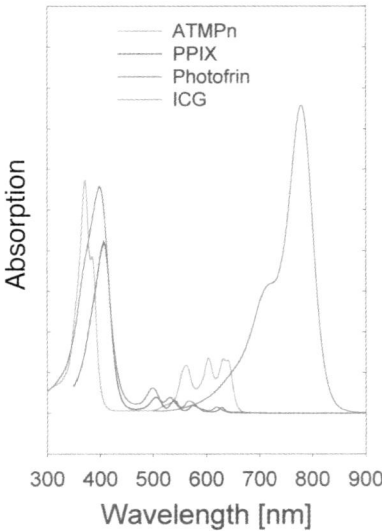

Figure 2. The absorption spectra of different sensitizers in dimethylsulfoxide (ATMPn, Photofrin®, PPIX) and water (Indocyanine green).

where N is Avogadro's constant $(6.022 \times 10^{23} \text{ mol}^{-1})$. In Fig. 2 the absorption spectra of some sensitizers are shown.

6.3.1 PDT

The absorption spectrum depends on the solvent used. For example, there is a shift of the absorption spectrum of Protoporphyrin IX by 6 nm using human plasma instead of ethanol as a solvent [4]. This effect is not crucial when using an incoherent light source due to its broadband emission which usually covers the shift of the absorption spectrum. However, the shift has to be considered when using a laser. The wavelength of the laser must be adjusted precisely to a maximal absorption of the sensitizer in organic tissue. Regarding Photofrin® there are at least four maxima of absorption in the long-wave range of the visible spectrum. Due to the broadband emission of a ILS the energy absorbed per cm^2 (E_{abs}) is given by

$$E_{abs} = \Delta t \cdot \int_\lambda \alpha(\lambda) \cdot S(\lambda) \cdot d\lambda$$

while $S(\lambda)$ is the emission spectrum of the light source and Δt the exposure time. In case of using a laser the spectral intensity is approximated by $S(\lambda) = S(\lambda_{em}) = I_L$, where I_L is the intensity of the laser at its wavelength λ_{em}. The energy absorbed per cm^2 is

$$E_{abs} = \Delta t \cdot \alpha(\lambda_{em}) \cdot I_L$$

Using either laser or ILS the calculation of the energy absorbed in the sensitizer is different and the amount of ROS might depend on the light source used. However, as shown by investigations *in vitro* [2] and *in vivo* [5] incoherent light sources and lasers are equivalent regarding the efficacy of PDT. Therefore, it is not necessary to calculate E_{abs}. The light dose applied for PDT is calculated by using the fluence on the skin surface.

6.3.2 Fluorescence diagnosis

Since the radiative decay of a sensitizer molecule usually starts from the first excited singlet state the fluorescence depends not on the wavelength of the excitation light. It is an essential prerequisite in the context of FD to separate the excitation light from the fluorescence within the lesion. 5-aminolevulenic acid induced protoporphyrin IX (PPIX) is frequently used for FD [6,7]. The fluorescence quantum yield of PPIX is low ($\phi_F \sim 0.18$) [8] as compared to fluorescent dyes such as rhodamines ($\phi_F \sim 0.98$) [9]. Since all porphyrins show a high absorption at 400 nm (Soret-band) light in the blue part of the visible spectrum is used for the excitation of fluorescence (see Fig. 2). By using optical filters the red fluorescence (at 640 nm) can be easily separated from the blue excitation light.

6.4 Light dose

6.4.1 PDT

Regarding the light dose in PDT the basic variables are the intensity, the exposure time and the fluence of the light. The fluence E_F (J/cm^2) is described by

$$E_F = \frac{P \cdot t}{A} = I_L \cdot t$$

where P (W) is the optical power and I_L (W/cm^2) is the intensity of the light source, t (s) is the exposure time and A (cm^2) the area irradiated.

The fluence and the intensity used is different for oncologic and non-oncologic indications. Basically, the fluence should be less than 150 J/cm^2 and the intensity less than 200 mW/cm^2 in order to avoid photothermal effects. After defining the fluence and the intensity the exposure time is calculated by

$$t = \frac{E_F}{I}$$

according to the intensity measured on the lesion using an especially suited device. The intensity of the light source measured on the skin surface should be as homogeneous as possible to provide equal therapeutic conditions in the area irradiated (Fig. 3).

6.4.2 Fluorescence diagnosis

Performing fluorescence diagnosis the intensity of light is usually less than 1 mW/cm^2 and the exposure time is as short as possible to avoid a fluence in the therapeutic range.

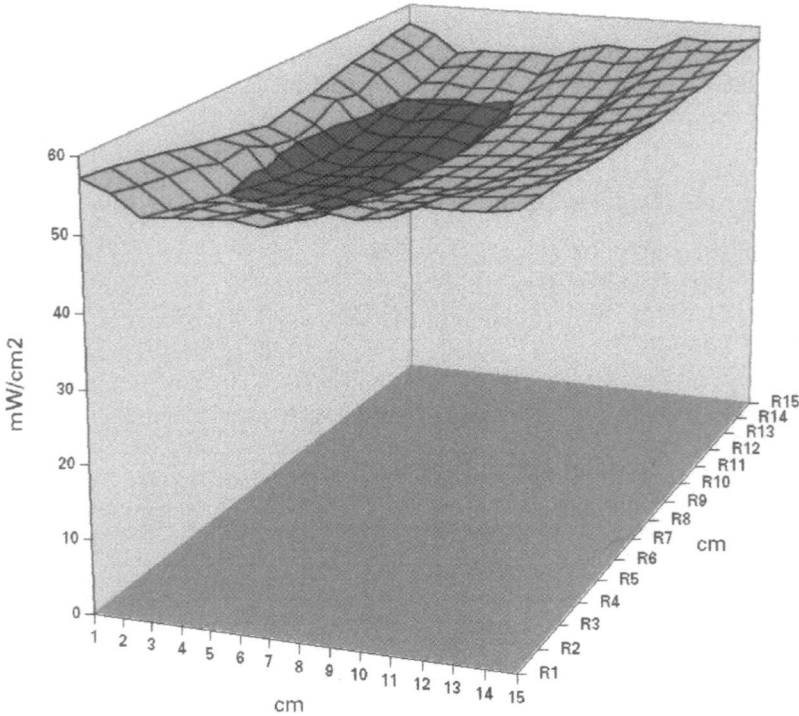

Figure 3. The intensity of the light source should be as homogeneous as possible measured on the skin surface to provide equal therapeutic conditions in the area irradiated.

Due to the high sensitivity of CCD-cameras detecting the fluorescence it is possible to keep the fluence of the excitation light at a low level.

6.5 Penetration of light into skin

The treatment of skin disorders using PDT requires the penetration of light into the skin in order to photoactivate the sensitizer either in the epidermis or in the dermis. However, due to scattering and absorption inside the skin the intensity of light decreases with penetration depth. At present, a rigorous theory is far from being available, because skin is irregularly shaped, has hair follicles and glands, is inhomogeneous, multilayered and has anisotropic physical properties [10]. Nevertheless, wavelength dependent coefficients of scattering and absorption can be determined for the epidermis and the dermis. The main chromophors absorbing light are proteins (UV), melanin (UV, visible) and hemoglobin (visible), the absorption spectra of melanin and hemoglobin are shown in Fig. 4.

6.5.1 PDT

Comprising the effects of scattering and absorption the penetration of light into skin increases with increasing wavelength. When using light between 400 and 600 nm the

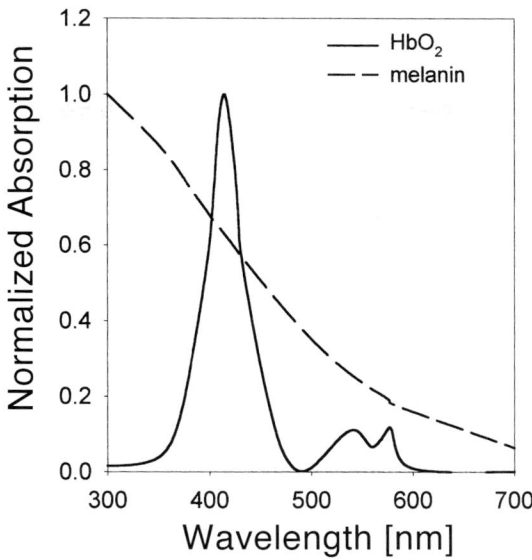

Figure 4. The absorption spectra of chromophors in the skin such as oxy-hemoglobin (HbO2) and melanin.

fluence inside the skin is clearly affected by the absorption of light in melanin and hemoglobin (Fig. 5). This absorption decreases significantly at wavelengths longer than 600 nm and consequently, the penetration of light increases and is maximal at 1000 nm.

However, a compromise must be found regarding the penetration depth of light and the absorption spectrum of the sensitizer used. With respect to porphyrins (e.g. Photofrin, PPIX) there is still an effective absorption of light for wavelengths above 600 nm. Consequently, lasers were adjusted to 630 nm (Photofrin) or 635 nm (PPIX) or incoherent light sources emit in the red part of the visible spectrum [4,11]. In case the absorption spectrum of the sensitizer is in the near infrared range (ICG, Lu-Tex) the conditions of photoactivation are optimal [12,13].

6.5.2 Fluorescence diagnosis

When using blue light for the excitation of fluorescence the penetration depth of the light into skin is limited to a few tenth of a millimeter, which is usually sufficient for the discovery of diseased skin. However, this fluorescence yields hardly any information from deeper structures inside the diseased tissue. In order to extend the range of fluorescence diagnosis in the skin sensitizers are necessary which absorb and emit light in the infrared range of the spectrum. When using laser light for excitation the excitation light and the fluorescence can be easily separated due to the narrow spectral width of the laser light.

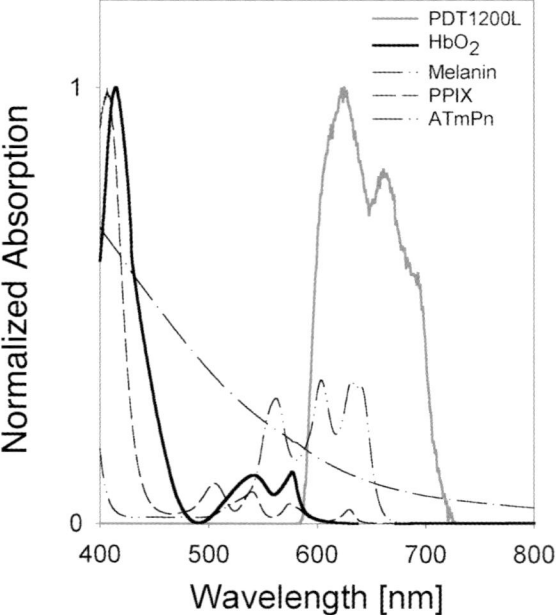

Figure 5. The absorption spectra of chromophors in the skin such as oxy-hemoglobin and melanin. For comparison the emission spectrum of a incoherent light source (PDT 1200) is added.

6.6 Laser systems

Laser is an acronym for "Light amplification by stimulated emission of radiation". In contrast to incoherent light sources (spontaneous emission) laser light is produced by the stimulated emission of atoms or molecules inside an active medium showing a population inversion. After absorption of energy there are more atoms or molecules in an upper level (upper laser level) than in the groundstate. The population inversion inside the active medium (solid state, gas, semiconductor, dye, etc.) is generated by a pump source (electrical discharge, flashlamp, electrical current etc.). Due to stimulated emission the laserlight is monochromatic, constant in phase (coherent) and direction of propagation. The amplification is enhanced by using an optical resonator.

The laser consists of a pump source providing the activation energy for the optical active medium and an optical resonator of at least two mirrors. The laser light is transmitted through one of the mirrors showing a reflection less than 100% (outcoupling mirror) and is available for any therapeutic or diagnostic procedure. Depending on the pump source (continuous, flash lamp) the laser show either continuous wave (cw) or pulsed emission.

As already mentioned above the wavelength of the light source used for either PDT or FD must be adjusted to the absorption spectrum of the sensitizer applied. The spectral width of the laser is determined by the spectral width of the laser transition in the active medium.

The light is delivered to the skin using optical fibers and a lens at the fiber tip to enable a homogeneous irradiation of the skin surface. The diameter of the area irradiated is adjustable by using different lenses or a lens system.

6.6.1 PDT

Dye laser. Using gas or solid state as active medium (except Ti : sapphire) the wavelength of the laser is usually fixed to sharp spectral lines (e.g. Argon: 488/514 nm). Therefore, these lasers are hardly used for PDT. In contrast, laser dyes show a broadband emission. When using a dye solution as active medium in the laser the emission wavelength is tunable up to 50 nm for one dye. Dye lasers are pumped by argon lasers or flash lamps yielding cw or pulsed emission, respectively. The pulse duration of the laser seems to play no major role regarding the efficacy of PDT [5]. However, cw dye lasers are preferably used for PDT providing an optical power of up to 7 watts which is sufficient to irradiate an area of about 45 cm^2 ($I_L = 150$ mW/cm^2) and 140 cm^2 ($I_L = 50$ cm^2). The disadvantages of the dye laser system are the difficult running of the system and the high purchase costs which prevents the spread of such a system.

Diode laser. The active medium of a diode laser is a semiconductor (Fig. 6) and the emission wavelength depends on the bandgap of the material used to form the laser structure (Table 1). For wavelengths shorter than 1 μm gallium aluminium (GaAlAs) or gallium indium arsenic phosphide (GaInAsP) are frequently used as active layer [14]. The wavelength of a diode laser is tunable by using different materials for the semiconductor. However, the wavelength is fixed after finishing the production of the laser. The optical output of diode lasers decreases with decreasing wavelength from hundreds of watts in the infrared to about 3 watts at 630 nm. The latter leads to a limited size of the irradiation area. The purchase costs for diode lasers are slightly less as compared to dye laser systems. There are different companies providing lasers for PDT (Table 2).

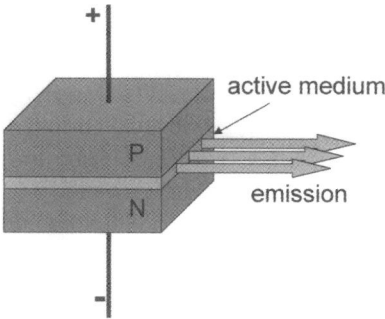

Figure 6. All diode lasers are based on light emitting diodes. The underlying structure is a p-n junction. Applying an electrical current to the diode light is emitted by an p-type active medium. The wavelength of the emission is correlated to the type of diode material used for the active layer.

Table 1. Materials for diode-laser sources (taken from [14])

Active medium	λ_{em}
Al-Ga-In-P	560–660
Ga-In-As-P	600–890
Al-Ga-As	670–890
Ga-In-As-P	900–1800

6.6.2 Fluorescence diagnosis

The argon laser exhibits some spectral lines in the ultraviolet or blue part of the light spectrum. In contrast to other solid state lasers the emission wavelength of the Ti: sapphire laser is adjustable ranging from about 800 nm to 1100 nm. When using frequency doubling light is generated from 400 nm to 550 nm. Both the argon laser and the short wave part of the frequency doubled Ti: sapphire laser can be applied for fluorescence excitation of porphyrins (e.g. PPIX, Photofrin) at 400 nm (Soret-band), however, these lasers are costly.

6.7 Non-coherent light sources

Light is emitted by either solid states (incandescent lamp, halogen bulb etc.) or discharge lamps at a different gas pressure (mercury, xenon, metal-halogen). The emission of these lamp systems $S(\lambda)$ is ranging from the ultraviolet to the infrared part of the spectrum (white light). When used for PDT or FD the major part of the infrared spectrum ($\lambda > 900$ nm) is blocked in any case to avoid thermal effects in the skin. The emission is additionally narrowed to the absorption spectrum of the sensitizer used by optical filters (Fig. 7). The skin surface is directly exposed to the light source or the light is delivered by fibers.

6.7.1 PDT

In view of the intensities of light used for PDT (e.g. 200 mW/cm^2) and the areas irradiated (e.g. 100 cm^2) an incoherent light source must provide an optical power of up

Table 2. Manufacturer of diode lasers

Manufacturer	
Asclepion Meditec AG Prüssingstr. 41 07739 Jena Germany	Coherent, Inc. Corporate Headquarters 5100 Patrick Henry Drive, Santa Clara, CA 95054
Carl Zeiss Carl Zeiss Str. 4-54 73447 Oberkochen, Germany	Diomed Limited, Cambridge Research Park, Ely Road, Cambridge CB5 9TE, UK

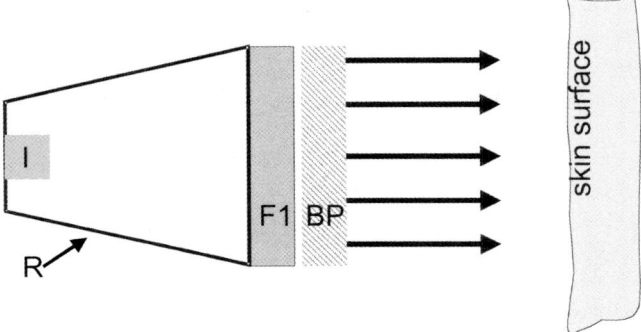

Figure 7. A typical setup of an incoherent light source. The infrared part of the emission of an illuminant (I) centered in a reflector (R) is blocked by an optical filter (F1). After that the spectrum of the emission is narrowed (band pass filter, BP) to the absorption spectrum of the sensitizers used. The reflector R provides a homogeneous irradiation of the skin surface.

to 20 Watts in a spectral range correlated to the absorption of the sensitizer used. The illuminant used [15] in the light source should exhibit a high optical output in the visible part of the spectrum and a low emission in the infrared and the UV (Table 3). The illuminant is frequently combined with a reflector and the whole is acting as a spotlight (e.g. Fig. 7). The intensity is easily adjusted by changing the distance of the light source and the skin surface (Fig. 8). In order to ensure the correct light dose the intensity must be measured at the skin surface prior to each treatment.

Recently, monolithic arrays of light emitting diodes (LED) are available using gallium arsenide diodes which are designed to emit diffused monochromatic light. The

Table 3. The optical output of different illuminants in the visible part of the spectrum, the infrared and the UV. The values are given in percent of the total optical output ranging from 100 nm to 10 μm and the loss of energy by thermal convection or conductivity

Illuminant	UV (280–380 nm)	Visible (380–780 nm)	IR (780–1400 nm)
Incandescent lamp 100 W	0	9	36
Mercury HTQ	11.2	15	25
Xenon CSX	3	15	20
Metal-halogen MSR	6	32	26

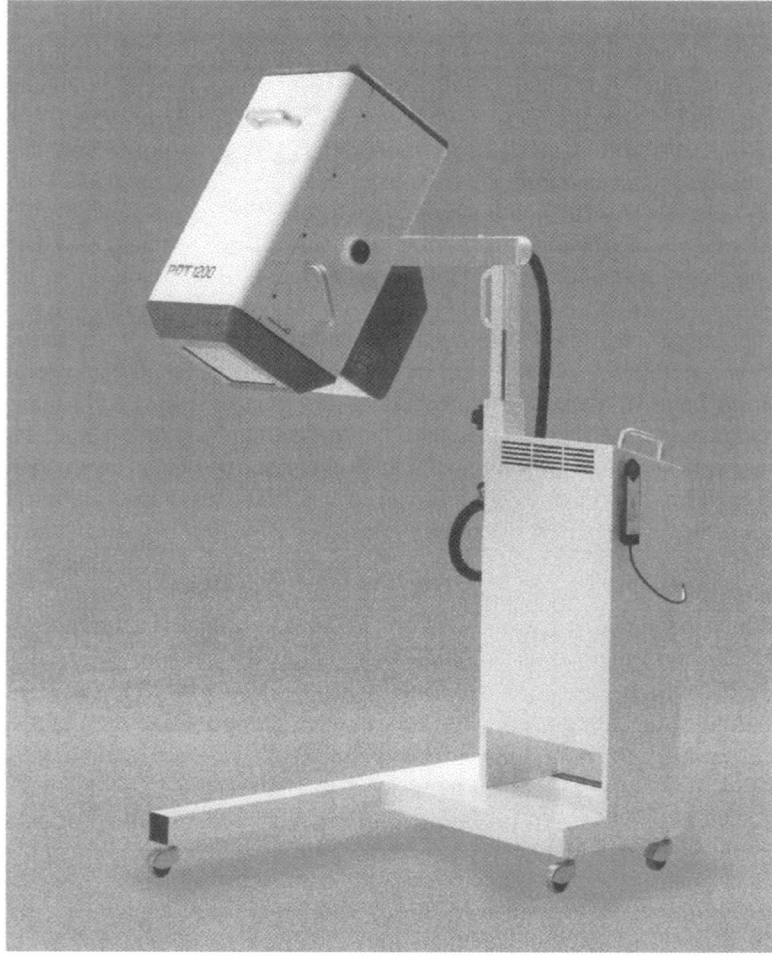

Figure 8. A typical incoherent light source used for PDT.

LED-based array has a bandwith of up to 30 nm [16]. There are different companies providing incoherent light sources for PDT (Table 4).

6.7.2 Fluorescence diagnosis

Due to the high absorption of porphyrins at 400 nm (Soret-band) and the good distinction of the red fluorescence generated usually blue light is used for FD. The light intensity necessary for FD is remarkably lower as compared to PDT. Due to a high emission at 400 nm usually xenon arc lamps are used (Fig. 9). The light is delivered using a fiber and lens system.

Table 4. Manufacturer of incoherent light sources for PDT and FD

Manufacturer	Product
Photocure ASA POBox 55, Montebello N-0310 Oslo, Norway	PDT CURElight
Saalmann Werrastr. 94 32049 Herford, Germany	PDT PDT-green light
Waldmann Lichttechnik Peter Henlein Str. 5 78056 Villingen-Schwenningen, Germany	PDT PDT 1200 L
Karl Storz GmbH & Co Mittelstr. 8 78532 Tuttlingen, Germany	FD D-light

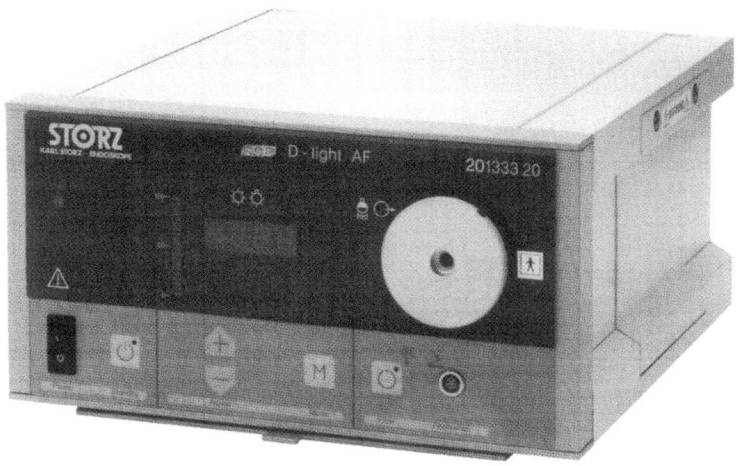

Figure 9. A typical incoherent light source used for fluorescence diagnosis. The light is delivered by a fiber and a lens system.

References

1. B.C. Wilson, M.S. Patterson (1986). The physics of photodynamic therapy. *Phys. Med. Biol.*, **4**, 327–360.
2. R.M. Szeimies, R. Hein, W. Bäumler, A. Heine, M. Landthaler (1994). A possible new incoherent lamp for photodynamic treatment of superficial skin lesions. *Acta. Derm. Venereol. (Stockh)*, **74**, 117–119.

3. J. Gahlen, J. Stern, H. Laubach, M. Pietschmann, C. Herfarth (1999). Improving diagnostic staging laparoscopy using intraperitoneal lavage of delta-aminolevulinic acid (ALA) for laparoscopic fluorescence diagnosis, *Surgery*, **126**, 469–473.

4. R.M. Szeimies, C. Abels, C. Fritsch, S. Karrer, P. Steinbach, W. Bäumler, G. Goerz, A.E. Goetz, M. Landthaler (1995). Wavelength dependency of photodynamic effects after sensitization with 5-aminolevulinic acid in vitro and in vivo. *J. Invest. Dermatol.*, **105**, 672–677.

5. S. Karrer, W. Bäumler, C. Abels, U. Hohenleutner, M. Landthaler, R.M. Szeimies (1999). Long-pulse dye laser for photodynamic therapy: investigations in vitro and in vivo. *Lasers Surg. Med.*, **25**, 51–59.

6. P. Hillemanns, H. Weingandt, H. Stepp, R. Baumgartner, W. Xiang, M. Korell (1999). Assessment of 5-aminolevulinic acid-induced porphyrin fluorescence in patients with peritoneal endometriosis. *Am. J. Obstet. Gynecol.*, **183**, 52–57.

7. G. Ackermann, C. Abels, W. Bäumler, S. Langer, M. Landthaler, E.W. Lang, R.M. Szeimies (1998). Simulations on the selectivity of 5-aminolevulinic acid-induced fluorescence in vivo. *J. Photochem. Photobiol. B: Biol*, **47**, 121–128.

8. G. Ackermann (2001). Fluoreszenzdiagnostik in der Dermatologie, Ph.D. thesis, University of Regensburg, Germany.

9. A. Penzkofer, W. Leupacher (1987). Fluorescence behaviour of highly concentrated Rhodamine 6G solutions, *J. Lumin.*, **37**, 61–72.

10. M. van Gemert, S. Jacques, H. Sterenborg, W. Star (1989). Skin optics. *IEEE Transactions on Biomedical Engineering*, **36**, 1146–1154.

11. P. Calzavara-Pinton, R.M. Szeimies, B. Ortel, C. Zane (1996). Photodynamic therapy with systemic administration of photosensitizers in dermatology. *J. Photochem. Photobiol. B.*, **36**, 225–231.

12. M.J. Hammer-Wilson, C.H. Sun, M. Ghahramanlou, M.W. Berns (1998). In vitro and in vivo comparison of argon-pumped and diode lasers for photodynamic therapy using second-generation photosensitizers. *Lasers Surg. Med.*, **23**, 274–280.

13. W. Bäumler, C. Abels, S. Karrer, T. Weiß, H. Messmann, M. Landthaler, R.M. Szeimies (1999). Photooxidative killing of human colonic cancer cells using indocyanine green and infrared light, *Br. J. Cancer*, **80**, 360–363.

14. E.J. Lerner (1998). Diode lasers light up disks, communications, and printers. *Laser Focus World*. **34**(1), 123–132.

15. Basics of optical radiation (1993). Philips Light, 14–15.

16. R. Ignatius, M. Ignatius (1998). Diode make cancer treatments cost effective. *Laser Focus World*. **34**(7), 139–143.

Part II:

Photosensitizers

Photodynamic Therapy and Fluorescence Diagnosis in Dermatology
P.-G. Calzavara-Pinton, R.-M. Szeimies and B. Ortel, editors.

Chapter 7

Photosensitizers – systemic sensitization

Cristina Zane, Giuseppe De Panfilis and Piergiacomo Calzavara-Pinton

Table of contents

Abstract

Photodynamic therapy (PDT) involves the sequential administration of photosensitizing drugs and light to patients. In Dermatology, photosensitizers have been used both topically and systemically, following intravenous injection or oral administration. The first and most widely used systemic photosensitizing drugs are Hematoporphyrin Derivative (HPD) and porfimer sodium (Photofrin®, QLT Phototherapeutics Inc., Vancouver, BC). They were found effective for the treatment of non-melanoma skin cancers but the prolonged generalised skin photosensitivity confines their use to a small number of selected patients with widespread and surgically inoperable lesions. However, new PS, which are more effective and clear rapidly from the skin with a short duration of photosensitivity, are now available. This group of second generation PS includes phthalocyanines, benzoporphyrin derivatives, tin etiopurpurin, chlorins, texaphyrins and others. At the same time, new, simple and efficient light systems have been developed. These progresses and results of ongoing well designed clinical trials argue in favour of further development of PDT with systemic delivery of the drug as a viable therapeutic modality in the near future. Finally, better understanding of the mechanisms underlying PDT-induced apoptosis and modulation of immune responses might open new exciting perspectives of systemic PDT for several new therapeutic applications, namely inflammatory and infectious skin diseases.

7.1 Introduction

Since the 1970s, photodynamic therapy (PDT) with systemic delivery of hematoporphyrin-derived photosensitizers (PS), i.e. Hematoporphirin Derivative (HPD) or Photofrin® (QLT Phototherapeutics Inc., Vancouver, BC), has been used successfully for the treatment of non-melanoma skin cancer (NMSC) and skin metastasis of breast carcinoma. Unfortunately, prolonged photosensitivity confines its use to very selected patients in highly specialized centers. However, "second generation" PS with high efficiency and mild and transient adverse effects are now under clinical investigation triggering a renewed and growing interest in systemic PDT as an effective and safe treatment modality of skin cancers.

In addition, the number of its possible clinical indications expanded beside oncology and now encompasses several inflammatory [1] and infective [2] skin diseases, particularly if widespread body areas are affected. These perspectives were a consequence of experimental findings demonstrating that PDT can promote apoptotic cell death and modulate several immune activities of the skin. Several PS can achieve these effects with combinations of light and drug doses much smaller than the doses needed for the induction of acute necrosis of tumor cells. The setting up of treatment regimens that can be fractionated in subsequent and repeated treatment sessions is allowed and the risk of adverse effects is minimized.

7.2 Systemic sensitization

A systemic PS should have high toxicity for the target (basic requirement for efficacy) and low or no toxicity for all healthy body organs.

Efficacy and toxicity are dependent upon the photophysical and photochemical properties of the PS. The "ideal" PS should have chemical purity, high quantum yields, quick tissue accumulation, short half-life and rapid clearance from normal tissues, activation at long wavelengths with optimal tissue penetration, and lack of dark toxicity and mutagenesis. In addition, the features of uptake as well as intracellular localization influence deeply the biological effects of the PS.

7.2.1 Photochemistry

The first step of photosensitizating mechanism is the absorption of a light photon by the PS in a ground state. Following excitation, the PS is promoted to the extremely unstable singlet state. The singlet state either decays back to the ground state, resulting in the emission of light in form of fluorescence, or undergoes intersystem crossover to the more stable triplet excited state by electron spin conversion.

The quantum yield of singlet molecular oxygen (ϕ_Δ) formation is one of the most important determinant for the activity of the photosensitizing drug and is governed by both the lifetime and yield of the triplet of the PS (ϕ_T). The formation of singlet oxygen requires a minimum energy level of 22.5 kcal mol^{-1} for the triplet PS, i.e. the energy of the $^1\Delta_g$ state of oxygen, unless endoergonic energy transfer takes place. This energy corresponds to a wavelength of 1270 nm, as deduced from the luminescence decay of $^1O_2$ [3]. However, in selecting the longest wavelength absorption maximum of a dye, we must take into account the energy gap between the lowest excited singlet and triplet states [3].

However, ϕ_Δ is not an exact measure of in vivo photodynamic efficiency [4]. The lack of correlation was regarded as being partly related to different degrees of aggregation, photodegradation, cellular uptake and binding to specific subcellular sites of the PS as well as to a different pH at the respective intracellular site [5]. In addition, type I photosensitization pathways, involving the generation of radical species via electron transfer between the photo-excited PS and suitable substrates, cannot be ruled out.

Finally, additional information on the photosensitizing activity of a compound may come from two additional indexes: the relative quantum yield (ϕ_A) for inactivation of cells per absorbed quantum of light and the relative quantum yield (ϕ_F) for inactivation of cells per absorbed quantum of emitted light from the PS.

7.2.2 Pharmacokinetics

The delay between drug delivery and its maximal accumulation within the target tissue determines the time point of light application and varies considerably between different types of cancer and different PS. If it is short, the treatment can be carried out on the same day as drug infusion and reduces the risk of accidental burns before treatment.

Short half-life and rapid clearance from normal tissues are needed in order to reduce the severity and duration of photosensitivity of healthy skin.

7.2.3 Localization in the target tissues.

The past 20 years have seen considerable advances in our understanding of the mechanisms by which PDT leads to the ablation of tumors largely sparing normal surrounding tissues. Selectivity has been attributed to two underlying pharmacokinetic characteristics: the tendency of several PS to be taken up to a greater extent and retained longer by hyperproliferative cell populations in comparison with the resting cells, and the cell characteristic accelerated uptake by neo-vascular endothelial cells. In addition, biological characteristics of cancer tissues, i.e. permeability and number of blood vessels, as well as poor lymphatic drainage, may contribute to the retention of the drug.

The most successful PS tested clinically have been found to exert both a direct tumoricidal effect as well as vascular damages leading to vascular occlusion [6]. In addition, release of histamine, prostaglandin D2 and platelet activating factor by mast cells and release of tumor necrosis factor by macrophages [7] can contribute significantly to PDT-induced tumor regression.

However, in spite of good tumor localizing properties of the PS, the photodynamic response of a tumor may be poor, hampered by a predominant localization in the stroma of the cancer. This lack of in vivo efficacy was reported for several hydrophilic dyes. In contrast, lipophilic compounds, such as Photofrin®, $TPPS_4$ and $AlPcS_4$, are believed to be mainly taken up by tumor cells [8,9].

Different approaches have been used to increase selectivity and extent of cellular uptake. Hydrophobic drugs, e.g. purpurins, phthalocyanines and naphthalocyanines, can be entrapped in the phospholipid bilayer of liposomes, solubilized in an oil emulsion prepared with the ethoxy castor oil [10,11], or bound covalently to polystirene microspheres [12]. A different approach takes advantage of the fact that malignant cells possess a significantly larger number of surface receptors for low-density lipoproteins (LDL), which are natural carriers of systemically injected hydrophobic dyes in the serum. LDL and the PS possibly embedded in the lipid moiety of the protein are taken into malignant cells through a receptor-mediated endocytic process. The recognition of the LDL molecule by the receptor involves some amino acid residues of apoprotein E, and hence there is no interference from the lipid-bound dye. The PS taken into the cell via an endocytic process is released inside the tumour cell and, on photoecitation, mainly destroys the cell membranes. These findings led to different strategies, including precomplexation of the PS with LDL [13], induction of the expression of LDL receptors [14] as well as direct PS targeting using covalently bound LDL as carrier [13].

Finally, the selectivity of tumor targeting can be improved by coupling covalently both porphyrins and chlorins with monoclonal antibodies directed against antigens specifically present at the surface of neoplastic cells. Antibody-bound PS fully retain their photophysical properties, including a good efficiency of $^1O_2$ generation. This technique allowed reducing the amount of PS needed for the achievement of a therapeutic effect and reduced skin phototoxicity [15,16]. However additional investigations are needed in order to enhance the number of PS molecules that can be associated with the antibody without impairing its ability to interact with the antigens in the malignant cells. In addition, since the antibody (and the bound PS) will remain outside the cell, we must assess the efficiency with which $^1O_2$ generated in close proximity, but externally, to the cell can attack endocellular targets [17,18].

7.2.4 Subcellular target sites

Lipophilicity/hydrophilicity, electric charge, non-specific protein binding and size of PS influence deeply the cellular uptake and the subcellular distribution of the PS [19].

The tumor uptake for lipophilic dyes, such as ATMPn [20] and porphyrins [21], usually enhances with increasing lipophilicity [22] and the low pH value of the interstitial fluid in cancers still increases their lipophilic character and uptake [23]. Following absorption, primary targets of lipophilic PS are mitochondria.

Photoactivation of PS located in the mitochondria inhibits ATP production and damage the respiratory chain complexes III, IV and I.

Most of the hydrophilic PS, such as $TPPS_n$ and $AlPcS_4$, which are taken up by pinocytosis, aggregates which cannot diffuse through the cell membrane and lipophilic PS which are administered via the LDL-endocytic pathway, localize preferentially in extranuclear granules, especially lysosomes [24].

Lysosomally located PS may cause the photochemical inactivation of cells by the release of lysosomal hydrolases. Alternatively, the PS may be released from the lysosomes during light exposure leading to the photochemical damage of other extralysosomal targets. In addition, cells may be damaged by a fraction of the PS located extralysosomally before light exposure.

Several PS may target endoplasmic reticulum and Golgi apparatus but the possible importance of photochemical damage to their functions for the inactivation of cells is still largely unclear. Finally, DNA damages have been reported and they cannot be excluded as a partial cause for the cell killing effect of PDT [25].

The pattern of PS localization in cell membranes and organelles gives rise to the type of damage since $^1O_2$, the main PDT induced cytotoxic agent, can diffuse only approximately 0.01–0.02 μm in cells during its very short lifetime ($\tau \approx 10^{-6}$–10^{-9} s) [26]. This is clearly visualized when cells loaded with either Photofrin® (extralysosomally localized) or phthalocyanines (lysosomally located) are given cytotoxic light exposures: in the former case the lysosomes remain intact whereas in the latter case they are permealized [27]. Furthermore, $^1O_2$ may hardly penetrate biological membranes and $^1O_2$ generated in the medium outside cells is inefficient in inactivating cells. However, the localization of the PS may be cell line dependent since porphyrins are found in lysosomes in some cell-lines (e.g. cultured human skin fibroblasts) and not in others (e.g. Hep-2 and NHIK) [28,29].

7.2.5 Light irradiation

The PS may be activated by all wavelengths corresponding to its absorption spectrum. However, the search for second generation PS is mainly focused compounds with high absorption extinction coefficients in the red and far-red regions because, as a general rule, longest wavelengths of the visible spectrum have optimal penetration depth into the mammalian tissues [30]. In this waveband, the penetration depth increases by a factor of about 2 going from 630 to 800 nm [31]. Therefore, an increase of the depth of the tumor necrosis is expected by using PS absorbing near 800 nm. In addition, far red absorbing PS should allow the treatment of highly pigmented tumors, such as melanoma, that are unaffected by treatment with HPD or Photofrin® and 630-nm light. An additional major advantage of PS absorbing in the far red and near infrared regions is the possibility of using the new simple and inexpensive diode lasers that are more reliable and less expensive than solid state lasers currently employed with HPD or Photofrin®.

7.2.6 Adverse effects

Skin phototoxicity is the only relevant adverse effect known so far. In addition, during light exposures, the patients may experience burning pain, stinging, or itching restricted to the irradiated area. These symptoms may require local anesthesia or premedication with benzodiazepines. Short-term toxicity to other body organs in the absence of light exposures is negligible for most PS. The occurrence of unexpected relevant long-term adverse effects, including dark toxicity and mutagenesis, is usually ruled out very carefully because it could seriously impair future trials, particularly for non-oncology indications. Finally, chemical purity of the compound is needed in order to avoid unwanted toxicity by chemical impurities.

7.3 Photosensitizers for systemic PDT

7.3.1 Porphyrins

HPD was the first systematically studied PS for clinical PDT. It is prepared from hematoporphyrin and is a very complex mixture of several components. Approximately 50% of HPD contain readily identifiable components, e.g. hematoporphyrins and protoporphyrins, with a low photosensitizing activity in vivo. The remaining part can be separated by size exclusion gel chromatography. It is a mixture of oligomeric porphyrins that are linked primarily by ether bonds and some ester bonds and aggregates. The purified commercial compound, porfimer sodium (Photofrin®) is a lyophilized and concentrated form of monomeric (hematoporphyrin, protoporphyrin and mono-hydroxyethyl-vinyl-deuteroporphyrin) and oligomeric (dimer to hexamer derivatives of hematorphyrin units linked via ether or esters bonds) porphyrins. Both HPD and Photofrin® accumulate selectively and/or are retained longer by several types of tumor cells or tumor tissue in comparison to normal surrounding tissues. The ratios of porphyrin concentration between tumors and peritumoral tissues of different body organs is variable between 15:1 in the brain and 1:1 in the muscle or skin at 24–48 hours after injection of Photofrin®. Cellular uptake is strongly dependent upon hydrophobicity. Both HPD and Photofrin® localize in membrane structures, in agreement with the fact that their diffusion constant in cells is close to that of lipids [22]. The stability of Photofrin® in tissues is questionable. Ester bonds linking hematopor-phyrin units are susceptible to esterase-catalyzed hydrolysis in body fluids. In addition, the non-covalent aggregates can be split into monomers in subcellular loci having a low dielectric constant, such as lipid regions of the cytoplasmic, mithocondrial and lysosomal membranes. These processes would lead to a population of porphyrins characterized by widely different photosensitizing activity [32]. A single intravenous bolus push of Photofrin® or HPD at doses of 0.5 to 2.0 and 3.0 to 5.0 mg/Kg of body weight, respectively, leads to maximal tumor-to-normal cell concentration ratios after 24 to 72 hours. One or more irradiation of the target area is delivered in this interval [32].

Although absorption is stronger at shorter wavelengths, HPD or Photofrin® sensitized skin is usually irradiated by 100–200 J/cm^2 of red light ranging from 625 to 633 nm in order to obtain a deeper penetration into tissues. Photofrin® has a small molar extinction coefficient around 630 nm: e ~ 3200 M^{-1} cm^{-1} in a medium with 2% DMSO in serum for the monomer, slightly lower for the aggregates. The triplet quantum yield is 0.6 for the monomeric species and 0.2 for the aggregated derivatives. The singlet oxygen quantum yields are 0.3 for the monomeric species and 0.1 for the aggregated derivatives [32].

Normal skin may retain detectable amounts of porphyrins for several weeks and prolonged skin photosensitivity is the major adverse effect of both HPD and Photofrin®. In two large clinical trials enrolling a large number of patients receiving Photofrin®, 19/110 (17%) and 20/90 (22%), respectively, reported a sun-related skin reaction [33]. Inflammatory reactions range from mild redness to painful blistering and severe edema of the skin and require careful photoprotective measures during 4 to 6 weeks after the treatment. Because of this adverse effect, most dermatologists consider PDT with HPD or Photofrin® as a reasonable treatment option only for patients suffering from multiple NMSC such as in chronic arsenicism and Gorlin's Syndrome. In addition, it is useful for patients with extensive skin cancers that are not suitable for surgery or other treatments, including PDT with topical application of PS.

7.3.2 δ-Aminolevulinic acid (ALA)

δ-ALA is not a PS "per se". It is a pro-drug that can be metabolized to photosensitizing protoporphyrin IX (PpIX) in the metabolic pathway to heme. PpIX may accumulate intracellularly because of the limited capacity of ferrochelatase to metabolize it to heme by introducing an iron molecule [34].

ALA plasma concentration peaks at approximately 60 minutes following a single oral administration.

Microscopic examination of superficial, nodular and morpheiform basal cell carcinoma showed full thickness PpIX fluorescence after oral administration of single doses of 10, 20 or 40 mg/Kg ALA [35], although tissue levels were often unpredictable. The PpIX fluorescence peaks in tumors 1 to 3 hours after ALA ingestion [35].

Significant plasma levels of PpIX may be detected in patients after oral doses of ALA [36] and PpIX concentrations in plasma and tumors correlate with the intensity of tumor surface fluorescence that can monitored with spectrophotofluorimetric measurements.

PpIX is activated by wavelengths ranging from UVA to the red region (with a relative peak at 630–633 nm) with a maximum in the Soret band.

PDT with oral administration of ALA at doses ranging from 30 to 60 mg/kg of body weight [37,38,39] is under investigation for the management of gastrointestinal, lung and cerebral tumors as well as nodular and infiltrating NMSC and non-oncological applications such as psoriasis.

Adverse effects of systemic ALA administration are transient rises of serum hepatic enzyme levels, occasional nausea and vomiting during the first 12 hours and mild generalized skin phototoxicity lasting less than 48 hours.

7.3.3 Phthalocyanines (Pc)

Pc have a porphyrin-like chemical structure characterized by the condensation of benzene rings with pyrrole moieties. The presence of zinc (II) and silicon (IV) ions, both of which can give hexacoordination through d^2sp^3 hybrid orbitals, guarantees a satisfactory yield of $^1O_2$ generation [36]. Pc are manufactured as pure compounds. Tumor-to-tissue ratios are high and are reached 1 to 3 hours after intravenous administration. Their absorption in the red part of the spectrum ($\lambda_{max} \approx 675$ nm) is strong allowing deeper penetration into tissues by the activating light as compared with porphyrins. Low accumulation levels in normal skin and rapid drug elimination result in minimal skin photosensitivity and toxicity in the absence of light is negligible.

A zinc phthalocyanine (ZnPc) (QLT Inc, Vancouver, BC), with a very high molar extinction coefficient (260.000 $M^{-1}\,cm^{-1}$) and high triplet (0.67) as well as singlet oxygen (0.53) quantum yields in organic solvents, is currently under investigation for the treatment of NMSC [32].

7.3.4 Chlorin derivatives

Chlorins are a heterogeneous group of porphyrin- or chlorophyll-derived compounds. Chlorins differ from porphyrins by the partial hydrogenation of one pyrrole ring and exhibit an intense (extinction coefficient above $10^5\,M^{-1}\,cm^{-1}$) absorption band in the 650–700 nm [32]. This group includes several PS currently under investigation for clinical uses.

7.3.4.1 Benzoporphyrin derivative-monoacid ring A (BPD-MA)
BPD-MA (Verteporfin, QLT Inc, Vancouver, BC) is a semi-synthetic porphyrin derived from protoporphyrin with maximal photoactivation peaks at both 630 and 690 nm [40]. The molar extinction coefficient is $e = 13000\,M^{-1}\,cm^{-1}$ in phosphate buffer. It is an example of the class of PS with poor water solubility that can be formulated successfully, either in liposomes or emulsions.

Optimal photodynamic efficacy is achieved 30 to 150 minutes after intravenous delivery of BPD-MA at doses ranging from 0.15 to 0.5 mg/Kg of body weight [41]. BPD-MA was found effective for the treatment of macular degeneration, choroidal melanoma, atherosclerotic plaques and psoriasis. Sun exposures must be avoided for 4–7 days corresponding to the duration of persistent skin photosensitivity.

Several pre-clinical studies have also investigated the effects of PDT with BPD-MA delivered according to an innovative approach, called transdermal PDT. Indeed, BPD-MA in the circulation can be activated without eliciting skin inflammation provided that light exposure is carried out before the drug has distributed to the skin. This treatment can modify certain immune responses and was tested successfully in an adjuvant-induced autoimmune arthritis model in mice [42].

7.3.4.2 Mono-l-asparthyl-chlorin e6 (NPe6)
NPe_6 has an absorption peak at 654 nm and $e = 40000\,M^{-1}\,cm^{-1}$ in phosphate buffer with pH 7.4. NPe_6 enters into the cells via endocytosis and accumulates predominantly in lysosomes. It is delivered intravenously at doses ranging from 0.5 to 3.0 mg/Kg of body weight. The optimal interval before irradiation ranges from 4 to 6 hours but tumor response can be achieved up to 72 hours post injection [43]. Generalized skin photosensitivity peaks within 96 hours and lasts 1–4 weeks. Strict avoidance of bright

sunlight is needed for 4–5 days and then a careful self-administered weekly phototesting must be repeated until the reaction subsides.

7.3.4.3 Tin ethyl etiopurpurin (SnET$_2$)

SnET$_2$ (Puryltin®, Miravant Medical Technologies, CA) is a synthetic chlorin analogue with maximal excitation at $\lambda \sim 665$ nm with $e = 27600$ M^{-1} cm^{-1} in benzene as a medium. Light doses from 100 to 300 J/cm^2 are delivered 24–72 hours after the infusion of 0.8 to 1.6 mg/Kg of body weight. SnET$_2$ is cleared from the skin within a few days after the treatment and, at the lowest therapeutic doses, generalized photosensitivity is mild. Exposures to sunlight must be avoided for 48 hours. Phase I/II trials to date with SnET$_2$ include NMSC, breast adenocarcinomas metastatic to the chest wall, and cutaneous Kaposi's sarcoma in AIDS patients [43,44].

7.3.4.4 Texaphyrins

Texaphyrins are expanded ring porphyrin analogues containing a central lanthanide, which both stabilizes the ring structure and causes a substantial red shift in the absorbance spectrum. They are synthetic water-soluble compounds that concentrate preferentially in malignant tissues. The lack of significant persistent skin phototoxicity is the most outstanding characteristic of these compounds and sunlight exposures must be avoided only for 24 hours. A texaphyrin containing lutetium (Lu-Tex) (Lutrin®, Pharmacyclics Inc., Ca) is a highly fluorescent dye that possesses a lowest energy maximum at $\lambda_{max} = 732$ nm in the far-red portion of the visible spectrum, where blood and bodily tissues are most transparent. Being diamagnetic and containing a bona fide heavy atom (lutetium), it produces singlet oxygen in good quantum yield (between 10 and 70% depending on condition [45]) when irradiated at 732 nm. Lu-tex is administered systemically at doses ranging from 0.6 to 7.2 mg/Kg of body weight. Quick accumulation in neoplastic tissues allows irradiation as early as 2 to 4 hours after drug administration. Lu-tex has shown efficacy against metastatic tumors to the skin and subcutaneous tissues and recurrent breast cancer to the chest wall [46].

7.3.4.5 Mesotetrahydroxyphenyl chlorin (m-THPC)

m-THPC (Temoporfin, Foscan®; Scotia Pharmaceuticals, Stirling, UK) has an absorption peak at 652 nm with a molar extinction coefficient (e) of 22400 M^{-1} cm^{-1}. Although it is poorly soluble in water, it was originally formulated as a powder that had to be reconstituted in an aqueous solution. Afterwards, a solvent was added so that it is more soluble and less painful to administer.

However, interactions between the formulated drug and plasma and/or tissue components may play a role in its biodistribution.

The ideal irradiation period is between 4–8 days following injection of Foscan® at doses ranging from 0.1 to 0.15 mg/Kg of body weight. Skin is irradiated with 20 Joules/cm^2 of 652-nm laser light at an intensity of 100 mW/cm^2.

Foscan® was found highly effective for the treatment of primary NMSC of the head and neck [47]. Photosensitivity lasts up to 6 weeks but strict avoidance of sunlight is usually needed only for two weeks.

Photosensitivity reactions have been reported in 22 out of 957 (2.3%) healthy volunteers and patients exposed to Foscan® [48].

Recently, 6 patients developed partial thickness burns of the forearms after minimal exposure to light. Healing was slow with prominent scarring. This severe side effect was most likely caused by leakage of the drug from the vein at the time of infusion [49].

7.3.5 Other photosensitizers

Nine-acetoxy-2, 7,12,17-tetrakis-(β-methoxyethyl)-porphycene is a synthetic electronic isomer of porphine. It is an efficient generator of singlet oxygen, has high fluorescence yield, and show a 10-fold increase in light absorption at 630 nm compared with HPD [50]. Bacteriochlorins display a significant absorption around 780 nm with a high extinction coefficient $(150000 \text{ M}^{-1} \text{ cm}^{-1})$ and might be particularly useful for the PDT of pigmented tumor [32]. Further effective PS include, anthraquinones, cationic porphyrins, hypericin and its analogs, rhodamine 123 and xanthene dyes. They have shown interesting properties under experimental conditions but their relevance for PDT in Dermatology is yet to be evaluated.

7.4 Conclusions

A great number of clinical and experimental investigations have reported that HPD and Photofrin have great efficacy against NMSC. Unfortunately, although systemic toxicity on internal organs is negligible, skin phototoxicity is prolonged up to 2 months, limiting its use to selected skin conditions. In addition, although Photofrin® is at present the only systemic PS available for clinical use in several European countries as well as USA, Canada and Japan, dermatological diseases are not included among the registered therapeutic indications. However, trials with HPD and Photofrin established PDT as a promising clinical modality in Dermatology. In order to make systemic PDT a practicable, safe and effective therapy, ongoing experimental and clinical investigations are developing new synthetic, chemically pure PS with high quantum yields, absorption peaks in the range of 650–800 nm, and short serum half lives that cause limited generalized photosensitivity. Hopefully, these requirements appear to be largely met by second generation PS, such as BPD-MA, NPe6, Pc, Lu-Tex, BPD-MA and $SnEt_2$. We can assume that, when these compounds will be developed in clinical trials, systemic PDT will enter in the therapeutic armamentarium against different types of skin cancer and several inflammatory diseases.

References

1. J.D. Spikes, G. Jori (1987). Photodynamic therapy of tumors and other diseases using porphyrins. *Laser Med. Sci.*, **2**, 3–15.
2. Z. Malik, H. Ladan, B. Ehrenber, Y. Nitzan (1992). Bacterial and viral inactivation. In: B.W. Henderson, T.J. Dougherty (Eds), *Photodynamic therapy. Basic principles and applications* (pp. 97–113). Marcel Dekker Inc., New York.
3. V. Cuomo, G. Jori, B. Rihter, M.E. Kenney, M.A.J. Rodgers (1990). Liposome-derived Si(IV)-naphtalocyanine as a photodynamic sensitiser for experimental tumours: pharmacokinetic and phototherapeutic studies. *Br.J. Cancer*, **62**, 966–970.

4. S.B. Kimel, J. Tromberg, W.G. Roberts, M.W. Berns (1989). Synglet oxygen generation of porphyrins, chlorins, and phthalocyanines. *Photochem. Photobiol.*, **50**, 175–183.

5. H. Kostron, D. Bellnier, C.W. Lin, M.R. Schwartz, R. Martuza (1986). Distribution, retention, and phototoxicity of hematoporphyrin derivative in a rat glioma. Intraneoplastic versus intraperitoneal injection. *J. Neurosurg.*, **64**, 768–774.

6. J.G. Levy, M. Obochi (1996). New applications in photodynamic therapy: Introduction. *Photochem. Photobiol.*, **64**, 737–739.

7. S. Evans, W. Matthews, R. Perry, D. Fraker, J. Norton, H.I. Pass (1990). Effect of photodynamic therapy on tumor necrosis factor production by murine macrophages, *J. Natl. Cancer Inst.*, **82**, 34–39.

8. Q. Peng, J. Moan, G. Farrants, H.E. Danielsen, C. Rimington (1991). Localization of potent photosensitizers in human tumor LOX by means of laser scanning microscopy. *Cancer Lett.*, **58**, 17–27.

9. D. Kessel, P. Thompson, K. Saatio, K.D. Nantwi (1987). Tumor localization and photosensitization by sulfonat derivatives of tetraphenylporphine. *Photochem. Photobiol.*, **45**, 787–790.

10. E. Reddi, S. Cernuschi, R. Biolo, G. Jori (1990). Liposome – or LDL administered Zn (II) – phthalocyanine as a photodynamic agent for tumors. III. Effect of cholesterol on pharmacokinetic and phototherapeutic properties. *Lasers Med. Sci.*, **5**, 339–343.

11. E. Reddi, C. Zhou, R. Biolo, E. Menegaldo, G. Jori (1990). Liposoma – or LDL-administered Zn(II)-phthalocyanine as a photodynamic agent for tumors. I. Pharmacokinetic properties and phototherapeutic efficiency. *Br. J. Cancer*, **61**, 407–411.

12. S. Biade, J.C. Maziere, L. Mora, R. Santus, C. Maziere, M. Auclair, P.P. Moliere, L. Dubertret (1993). Lovostatin potentiates the photocytotoxic effect of photofrin II delivered to HT29 human colonic adenocarcinoma cells by low density lipoprotein. *Photochem. Photobiol.*, **57**, 371–375.

13. R. Bachor, C.R. Shea, R. Gillies, T. Hasan (1991). Photosensitized destruction of human bladder carcinoma cells terated with chlorin e6-conjugated microspheres. *Proc. Natl. Acad. Sci. USA*, **88**, 1580–1584.

14. U. Schmidt-Ehrfurt, W. Baumann, E. Gragoudas, T.J. Flotte, N.A. Michaud, R. Birngruber, T. Hasan (1994). Photodynamic therapy of experimental choroidal malanoma using lipoprotein-delivered benzoporphyrin. *Ophtlmology*, **101**, 89–99.

15. T. Hasan, A. Lin, D. Yarmush, D. Oseroff, M. Yarmush (1989). Monoclonal antiboby-chromophore conjugates as selective phototoxins. *J. Controlled Release*, **10**, 107–117.

16. D. Mew, C.K. Watt, G.H.N. Towers, J. Levy (1983). Photoimmunotherapy treatment of animal tumors with tumor specific monoclonal antibody-hematoporphyrin conjugates. *J. Immunol.*, **130**, 1473–1477.

17. A.M. Richter, B. Kelly, J. Chow, D.J. Lui, G.N.H. Towers, D. Dolphin, J.G. Levy (1987). Preliminary studies on a more effective phototoxic agent than hematoporphyrin. *J. Natl. Cancer Inst.*, **79**, 1327–1332.

18. T. Hasan, C.W. Lin, A. Lin (1989). Laser-induced selective cytotoxicity using monoclonal antibody-cromophore conjugates. *Prog. Clin. Biol. Res.*, **288**, 471–477.

19. Rück, H. Diddens (1996). Uptake and subcellular distribution of photosensitizing drugs in malignant cells. In: H. Hönigsmann, G. Jori, A.R. Young (Eds), *The fundamental bases of phototherapy*. OEMF, Milano.

20. R.M. Szeimies, S. Karrer, C. Abels, P. Steinbach, S. Fickweiler, H. Messmann, W. Bäumler, M. Landthaler (1996). 9-acetoxy-2,7,12,17-tetrakis-(methoxyethyl)-porphycene (ATMPn) a novel photosensitizer for photodynamic therapy: uptake kinetics and intracellular localization. *J. Photochem. Photobiol. B:Biol.*, **34**, 67–72.

21. K.W. Woodburn, N.J. Kaye, A.A. Reiss, D.R. Phillips (1992). Evaluationof porphyrin characteristics required for photodynamic therapy. *Photochem. Photobiol.*, **55**, 697–704.
22. J.R. Shulok, M.H. Wade, C.W. Liu (1990). Subcellular localization of hematoporphyrin derivative in bladder tumor cells in culture. *Photochem. Photobiol.*, **51**, 451–457.
23. R. Pottier, J.C. Kennedy (1990). The possible role of ionic species in selective biodistribution of photochemotherapeutic agents toward neoplastic tissue. *J. Photochem. Photobiol. B:Biol*, **8**, 1–16.
24. G.G. Miller, K. Brown, R.B. Moore, Z.J. Diwu, J. Liu, L. Huang, J.W. Lown, D.A. Begg, V. Chlumecky, J. Tulip, M.S. McPhee (1995). Uptake kinetics and intracellular localization of hypocrellin photosensitizers for photodynamic therapy: a confocal microscopy study. *Photochem. Photobiol.*, **61**, 632–638.
25. C, Prinzse, L.C. Penning, T.M.A.R. Dubbelmann, J. Van Steveninck (1992). Interaction of photodynamic tretament and either hyperthermia or ionizing radiation and of ionizing radiation and hyperthermia with respect to cell killing of L929 fibroblasts, chinese hamster ovary cells, and T24 human bladder cells. *Cancer Res.*, **52**, 117–120.
26. J. Moan, K. Berg (1991). The phtotodegradation of porphyrins in cells can be used to estimate the lifetime of singlet oxygen. *Photochem. Photobiol.*, **53**, 549–553.
27. J. Moan (1990). On the diffusion length of singlet oxygen in cells and tissues. *J. Photochem. Photobiol.*, **6**, 343–347.
28. P. Moliere, R. Santus, J.C. Maziere, M. Bazin, E. Kohen, K.M. Smith, L Dubertret (1991). Lysosomes as primary targets of photofrin II photosensitization in cultured human fibroblats: a kinetic, spectral and topographic investigation by microspectrofluorometry on single living cells. *J. Cell. Pharmacol.*, **2**, 143–151.
29. J. Moan, Q. Peng, J.F. Evensen, K. Berg, A. Western, C. Rimington (1987). Photosensitizing efficiencies, tumor and cellular uptake of different photosensitizing druugs relevant for photodynamic therapy of cancer. *Photochem. Photobiol.*, **46**, 713–721.
30. J.A. Parrish (1981). New concepts in therapeutic photomedicine, photochemistry and targeting and the therapeutic window. *J. Invest. Dermatol.*, **77**, 45–50.
31. B.C. Wilson, W.P. Jeeves, D.M. Lowe, G. Adam (1984). Light propagation in anima tissues in the wavelength range 375–825 nanometers. In: D.R. Doiron, C.J. Gomer (Eds), *Porphyrin Localization and treatment of tumors* (pp 115–132). R. Alan, Liss Inc., New York.
32. G. Jori (1992). Far-red-absorbing photosensitizers: their use in the photodynamic therapy of tumours. *J. Photochem. Photobiol. A:Chem.*, **62**, 371–378.
33. C.J. Lightdale, S.K. Heier, N.E. Marcon, J.S. McCaughan Jr, H. Gerdes, B.F. Overholt, M.V. Sivak Jr, G.V. Stiegmann, H.R. Nava (1995). Photodynamic therapy with porfimer sodium versus thermal ablation therapy with Nd:YAG laser for palliation of esophagealcancer: a multicenter randomized trial. *Gastrointest. Endosc.*, **42**(6), 507–512.
34. S. Sassa and A. Kappas (1981). Genetic metabolic, and biochemical aspects of the porphyrias. *Adv. Hum. Genet.*, **11**, 121–231.
35. W.D. Tope, E.V. Ross, N. Kollias, A. Martin, R. Gilles, R.R. Anderson (1998). Protoporphyrin IX fluorescence induced in basal cell carcinoma by oral delta-aminolevulinic acid. *Photochem. Photobiol.*, **67**, 249–255.
36. J. Webber, D. Kessel, D. Fromm (1997). On line fluorescence of human tissues after oral administration of 5- aminolevulinic acid. *J. Photochem. Photobiol. B:Biol*, **38**, 209–214.
37. C. Abels, G.E.H. Kuhnle, A.E. Goetz (1995). Photodynamic diagnosis and therapy of tumors with 5 aminolevulinic acid (ALA). Kinetics and efficacy of ALA induced porphyrins. *Arch. Dermatol. Res.*, **287**, 355.
38. L. Gossner, R. Sroka, E.G. Hahn, C. Ell (1994). Orale gabe von 5-aminolavulinsaure zur Photodynamischen therapie von gastrointestinalen tumoren. *Lasermedizin*, **10**, 110.

39. E.W. Grant, C. Hopper, A.J. MacRobert, P.M. Speight, S.G. Bown (1993). Photodynamic therapy of orla cancer: photosensitization with systemic aminolaevulinic acid. *Lancet*, **342**, 147.

40. E.M. Waterfield, M.E. Renke, C.B. Smits, M.D. Gervais, R.D. Bower, M.S. Stonefield, J.G. Levy (1994). Wavelength-dependent effects of benzoporphyrin derivative monoacid ring A in vivo and in vitro. *Photochem. Photobiol.*, **60**, 383–387.

41. H. Lui (1994). An overview of clinical experience with benzoporphyrin derivative for photodynamic therapy. Proc. 5th biennial IPA meeting, Amelia Island, USA.

42. R.K. Chowdhary, L.G. Ratkay, A.J. Canaan (1994). The use of transcutaneous photodynamic therapy in the prevention of adjiuvant -enhanced arthritis in MRL/lpr mica. *Clin. Immunol. Immunopathol.*, **72**, 255–263.

43. D.R. Doiron, N.J. Razum, R.M. Trommer (1994). Clinical evaluation of tin-ethyl etiopurpurin: SnET2. Proc. 5th biennial IPA meeting, Amelia Island, USA.

44. G.M. Garbo (1996). Purpurins and benzochlorins as sensitizers for photodynamic therapy. *J. Photochem. Photobiol. B:Biol.*, **34**, 109–116.

45. J.L. Sessler, W.C. Dow, D. O'Connor, A. Harriman, G. Hemmi, T.D. Mody, R.A. Miller, F. Qing, S. Springs, K. Woodburn, S.W. Young (1997). Biomedical applications of lanthanide (III) texaphyrins as potential PDT sensitizers. *J. Alloys Compd.*, **249**, 146–152.

46. J.L. Sessler, R.A. Miller (2000). Texaphyrins. New drugs with diverse clinical applications in radiation and photodynamic therapy. *Biochem. Pharmacol.*, **59**, 733–739.

47. A.C. Knubler, T. Haase, C. Staff, B. Kahle, M. Rheinwalde, J. Muhling (1999). Photodynamic therapy of primary nonmelanomatous skin tumors of the head and neck. *Laser Surg. Med.*, **25**, 60–68.

48. R. Bryce (2000). Burns after photodynamic therapy. *Br. Med. J.*, **320**, 1731.

49. S. Hettiaratchy, J. Clarke (2000). Burns after photodynamic therapy. *Br. J. Derm.*, **320**, 1245.

50. P.F. Aramendia, R.W. Redmond, S. Nonell, W. Schuster, S.E. Braslavsky, K. Scaffner, E. Vogel (1986). The photophysical properties of porphycenes: potential photodynamic therapy agents. *Photochem. Photobiol.*, **44**, 555–559.

Photodynamic Therapy and Fluorescence Diagnosis in Dermatology
P.-G. Calzavara-Pinton, R.-M. Szeimies and B. Ortel, editors.

Chapter 8

Basic principles of 5-aminolevulinic acid-based photodynamic therapy

Kristian Berg

Table of contents

Abstract

Treatment of cells and tissues with 5-aminolevulinic acid and derivatives thereof induces accumulation of intermediates of the heme synthesis pathway. Protoporphyrin IX is preferentially accumulated in neoplastic lesions and induces cytotoxic effects upon exposure to light. The photocytotoxic effect of the accumulated photosensitizers is exerted mainly through formation of singlet oxygen. This photodynamic action has shown great promise for treatment of many oncological and non-oncological diseases. One of the main applications of ALA-based PDT is in dermatology and in particular for treatment of solar keratosis and basal cell carcinoma with a topically administered prodrug. Several other treatment options are presently under evaluation. In this chapter the basic principles of 5-aminolevulinic acid-based photodynamic therapy will be reviewed.

8.1 Introduction

5-Aminoleulinic acid (ALA) is a prodrug for application in photodynamic therapy (PDT) and diagnosis (PDD) that have attracted great attention the last 10 years. In 1987 Malik and Lugaci [1] were the first to document ALA-based photosensitization of cells in culture. Treatment with 5-ALA induces accumulation of porphyrins, mainly protoporphyrin (PpIX), which sensitize cells to photoinactivation. The accumulation of PpIX is higher in neoplastic lesions than in most normal tissues, which is of great importance for the clinical applicability of the treatment. Another feature, which favors ALA-based PDT, is the short half-life of the formed photosensitizers as compared to the use of chemically synthesized photosensitizers. In addition to the use of 5-ALA in PDT and PDD derivatives of ALA, mainly 5-ALA esters, are under evaluated for treatment of dermatological diseases and cancer diagnosis [2–4]. The Food and Drug Administration in USA has recently approved the treatment of actinic keratosis with ALA-PDT and promising results have been reported for the use of 5-ALA and 5-ALA esters in treatment and diagnosis of both dermatological and non-dermatological diseases [2,4–11]. The basic mechanisms involved in the use of ALA-based PDT will be described in this chapter, while the clinical application in diagnosis (Part III) and therapy (Part V) will be described elsewhere.

8.2 Regulation of heme synthesis and degradation

8.2.1 Heme synthesis

The tetrapyrrole class of compounds includes the metallopigments heme (the prosthetic group of proteins like hemoglobin, cytochromes, catalase, peroxidase and tryptophane pyrrolase), vitamin B_{12}, chlorophyll, siroheme (in nitrite and sulphite reductases) and factor F_{430} (cofactor of methyl CoM reductase). All these compounds are synthesized with uroporphyrinogen III as a common intermediate and modified to permit coordination of different metals at the ring centre, i.e. Fe in heme and siroheme, Mg in chlorophyll, Co in Vitamin B_{12} and Ni in factor F_{430}.

The initial step in the heme synthesis pathway is the formation of 5-aminolevulinic acid (5-ALA) (Fig. 1). In mammals and photosynthetic bacteria, 5-ALA is formed from glycine and succinyl-CoA by the pyridoxal phosphate-requiring enzyme ALA synthase (ALAS, succinyl-CoA:glycine C-succinyltransferase, EC 2.3.1.37). In higher plants and many prokaryotic cells, 5-ALA is formed through the C-5 pathway using glutamate-tRNA as a precursor. In vertebrates, there are two ALA synthase isoenzymes, a housekeeping and an erythroid-specific isoenzyme (see Section 2.1.2). The former has been mapped to chromosome 3 (3p21.1) [12], while the erythroid specific enzyme is located on the distal subregion of Xp11.21 in humans [13]. The enzyme has been purified, characterized, sequenced and the genomic organization of the human erythroid ALA synthase reported [14,15]. The enzyme has been located to the matrix side of the inner mitochondrial membrane [16], loosely associated with the membrane [17]. The enzyme has the main regulatory function of the pathway, which will be described in some detail in Section 8.2.2.

The next enzyme in the pathway, 5-ALA dehydratase (ALAD) is located in the cytosol and induces the condensation of 2 molecules of 5-ALA to yield porphobilinogen (PBG) with the elimination of two water molecules. This enzyme comprises 8 subunits and requires zinc for optimum activity in bacterias and mammals and magnesium in

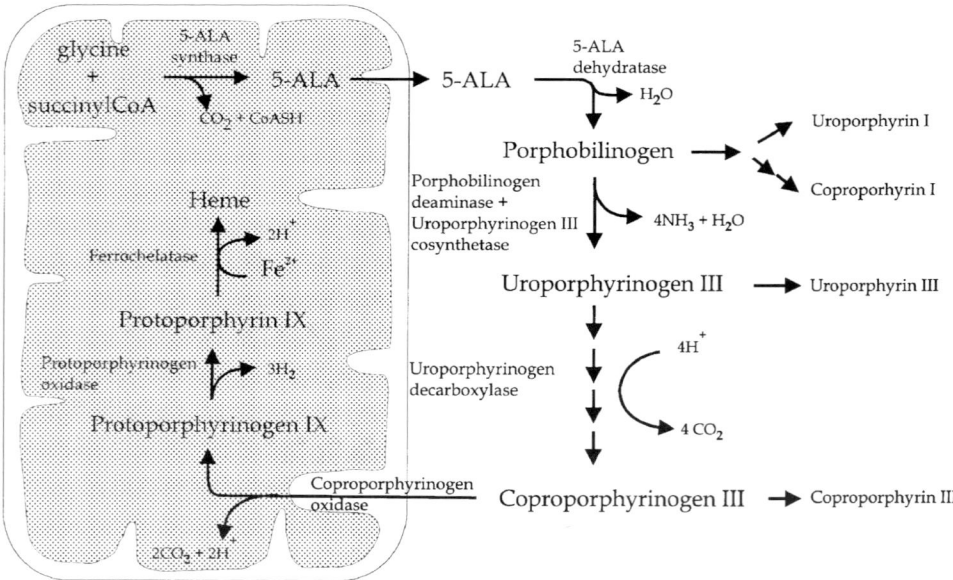

Figure 1. The heme biosynthetic pathway. The synthesis occurs as described in the text partly in the mitochondria as indicated on the left side of the figure and partly in the cytosol. Porphobilinogen, uroporphyrinogen and coproporphyrinogen may ondergo enzymatic or autooxidative conversion to photochemically active porphyrins as indicated on the right side of the figure.

plants. The zinc-binding enzyme is dependent on 4 or more oxygen-sensitive cysteines whereas aspartate occupies the same positions in the magnesium-binding enzyme. Zn^{2+} is necessary for maintaining the sulfhydryl groups in a reduced state [18]. Antioxidants such as ascorbic acid may be involved in the regulation of ALAD activity through redox reactions with sulfhydryl groups on ALAD [19]. Although PBG is subsequently used to form heme, a small fraction may be metabolized to other monopyrroles by porphobilinogen oxidase. ALAD is identical to the CF-2 component of the proteasome, which acts as an inhibitor of proteasome-induced protein degradation [20]. ALAD has also been reported to stimulate protein renaturation by heat shock protein 70 [21]. In patients with chronic renal failure, anemia and various abnormalities of the porphyrin metabolism have been reported. Red blood cells of patients undergoing hemodialysis have been shown to have diminished activity of ALAD, which may be due to a 56.2 kD inhibitory peptide isolated from plasma of patients with chronic uremia [22].

The concerted action of porphobilinogen deaminase (PBGD), coded for on chromosome 11 [23], and uroporphyrinogen III (co)syntase [24] condense in a head-to-tail manner 4 molecules of PBG and cyclize the tetrapyrrole chain to form uroporphyrinogen III. The PBGD alone forms only the symmetrical uroporphyrinogen I which will not proceed past the stage of coproporphyrinogen in the heme pathway (Fig. 1). Both enzymes are located in the cytosol of which PBGD deaminase is the rate limiting step.

A series of decarboxylations and oxidations have to take place before iron can be inserted into the tetrapyrrole ring. The first part of this process is performed in the cytosol by uroporphyrinogen decarboxylase (EC 4.1.37). This enzyme removes 4 acetic acid carboxyl groups from uroporphyrinogen to form the tetracarboxylic coproporphyrinogen. Ring D seems to be decarboxylated first, followed by ring A, B and C to form hepta-, hexa-, penta- and tetracarboxylic intermediates [25]. Uroporphyrinogen decarboxylase can use both the I and III series as substrates, although the reaction proceeds more rapidly with the III series [26]. The enzyme is coded for on chromosome 1, and the cDNA sequence and the structure of the gene have been determined [27,28]. Microsomes containing cytochrome P-450 activity catalyze uroporphyrinogen oxidation to uroporphyrin [29,30]. The hepatic P450 isoform CYP1A2 has been postulated to play a major role in uroporphyrinogen oxidation to uroporphyrin [31]. Administration of 5-ALA may cause uroporphyria in iron-treated mice [32].

At the stage of uroporphyrinogen III, pathways for formation of the different tetrapyrrole classes diverge. While the next step in formation of heme and chlorophyll is commited by uroporphyrinogen decarboxylase, siroheme, factor F_{430} and vitamin B_{12} is formed by a S-adenosyl methionine mediated C-methylation of uroporphyrinogen III to form precorrin-2 followed by as series of diverging reactions. Coproporphyrinogen III, to be used for heme synthesis, is now exposed to coproporphyrinogen oxidase which is situated in the intermembrane space of the mitochondria [33,34]. The enzyme decarboxylates and oxidize the propionic side chains in ring A and B to vinyl groups and protoporphyrinogen IX is formed. Two different types of coproporphyringen oxidase have been reported, one in aerobic organisms requiring oxygen for activity and another in anaerobic organisms [35]. In humans the enzyme is coded for on chromosome 9 and its cDNA sequence has been characterized [36]. The enzyme can only use the III and IV isomers of coproporphyrinogen as substrates [37].

The final step in the synthesis of PpIX is the oxidation of the tetrapyrrole ring by removal of 6 hydrogens from protoporphyrinogen IX, catalyzed by protoporphyrinogen oxidase (EC 1.3.3.4). The enzyme is embedded in the inner mitochondrial membrane with its active site on the matrix side of the membrane [38]. It is an oxygen-dependent enzyme with high substrate specificity [25]. Protoporphyrinogen IX may spontanously oxidize to protoporphyrin IX, but this is less likely to occur in the mitochondria where protoporphyrinogen IX is exposed to a rather anaerobic and reducing environment [25].

The tetrapyrrole structure is now ready for incorporation of iron, which is catalyzed by ferrochelatase (EC 4.99.1.1). Ferrochelatase, as protoporphyrin oxidase, is located in the inner mitochondrial membrane. Its substrate specificity is, however, low and mesoporphyrin and Zn are often used as substrates for measurements of ferrochelatase activity [39]. Thus, chelation of ferrous iron may induce incorporation of Zn in protoporphyrin in ALA-treated cells [40]. This will not be recognized when PpIX is isolated by strong acidic extractions methods since this removes Zn from PpIX [41]. It has recently been reported that ferrochelatase contains an [2Fe-2S] cluster at the carboxyl terminus, which may be of only structurally, or regulatory importance [42]. The source of ferrous iron is probably an intramitochondrial pool that is not associated with cytochromes or iron-sulfur proteins. Ferrochelatase-catalyzed heme synthesis in vitro is best accomplished in an anaerobic environment. This reduces autoxidation of Fe(II) to Fe(III) and, to some extent, prevents the breakdown of heme [43].

8.2.2 Regulation of the heme synthesis pathway

All the enzymes in the heme pathway act irreversibly. The pathway is partly regulated by substrate availability and feedback inhibition of ALAS. The concentrations of substrates and intermediates are usually far below the Michaelis constans of all the enzymes involved [44]. As an example ALA synthase has a low affinity for glycine, with a K_m of about 10 nM, while the glycine concentration in liver is even lower [25]. Of all the enzymes in the pathway ALAS has the lowest activity, followed by PBGD, while the other enzymes have much higher activities. In human erythroid cells ferrochelatase activity is also low, being only about 3-fold higher than ALAS [44].

A main regulatory step in the heme pathway is linked to ALAS activity (Fig. 2). Heme can inhibit the enzyme directly [45] as well as the transcription, translation and transport of the protein into mitochondria. The direct inhibition of the enzyme may however be of minor importance since the inhibition occurs only at around 10^{-5} M while the formation of ALAS is controlled at 10^{-7} M. A free heme pool at a concentration of about 10^{-7} M has been suggested to be involved in this regulation [46]. In liver 65% of the synthesized 5-ALA is used for production of cytochrome P-450 [47]. PpIX has been shown to inhibit cytochrome P-450 activity [48], while alcohols have been shown to stimulate ALAS and cytochrome P-450 [49]. Pyridine, a major constituent of tobacco smoke, stimulates expression of cytochrome P-450, ALAS and heme oxygenase (HO-1), indicating a coordinated regulation of heme metabolism and cytochrome P-450 expression [50].

The housekeeping ALAS (ALAS-N), expressed in all tissues, and the erythroid isoenzyme (ALAS-E) are regulated differently. ALAS-N transcription is regulated by 2 binding sites for nuclear respiratory factor 1 (NRF-1) in the promoter region, which is not found in the promoter of the ALAS-E gene [51]. The same sequence has been found in the promoters of mitochondiral respiratory chain proteins indicating a coordinated supply of heme and respiratory cytochromes. This is in accordance with the observed induction of ALAS through an NRF-1-dependent mechanism upon an increase in cellular metabolism [52]. At the transcriptional level ALAS-N mRNA stability has been shown to be reduced by heme (and Zn-mesoporphyrin) in conjunction with a labile protein [53–56]. ALAS is initially formed as an approximately 70 kD cytosolic precursor protein which is posttranslationally transported into the mitochondrial matrix. The aminoterminal end of preALAS is involved in the translocation process and is cleaved off in the mitochondrial matrix to form a 65 kD mature enzyme [57,58]. The translocation has been shown to be regulated, i.e. inhibited, by heme both in vivo and in vitro [59,60]. A conserved motif, termed the heme regulatory motif, in preALAS-E has been shown to bind hemin [61]. This binding was shown to inhibit translocation of

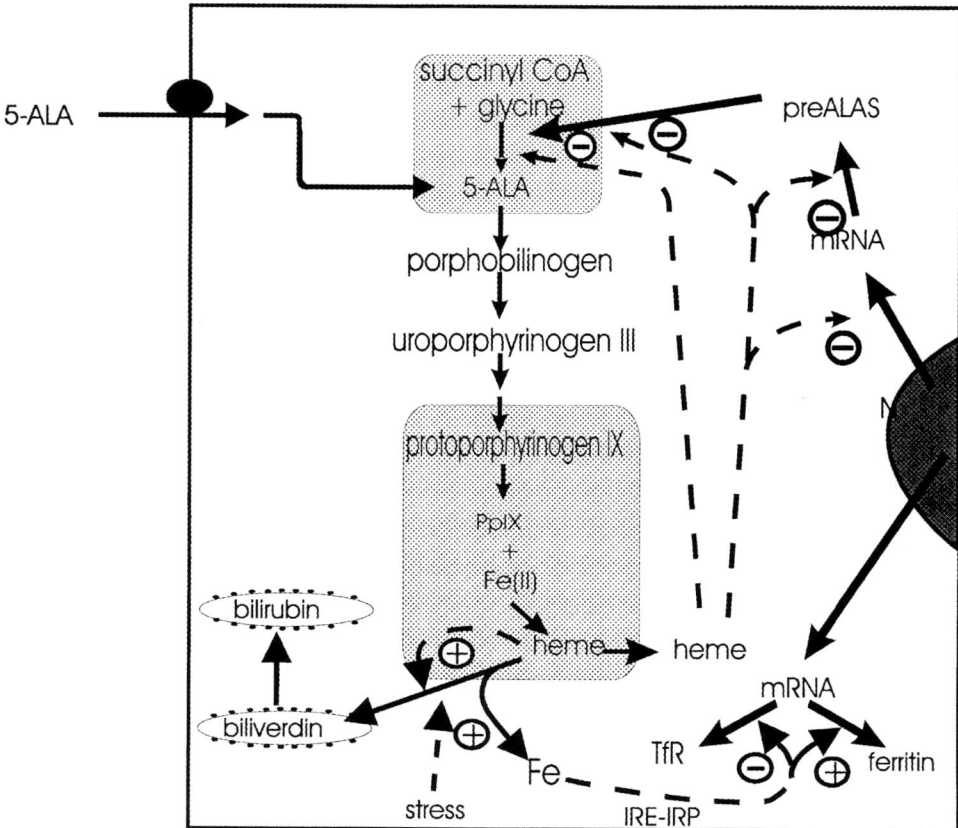

Figure 2. Regulation of heme synthesis. The synthesis steps are indicated by solid arrows and the regulatory steps by dashed arrows and a + or − sign depending on the effect of the regulatory molecule. Mitochondria are indicated as light grey boxes and the nucleus as a dark grey area. Bilirubin and biliverdin are located in endoplasmic reticulum. For abbreviations, see text.

preALAS-E into the mitochondria. ALAS has a short life-time which is necessary for rapid changes in the ALAS activity [62].

In contrast to the regulation of ALAS-N activity, ALAS-E expression is not directly related to the intracellular heme concentration [63,64]. The erythroid gene seems to be regulated by iron at the translational level. The ALAS-E mRNA contains a sequence in the 5'untranslated region which forms a stable haipin structure that shares structural similarities with ferritin and transferrin iron-responsive elements (IRE) [61]. A protein, iron-regulatory protein (IRP), can bind IRE and shut down translation. Iron inhibits the interaction between IRE and IRP and thereby triggers translation of ALAS-E mRNA [65]. Ferrochelatase has been shown to bind an iron-responsive element in ALAS-E mRNA and suggested thereby to regulate heme biosynthesis in differentiating erythrocytes (see Section 2.3) [66]. c-Myc appears to repress the expression of ALAS-E in murine erythroleukemia cells [67].

There are indications that the heme synthesis pathway is hormonally controlled in hormonally responsive tissues ([68,69] and references therein). 5-aminolevulinic acid treatment has also been shown to increase PBGD activity in R3230AC rat mammary adenocarcinoma [70].

8.2.3 Regulation of iron metabolism

Because iron is a substrate in the formation of heme, the regulation of iron metabolism has an impact on the accumulation of PpIX and other porphyrin intermediates (Figs. 1 and 2).

Iron regulates the expression of ferritin, transferrin receptor (TfR) and ALAS-E posttranscriptionally [71]. When intracellular iron levels are low, TfR biosynthesis increases, whereas the rate of mRNA translation of both ferritin and ALAS-E decreases. Intracellularly ferritin serves as a storage depot for iron and thereby as a defence mechanism against formation of free radicals through a Fenton-type reaction. Ferritin expression is regulated at the translational level. In iron-deficient cells ferritin mRNA is translationally inactive due to the binding of an IRP to the 5'untranslated region of the mRNA. Treatment with iron salts release IRP from IRE and ferritin mRNA is translated. Two different IRP proteins, IRP1 and IRP2, with different activities and regulations, exist [72–74]. IRP1 displays either aconitase or mRNA-binding activity, depending on the iron staus [75]. The activity of IRP1 is regulated by disassembly and reassembly of its [4Fe-4S] cluster with the disssembled form as the IRE-binding form [76]. The IRP1 protein level, in contrast to IRP2, does not change with the intracellular iron status [72]. When ferritin mRNA translation is stimulated by iron, synthesis of transferrin receptor (TfR) is inhibited. IRP binds also to TfR mRNA in a similar manner as for ferritin mRNA although on the 3'untranslated region. However, the IRP binding protects TfR mRNA from degradation. Once iron induces release of IRP from IRE TfR mRNA is degraded by nucleases. The regulation has been shown by Goessling and co-workers to occur through the rate of IRP degradation [77]. It is not clear whether iron itself or complexed with porphyrin, as heme or other compounds, binds to IRP [74,78]. Iron was found to be a poor stimulator of ferritin synthesis in the absence of porphyrin synthesis [77]. Heme and 5-ALA (when inducing accumulation of heme) reduce the number of transferrin receptors [78]. The level of IRP increases in cells growth-arrested by low serum, probably due to shortage in iron-loaded transferrin in the culture medium [79]. The oxidative properties of ALA, as described below, could have an important role as an intracellular regulator of iron metabolism by means of IRP1 activation [80]. The regulation of mRNA utilization to coordinate the synthesis of proteins for iron homeostasis and oxygen metabolism has recently been reviewed [81–85].

It should be noted that iron can be taken up into cells in a transferrin-independent manner [86].

8.2.4 Heme degradation

Heme oxygenase (HO), the rate-limiting enzyme in the degradation of heme, cleaves heme to form biliverdin (Fig. 1). Biliverdin is subsequently converted to bilirubin, which is then conjugated to glucuronic acid to increase the solubility of bilirubin. Biliverdin and solubilized bilirubin are both excretions products.

HO is a microsomal enzyme, which together with ALAS are considered to play the key roles in the regulation of heme turnover. Induction of HO activity is stimulated by heme as well as a large variety of compounds and oxidative stress [87]. Two different forms of HO, HO-1 and HO-2, have been identified in several mammalian species. The forms are encoded by different genes and there are large differences in their structure, regulation and tissue distribution [88]. HO-1 is termostable and resistant to inactivation by free radicals, and its activity is stimulated at the transcriptional level by several physical and chemical agents [89]. In contrast, HO-2 is easily inactivated by several stimuli and is not, except for a glucocorticoid response, induced by chemical stimuli. The activity of HO can be stimulated by $CoCl_2$ $SnCl_2$ and Co-protoporphyrin IX and inhibited by Sn protoporphyrin IX [78,90].

Although it has not been proven ALA-PDT is likely to stimulate heme oxygenase activity. This is supported by the observation that ALA-PDT and singlet oxygen stimulates heme oxygenase mRNA expression (G.Schnitzhofer et al., manuscript in preparation) [91]. It should also be noted that hematoporphyrin and Zn phthalocyanine induce heme oxygenase mRNA in the absence of light exposure [92]. Stimulation of heme oxygenase activity may have great impact on porphyrin synthesis and iron availability.

8.3 5-ALA stability

5-aminolevulinic acid is a relatively unstable compound, which may dimerise in solution [93]. Two molecules of 5-ALA react and form a pyrazin (2,5(β-carboxyethyl)pyrazine) as described in Fig. 3 [94,95]. Under anaerobic and alkaline conditions pseudo-PBG and 2,5(β-carboxyethyl)dihydropyrazine (DHPY) is formed in a ratio of about 2 : 1 and 2,5 (β-carboxyethyl)pyrazine is suggested to be formed from the dihydropyrazine [94,96]. The rate of DHPY formation is not influenced by the presence of oxygen, while oxygen stimulates the oxidation of DHPY to PY [96]. The amino group in its free base form is required for inducing degradation and the reaction rate therefore follows the relative amount of 5-ALA in its free base form. The pK for the amino group in 5-ALA has been shown to be between 8.05 and 8.9 by different authors [93,94,97]. The dimer has an absorption peak at 277 nm, which extends into the visible wavelength area and can be seen as a yellow to brown color depending on the amount of dimers formed. A 10% solution of 5-ALA becomes instantly light yellow at pH 7. No color is seen for 5 hours at pH 5.5, while it becomes brown after 24 h in solution [98]. At pH 2–2.5 5-ALA is a relatively stable compound in solution (> 30 days) [97–99] and no degradation is observed when 5-ALA is dissolved in strong acid and incubated for 6 weeks at 40°C [96]. The stability of 5-ALA in solution is a temperature dependent process that follows second order kinetics [96,97]. The iron chelator EDTA and antioxidants such as ascorbate and Na_2So_3, do not influence the degradation of 5-ALA [96,97].

8.4 Cellular uptake of ALA

In cells treated with 5-ALA accumulation of fluorescing porphyrins may be limited by the rate of 5-ALA transport across the plasma membrane. In cultured hepatocytes at optimal concentration of exogenously added 5-ALA for formation of porphyrins 2-propyl-2-isopropylacetamide stimulated uptake of 5-ALA without a subsequent

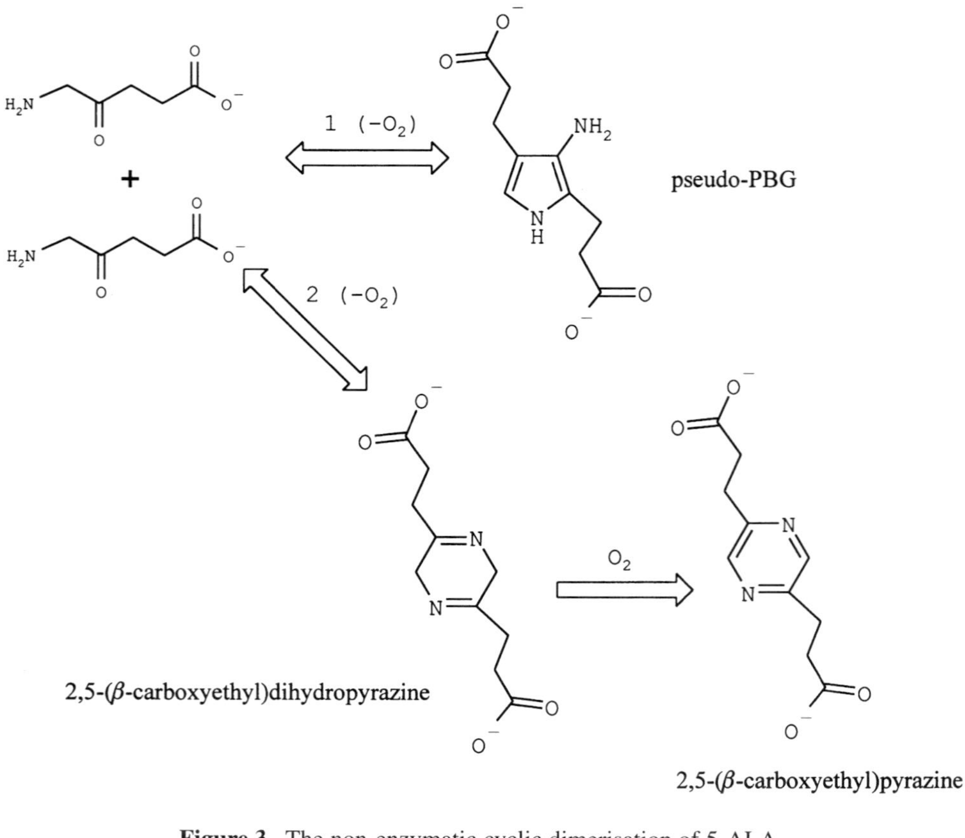

Figure 3. The non-enzymatic cyclic dimerisation of 5-ALA.

stimulation of porphyrin synthesis, indicating that at high concentrations of 5-ALA transport of 5-ALA across the plasma membrane is not a rate limiting step in the porphyrin synthesis [100]. To increase the rate of 5-ALA transport across membranes 5-ALA has been esterified with aliphatic alcohols by several authors [101–103]. Long chained (C_6-C_8) ALA-esters induces formation of PpIX at the maximum at a rate not more than 50% higher than that induced by ALA, depending on the cell line. However, the concentration needed to form similar amounts of PpIX is 30–150-fold lower with these esters than with unesterified 5-ALA [101,102]. These results indicate also that V_{max} for uptake of 5-ALA at least in some cell lines is sufficient to saturate the heme pathway while 5-ALA uptake may be rate limiting at suboptimal 5-ALA concentrations. This has been confirmed in adenocarcinoma cells [101]. Serum has been shown to attenuate 5-ALA uptake in mammalian epithelial cells (by 50% when 10% serum is added) [104].

5-aminolevulinic acid is taken up mainly by active transport mechanisms in adenocarcinoma, rabbit brain cerebral cortex and rat choroid plexus cells [105–108]. The uptake in neurons and glia cells are relatively low (~ 2–5 pmol/mg/min) and occur by passive diffusion [109], while uptake of 5-ALA has not been detected in synaptosomes and cerebral capillaries [110]. 5-aminolevulinic acid has on the other hand, been shown to be taken up in rat brain with some preference for implanted C6 gliomas [111]. In WiDr adenocarcinoma cells 5-ALAuptake occurs mainly through the Na^+/Cl^- – dependent β-amino acid and GABA transporters [107]. Accordingly, of all the compounds tested GABA was the most efficient (37%) inhibitor of 5-ALAuptake in rabbit brain cerebral cortex [105]. Uptake of 5-ALA through GABA transporters has also been documented in non-mammalian organisms [112–114]. In contrast, no uptake of 5-ALA through GABA transporters was found in rat choroid plexus as well as in CNCM-I-221 mammalian epithelial cells [104,108]. There are at least 5 different transporters (GAT-1, GAT-2, GAT-3, BGT-1 and TAUT) for β-amino acids and GABA, suggested to be grouped as a superfamily named system BETA [115]. Choroid plexus express GAT-2, but not GAT-1 and GAT-3 [116,117]. The GABA transporter GAT-1 expression is confined to the brain and retina, while GAT-3 and especially GAT-2 expression have been found in other tissues such as kidney, liver and heart [115]. BGT-1 expression has been observed in placenta, skeletal muscle, kidney, liver as well as brain while TAUT expression seems to be ubiquitous [115] including choroid plexus [118]. Uptake of 5-ALA in WiDr cells is therefore most likely through system BETA transporters that are different from GAT-1 and-2. It should be noted that GAT-3 has been detected in a layer of skin during late developmental stages [119], that GABA are formed in keratinocytes from putrescine [120] and that GABA has several functions outside the brain [121,122]. 5-ALA has also been shown to bind to GABA receptors [123], and due to the immunomodulatory role of GABA 5-ALA may also have a role in modulation of immunoresponses [124].

Transmembrane transport of 5-ALA has been shown by Daniel and coworkers to occur through the intestinal dipeptide transporter PEPT1 and the renal form PEPT2 expressed in oocytes and yeast cells [125]. PEPT2 was found by the same authors to be expressed in a variety of rabbit tissues and later Keep et al. found that 5-ALA transmembrane transport was through PEPT2 and probably two Na^+ and HCO_3^- dependent pathways [108]. The V_{max} (maximum rate of uptake) for 5-ALA uptake in choroid plexus and WiDr adenocarcinoma is of the same order while the K_m for the PEPT2 pathway in choroid plexus cells is about 40-fold lower than in the WiDr cells. This efficient and relatively high affinity uptake of 5-ALA in choroid plexus cells is not reflected in the in vivo ability of this organ to form PpIX [126], indicating possibly a low heme synthesis capacity or high conversion rate of PpIX to heme. C6 glioma cells in culture has been reported to be 20–30-fold less efficient in the uptake of 5-ALA than choroid plexus cells, but to accumulate efficiently PpIX [108,126]. 5-ALA, but not 5-ALA esters, enters pancreatic tumor cells through the PEPT1 transporter [127].

Amelanotic melanoma cells (A-Mel-3) grown in dorsal skin fold chambers of Syrian golden hamsters accumulate fluorescing porphyrins upon exposure to 5-ALA [128]. This porphyrin accumulation was shown to be substantially attenuated by a simultaneous application of as little as 20 μM glycine. This effect of glycine may be due to inhibition of 5-ALA uptake into the melanomas. A 50% inhibition of 5-ALA uptake was also observed in WiDr adenocarcinoma cells [107]. The indoles melatonin, serotonine and tryptophane reduce by 50–75% the accumulation of fluorescing porphyrins in a spontaneously immortalized human keratinocyte and a metastatic melanoma cell line, but not in a liver carcinoma cell line [129]. The cause for this influence of indoles is not clear, but may be due to influence on the transmembrane transport of 5-ALA or the porphyrin biosynthetic pathway. There is also evidence that 5-ALA may be taken up by transport systems for p-aminohippurate [130].

The uptake mechanisms for 5-ALA derivatives have so far only been studied in WiDr adenocarcinoma cells [131]. In this cell line ALA-methyl ester was found to be take up by active transport mainly through transporters of non-polar amino acids such as L-alanine, l-methionine, L-tryptophan and glycine, but not through system Gly, a specific transporter for glycine, or system BETA. ALA-hexyl ester is taken up by passive diffusion in WiDr cells (J. Howard et al., manuscript in preparation) which is in accordance with the observation that the ALA-hexyl ester induced formation of PpIX in the normal urothelium of porcine and human origin is not Na^+-dependent [132].

The cellular uptake mechanisms of 5-ALA and its derivatives may have several implications for the formation of photodynamically active porphyrins: (1) The increased efficacy of 5-ALA in PpIX formation by its esterification with long-chain aliphatic alcohols which transform the uptake mechanism from an active to a passive form strongly indicate that PpIX formation is rate limited by the transmembrane transport in neoplastic cells; (2) The composition of plasma membrane transporters varies highly from cell type to cell type [115], implicating a potential influence of the membrane transporter composisition on the specificity of the 5-ALA prodrug for different cell types; (3) The extracellular medium may contain inhibitors of 5-ALA or 5-ALA derivative uptake. The large (in the millimolar range) concentrations of amino acids such as taurine, glycine and alanine in tumors [133] and a tumor dependent and a likely tumor size dependent variations of such solutes should be expected to influence on the active transport mechanisms for 5-ALA and some derivatives thereof.

8.5 Cytotoxic effects of 5-ALA and PpIX

5-aminolevulinic acid has been shown to undergo iron-catalyzed aerobic formation of reactive oxygen species, like superoxide, hydrogen peroxide and hydroxyl radicals

[134–136]. Particularly, 5-ALA has been shown to induce damage to ALAD and glucose uptake in isolated rat cerebellum [137], Ca^{2+}-release from isolated mitochondria, uncoupling of State 4 respiration and mitochondrial swelling [138,139]. Ca^{2+} and thiol-groups seem to be involved in the mitochondrial swelling and uncoupling of respiration [140]. 5-ALA has also been shown to induce lipid peroxidation in cardiolipin-rich liposomes and cells in culture, to decrease intracellular glutathione (GSH) and increase GSSG levels, to induce single strand breaks in plasmid DNA, and to induce 8-hydroxy-2′-deoxyguanosine and 5-hydroxy-2′-deoxycytidine in DNA in vitro as well as in rat liver [136,141–145]. It is suggested that 5-ALA in its enolic form reacts with oxygen to form superoxide and 4,5-dioxovaleric acid (DOVA), the final oxidation product of ALA, as described in Fig. 4 [134,146]. The reactions are catalyzed by iron and 5-ALA will promote its own oxidation by release of iron from ferritin stimulated by the formed superoxide [147]. Ferritin has also been shown to stimulate

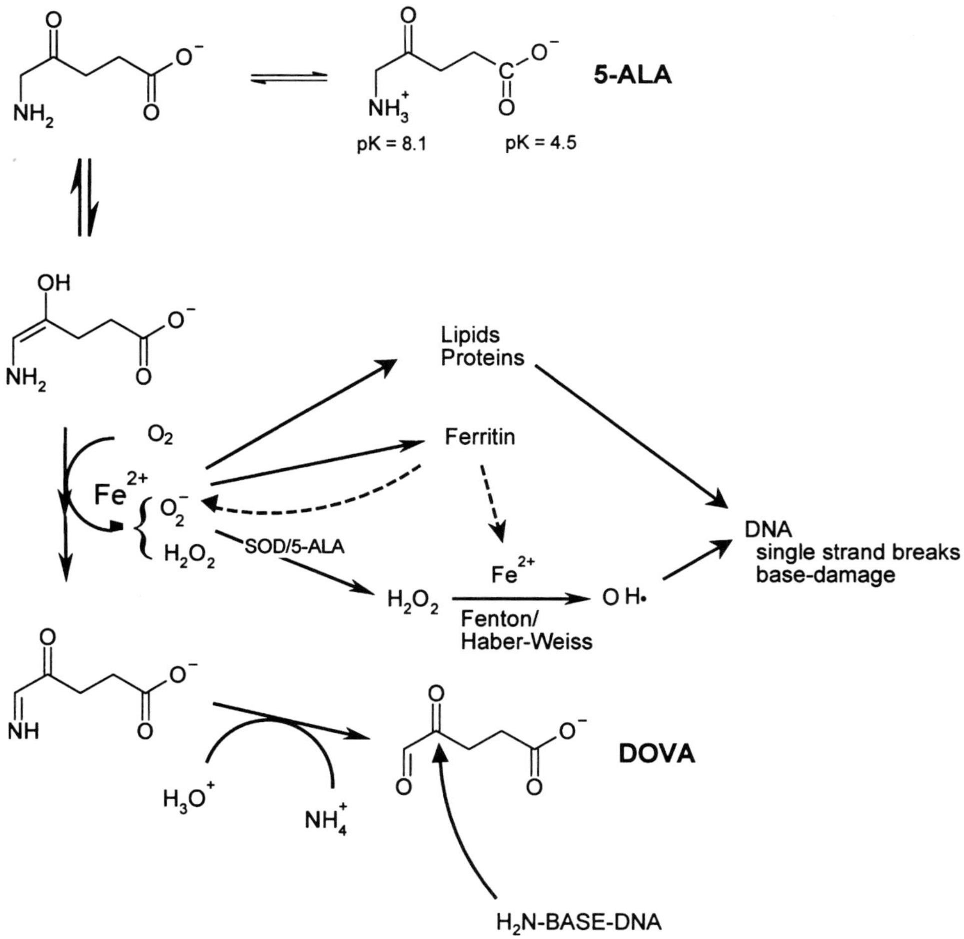

Figure 4. Proposed mechanisms of 5-ALA induced reactive oxygen species and damage to DNA.

ALA-induced cleavage of plasmid DNA and the enhancement of the formation of 8-oxo-7, 8-dihydro-2'-deoxyguanosine (8-oxodG) [144]. Hydrogen peroxide may be formed directly in an iron-catalyzed reaction with O_2, in an ALA-induced reaction with O2- or in a superoxide dismutase reaction (SOD or spontaneous) [134]. These effects may explain the cytotoxic effects of 5-ALA on cells in culture [101,148]. Several antioxidants, such as SOD, N-acetylcysteine, melatonin, catalase and dimethyl sulfoxide, have been shown to reduce the cytotoxic effects of 5-ALA [137,148–154].

The reactive oxygen species formed by 5-ALA are known carcinogens [155]. The final oxidation product of ALA, DOVA, is an aldehyde that has recently been shown to react primarily with the amine-group of guanine and form a Schiff base with the keto-group on DOVA [146,156]. Other aldehydes, such as acetaldehyde and malondialdehyde, which have been shown to form Schiff bases with DNA are known to be mutagenic [157,158]. Furthermore, 5-ALA or the products formed intracellularly induces an increase in the number of micronuclei and chromosomal abberations in primary hepatocytes [159]. Under pathological conditions, such as lead poisoning, and acute intermittent porphyria (AIP) and tyrosinosis the serum level and liver content of 5-5-ALA are raised [160]. There seems to be a correlation between 5-ALA accumulation and increase in hepatic cancers [161–163]. The plasma level of 5-ALA in AIP patients has been shown to be about 10 μM, which is 100-fold higher than normal plasma levels [160]. The plasma level of 5-5-ALA in patients treated with 40 mg/kg body weight 5-ALA orally was as high as 250 μM shortly after administration, but returned to baseline levels in less than 8 hours [164]. Similar plasma concentrations have been observed in rats treated orally or i.v. with 200 mg/kg ALA, but the baseline level was reach after 24 h [165,166]. Three hours after i.p. injection of ^{14}C-labeled 5-ALA (40 mg/kg body weight), its concentration raised 15 time in blood and is largerly taken up by liver (1300-fold), brain (10-fold) and heart (5-fold) [167]. Hamsters i.v. administered with 500 mg/kg body weight 5-ALA accumulated at the maximum 100 μM 5-ALA in the erythrocytes [168]. Wennberg and coworkers have used a microdialysis technique to measure the 5-ALA concentration of the fluid reaching the microdialysis catheter located 0.5 mm below the skin surface [169]. The 5-ALA concentration was found to increase to 3.1 mM 15 min after topical application of 20% 5-ALA dissolved in a cellulose gel to superficial basal cell carcinomas (BCC), while no measurable amounts was found in healthy skin. However, 5-ALA administered topically to small surface areas ($<72 \text{ cm}^2$) did not induce any increase in the red blood cell and urinary levels of 5-ALA and PpIX [170]. On the other hand, oxidation of liver DNA has been observed in rats repeatedly injected i.p. with 5-ALA (40 mg/kg body weight) [136]. Oxidative radical reactions have been observed in the liver, soleus muscle and brain of rats after 1–2 i.p. injections of ALA, while 5-ALA does not interfere with the plasma antioxidant capacity, does not promote oxidation of plasma elements nor bind to plasma proteins [166]. The lack of oxidative effects in the plasma in contrast to solid tissues was suggested to be due to the low content of free iron in the plasma. The antioxidants in plasma can even protect 5-ALA from oxidation [166]. In conclusion, there is a carcinogenic potential in the treatment with ALA, which probably requires repeated systemic treatments to be measurable [136].

The neuropathic effects of 5-ALA in AIP patients is also well recognized and may be due to binding to GABA receptors, decreasing glucose uptake and inhibition of Na^+,

K^+-ATPase activity [109,123,171], but is not documented to play any role in the ALA-based PDT. However, behavioral alterations have been observed in rats and mice, but not in hamster after systemic administration of 5-ALA [168,172–174].

As known from the inherited disorder erythropoietic protoporphyria PpIX may have toxic effects also in the dark. In these patients cholestatic hepatitis, fibrosis, cirrhosis and even terminal liver failure may occur. This dark toxicity is possibly caused by crystalline protoporphyrin deposits in the hepatobiliary structure and accumulation in the liver [25]. PpIX may also act as a pro-oxidant interfering with the cellular redox systems [175,176]. Similarly, heme may induce oxidative stress and TBARS formation [177]. It is not likely that the amounts of PpIX accumulating after topical application of 5-ALA induce any severe cytotoxic dark effects [178] and hemin and other porphyrins has been shown to protect against mouse skin carcinogenesis induced by 7,12-dimethylbenz[a]anthracene [179]. In contrast to the carcinogenic potential of 5-ALA itself, ALA-PDT has also been shown to have an anticarcinogenic potential [180].

8.6 ALA-based accumulation of porphyrins

8.6.1 Rate-limiting steps in photosensitizer accumulation

The rate of ALA-induced porphyrin synthesis has been shown to be higher in malignant and premalignant cells and tissues than in their normal counterparts [3,10,181–184]. This has been shown e.g. by comparing porphyrin synthesis in breast cancer tissues with the synthesis in the normal tissues [185]. The tumor tissues from 7 patients were found to synthesize on the average 20-fold (range 8–2410) more porphyrins than the normal breast tissue. The enzymatic activity of ALAD, PBGD and uroporphyrinogen decarboxylase was increased accordingly. Malignant urothelial cells have been found to accumulate more PpIX and to be more sensitive to photoinactivation than cells derived from normal urothelium [186,187]. Similar results have been obtained from cells of cervical origin [188], explants of BCC and keratoacanthomas as compared to normal skin [181], amelanotic melanoma after systemic administration [168] and other origins [189–193]. The difference between malignant and premalignant cells versus normal cells from the same origin in the accumulation of photosensitizers after treatment with 5-ALA range from 2-fold to more than 10-fold. PpIX synthesis in murine splenocytes treated with 5-ALA and 1,10-phenantroline was found to be increased 10-fold by treatment with the mitogenic lectin Concanavalin A [194]. Similarly, peripherial lymphocytes need activation in order to accumulate substantial amounts of PpIX [195,196]. Transfection of fibroblasts with several oncogens, e.g. c-myc, IGF-1 receptor, v-fos, v-raf, v-Ki-ras, v-abl or polyomavirus middle T antigen, triggers a several-fold increase in the maximum rate of ALA-induced PpIX accumulation [189], indicating a close correlation between the ability to generate PpIX from 5-ALA (5-ALA phenotype) and malignant transformation. It was also found in the same study a correlation between the 5-ALA phenotype and the ability of the cells to form colonies in an anchorage-independent manner [189]. In contrast, photochemical treatment with 5-ALA induced similar cell killing effects on normal human keratinocytes, spontaneously transformed human keratinocytes and on squamous carcinoma cells [197]. However, epidermal growth factor-α or interferon-γ enhanced the sensitivity of normal keratinocytes to ALA-induced photoinactivation. Accumulation of PpIX in 5-ALA treated cells does not seem to be dependent on transcriptional or translational activation of involved enzymes [104].

An important prognostic factor for the use of ALA-PDT could be the influence of cell differentiation on the porphyrin accumulation and photocytotoxicity. Ortel and coworkers have suggested on the basis of their studies with mouse keratinocytes that more differentiated tumors may be the better targets of ALA-PDT [198]. This is in accordance with the higher PpIX formation and cytotoxicity induced in well differentiated as compared to poorly differentiated urothelial transitional cell canrcinomas [186,187]. Similar conclusions may be drawn from studies of mammary ductal carcinomas [185]. In contrast, Steinbach and co-workers found no correlation between rate of PpIX formation and degree of differentiation although they found the normal counterpart to produce substantially lower quantities of PpIX [199]. Highly metastatic lymphomas have been found to produce more fluorescing porphyrins than a low metastatic counterpart [200]. The ability to accumulate PpIX was much lower (60-fold at 100 μg/ml) in a benign brain tumor, meningioma, cell line than in a glioma cell line [201]. It may be concluded that the changes in 5-ALA phenotype with changes in the state of differentiation are cell-type specific [202]. There is also a large variation in the generation of PpIX from 5-ALA in cells of the same histological origin, stage and grade [203]. It has been suggested that the mitochondrial content of cells is a good prognostic factor [204].

The regulation of the heme synthesis pathway in cells treated with 5-ALA or 5-ALA derivatives has been extensively studied. However, the rate-limiting step(s) in the formation of PpIX has still not been fully revealed. Yoshida ascites hepatoma cell lines have been shown to exhibit decreased ALAS activity and increased ALAD activity as compared to epithelial cell lines of normal rat liver origin [205]. However, the activities of these enzymes as well as PBGD and ferrochelatase and the K_m values for their respective substrates varied widely from one cell line to another, a finding suggesting that specific regulatory mechanisms for porphyrin metabolism might operate in each cell type.

The following summarize the current knowledge:

1. The rate of PpIX accumulation does not seem to correlate with the growth rate of the cells [187,190].
2. In most cell lines, including normal human cell lines, the rate of PpIX accumulation increases with increasing concentration of 5-ALA up to about 1 mM (range 0.3–1.7 mM) above which the rate of PpIX accumulation is independent of the 5-ALA concentration [187,206–211]. This rate limitation in PpIX accumulation is not caused by the uptake of 5-ALA over the plasma membrane since the K_m for the uptake of 5-ALA is much higher than 1 mM [107,211,212] and the maximum rate of PpIX accumulation with 5-ALA hexyl ester, penetrating the plasma membrane by passive diffusion, is similar to that of 5-ALA [101,102,132].
3. In most cell lines the main photosensitizer accumulated in ALA-treated cells is PpIX, indicating that the enzymes converting Uro III to PpIX are not rate limiting in the

pathway [101,194,200,207–209,213–217]. In some cases, however, like in HaCaT keratinocytes [218], Ehrlich acites carcinomas [219], normal and neoplastic breast tissues [220] and primary neural tissue cultures [221], other porphyrin intermediates were also observed. The accumulation of fluorescing porphyrins different from PpIX seems to be higher in normal cells than in cancereous cells [181,220,221]. Skin explants of normal skin and the squamous cell carcinoma-like keratoacanthomas, by some viewed as non-malignant, accumulate more coproporphyrin and uroporphyrin relative to PpIX than basal cell carcinoma explants [181]. It should be noted that the different extraction and analysis methods for porphyrins accumulated in biological samples may influence on the quantification of these compounds [222].

4. In most malignant cell lines PpIX is the predominant photosensitizer accumulating after treatment with 5-ALA or 5-ALA esters, indicating that the rate limiting steps in the synthesis pathway from 5-ALA to heme is one or more of the three first enzymes in the pathway before formation of porphyrins and the conversion of PpIX to heme. An obvious conclusion is that both an activation of one or more rate limiting steps in the heme pathway and a limited conversion of PpIX to heme is necessary in order to obtain a high rate of PpIX accumulation.

5. Ferrochelatase activity is high in non-erythroid cells relative to many of the other heme synthesis pathway enzymes [44] and higher [223,224] or lower [225,226] in malignant tissues than in their normal counterpart. Ferrochelatase activity may be sufficiently high to convert PpIX into heme under conditions of high intracellular concentrations of ALA. As exemplified in Table 1 the ferrochelatase activity is in most cases more that 10-fold higher than that of PBGD, the enzyme with lowest intracellular activity. In tissue samples from patients with Barrett's oesophagus and adenocarcinoma of the oesophagus as well as squamous epithelium and gastric cardia there is a week inverse relationship between ferrochelatase activity and PpIX accumulation ($r^2 = 0.516$) [223].

6. In differentiating Friend erythroleucemia cells heme synthesis does not seem to be limited by ALAS activity, but by the availability of iron to be incorporated into PpIX [227]. In these cells 5-ALA did not stimulate iron incorporations into heme when iron was given as Fe-transferrin, while salicylaldehyde isonicotinoyl hydrazone chelated iron (Fe-SIH) did. Fe-SIH increases the availability of free iron and indicates that uptake of iron through transferrin/transferrin receptors may limit the formation of heme. Even under such conditions heme is formed and reducing the content of available iron may enhance the accumulation of photosensitizers in ALA-treated cells [228]. This is in accordance with the good correlation between PpIX accumulation and transferrin-receptor expression in activated lymphocytes [195]. Activated, replicating lymphocytes have low intracellular iron levels and therefore an increased expression of transferrin-receptor (CD71/TfR). A similar correlation between PpIX accumulation and CD71 expression was not seen in cells of urothelial origin [187]. Some malignant cell lines may have elevated TfR expression that is unrelated to their intracellular iron stores [229]. Thus, CD71/TfR expression may not always be a good marker of ALA-PDT sensitive cells, while it has been suggested that IRP activation is a better marker since it more directly correlates with the level of the intracellular pool of available iron [230]. Combination of ALA-treatment with the iron chelators EDTA and desferrioxamine (DEF) has been shown to increase the

Table 1. Enzyme activities and rate of 5-ALA uptake in vitro and in vivo. The activities are converted to rates of PpIX equivalent formation

Cell line/Tissue	ALA	ALAD	PBGD	FC	PpIX	References
MCF-7[1]	36	1.34×10^6	6.4	978	15	(204)
EMT-6[1]	34	1.39×10^6	6.06	1074	7.8	(204)
R3230[1]	12.5	256×10^3	4.74	64	6.7	(204)
H-MESO-1[1]	16.3	140×10^3	1.7	65	3.1	(204)
Squamous epithelium – human oesophagus[2]			20/23	391/444	92*	(223,224)
Adenocarcinoma – human oesophagus[2]			39/55	532/582	112*	(223,224)
Rat oesophagus[2]			27–38	800–1040	30–43**	(237)
Rat liver[2]		7.2×10^3	650			(176)
			78.8	2470		(343)
Normal lymphocytes[2]		1350	13,2	610		(225)
Lymphomas[2]		700	39,4	353		(225)
CLL[§2]		638	62,1	396		(225)
Rat colon adenocarcinoma[2]		49,3		840		(343)

[1] pmol PpIX equivalents/105 cells × 3 h
[2] pmol PpIX equivalents/mg protein x h from tissues
* 60 mg/kg administered orally 6.7 h prior to PpIX measurements
** 200 mg/kg administered orally 3–7 h prior to PpIX measurements
§ chronic lynphocytic leukemia

rate of accumulation of photosensitizing porphyrins or to increase the sensitivity of the cells to photoinavtivation [104,206,207,213,215,231,232]. Iron chelators increase the accumulation of PpIX even in the absence of an exogenous iron source [213]. At similar concentrations DEF seems to be somewhat more efficient in accumulating PpIX than EDTA [213]. It has been suggested that the low availability of iron in tumor tissues is a major cause of PpIX accumulation [195,230]. In cells in culture the extracellular medium may have great influence on the free intracellular iron and therefore the rate of iron incorporation into PpIX. The iron-chelator induced increase in PpIX accumulation is higher in serum (and thereby iron) containing medium [187,206–208] and at suboptimal 5-ALA concentrations [213]. Surprisingly, DEF showed no potentiation of 5-ALA hexyl ester induced PpIX accumulation in porcine and human urothelium, although this was the case when 5-ALA was used as a prodrug [132].

7. Of the enzymes catalyzing the steps between 5-ALA and heme PBGD has the lowest endogenous activity [44,233] (Table 1). This is also supported by an accumulation of PBG in ALA-treated rat cerebellum and cerebral cortex particles in a concentration-dependent manner similar to that of porphyrin accumulation [211,212]. PBGD activity, although it seems to be somewhat increased in neoplastic tissues, may therefore be a rate-limiting step in the formation of porphyrins from ALA. There is also some indirect evidence for such a role of PBGD [70,223,224,234]. However, the important experiment where the PBGD activity was raised by transfection with a plasmid encoding PBGD did not influence on the rate of ALA-induced PpIX accumulation [235]. It s still striking that the ratio in PBGD activity between the 2 cell lines used for transfection as well as that of the R3230 cells reflected well the ratio between the maximum rate of porphyrin accumulation in these cell lines [70,210,235], although this correlation is not always that clear [187]. PpIX has been shown to bind and inhibit the activities of partly purified PBGD and ALAD in the absence and presence of light [236].

Gibson and coworkers have measured the rate of 5-ALA uptake, enzymatic activities of PBGD, ALAD and ferrochelatase as well as the rate of PpIX accumulation in 4 different cells lines [204]. These activities may be converted to PpIX equivalents as described in Table 1. The precausion should be taken that the enzymatic activities reflect the absolute activities in the cells since the measurements are performed on cell extracts. However, the results clearly indicate that even with a high K_m ALAD is not likely to be a rate-limiting step in the formation of PpIX. It should be noted that iron chelators, such as EDTA, may also bind Zn^{2+} and thereby inhibit ALAD, which contains labile and essential Zn^{2+}[18]. As seen from Table 1 the rate of 5-ALA uptake is 1.9–5.2 fold higher than what is observed accumulating as PpIX, while PBGD activity seems to be lower than what is needed for the observed rate of PpIX accumulation. Another striking observation is that the ferrochelatase activity was found to be 10–140 fold in access of the rate of PpIX accumulation, indicating that a low FC activity per se is not the cause of PpIX accumulation. The ferrochelatase activity intracellularly may as pointed out above be limited by the iron availability. These results are in accordance with enzymatic activities in oesophageal cancereous diseases [224]. The large accumulation of PBG as compared to 5-ALA and uroporphyrin, the low PBGD and high ferrochelatase activities suggests that PBGD influence on the rate of PpIX accumulation in cancereous tissues [224]. Similar conclusions may be drawn from studies with normal rat oesophagus [237] (Table 1). ALA-PDT could under specific conditions increase the PBGD:ferrochelatase activity, mainly due to a lowered ferrochelatase activity, without a concomitant increased rate of PpIX accumulation after the exposure to light [237], also indicating that the ferrochelatase activity is not influencing the rate of PpIX accumulation. The intracellular concentration of free iron may vary largely between cell lines and the incubation conditions.

An interesting, yet unexplained, phenomenon is the cell density dependent rate of porphyrin formation from 5-ALA [199,238,239]. This is probably not due to changes in cell cycle distribution since PpIX accumulation is almost independent of cell cycle distribution [187,240]. We found PpIX accumulation (per mg of cellular protein) to be linear with the amount of mg protein per dish in exponentially growing cells [238]. This indicates a cell-cell dependency of the metabolic activities [241–244]. Washbrook et al. [104] showed that the uptake of 5-ALA and the synthesis of porphyrins were more rapid in less dense cell cultures than in nearly confluent cultures. Steinbach and co-workers observed that PpIX formation could both increase and decrease in changing from a two to a three dimentional cell-cell interaction, depending on the cell line [199]. The cell density dependency on PpIX accumulation does not seem to occur in normal cells [104,187,199].

When cells are treated with 5-ALA in the presence of serum a large fraction of the PpIX formed will leave the cells [245,246]. The lipoproteins in the serum are not necessary for this loss of PpIX since albumin can efficiently extract PpIX from the cells [199].

The rate of ALA-induced PpIX formation is highly pH dependent with a maximum at pH 7.4 and less than 10% of that activity at pH 6.0 [247,248]. This may be of great importance since the pH of normal tissues, except the skin, is in the range from 7.1 to 7.4, while in tumors the pH ranges from 5.8 to 7.2 [249].

8.6.2 Accumulation of photosensitizers by means of 5-ALA derivatives

A large number of 5-ALA derivatives has been synthesized in order to develop compounds that penetrates the plasma membrane of the target cells and diffuse through epidermal layers more easily than 5-ALA itself (Table 2). The derivatives become more lipophilic depending on the groups added [102]. 5-ALA has been derivatized on both the carboxyl and the amino groups. Most of the results so far indicate that derivatization of the amino group on 5-ALA reduces its ability to form PpIX. This may be due to a reduced ability of these compounds to penetrate the plasma membrane or by a limited ability of the cells to remove the N-substituents. In contrast, many of the 5-ALA esters are more efficient in inducing porphyrin accumulation than 5-ALA itself. A prerequisite for improved PpIX accumulation seems to be to avoid branched chains on C1 of the ester group. This is seen by the higher rate of PpIX formation induced by 5-ALA cyclopentylmethyl ester as compared to 5-ALA cyclopentyl ester as well as by comparing 5-ALA isobutyl and isopentyl ester with 5-ALA isopropyl ester [127]. Short-chain alkanes are less efficient in inducing PpIX formation than the long-chain alkanes

Table 2. Overview of 5-ALA derivatives evaluated for the ability to form photosensitizing compounds

The efficacy in forming photosensitizing compounds is described semiquantitatively by comparing the concentrations of the derivative and 5-ALA needed to form the same amounts of photosensitizer: ++, a substantially lower concentration of the derivative is needed; +, the derivative is only moderately more efficient than 5-ALA in forming photosensitizers; 0, the derivative and 5-ALA is equally efficient; – the derivative is less effcient; –, the derivative forms no or only very little photosensitizer

Structure of 5-ALA-derivatives	Efficacy compared to 5-ALA	Experimental setup	References
5-ALA methyl→octyl ester	++ when n > 3–4 – when n < 3–4 +/++ when n = 1–4 and very short incubation period or mouse skin	Cells in culture Skin explants Mouse skin	(102,103, 127,250)
n=0,2-5 N-acetyl→N-heptyl 5-ALA	++ when n = 0 in cells in culture, – in mice skin 0 when n = 2 or 3 – when n = 4 – when n = 5 (rat) – when n = 5 (human)	Cells in culture, mice skin, skin explants	(103,250)
Carbobenzyloxyglycine-5-ALA ethyl or hexyl ester	– when n = 1 –/– when n = 5 (rat) – when n = 5 (human)	Skin explants	(250)
Carbobenzoyloxo-D-phenylalanyl-5-ALA-ethyl ester	+ for rat, 0/– for human	Skin explants	(250)
N-phthalylimido-5ALA-glucosamine-tetraacetate	0	Skin explants	(250)

Table 2. Continued

Structure of 5-ALA-derivatives	Efficacy compared to 5-ALA	Experimental setup	References
 N,N-dimethyl-5-ALA	–/– in rat – in human	Skin explants	(250)
 N-acetyl-5-ALA	–/– in rat – in human	Skin explants	(250)
 N-phthalimido-5-ALA	0	Skin explants	(250)
 n=0-7 R,S-ALA-2-(hydroxymethyl)tetrahydrofuralnyl ester	++	Cells in culture	(103,341)
 ALA-isopropyl ester	–	Cells in culture	(127)
 ALA isobutyl ester	++	Cells in culture	(127)

Table 2. Continued

Structure of 5-ALA-derivatives	Efficacy compared to 5-ALA	Experimental setup	References
ALA isopentyl ester	++	Cells in culture	(127)
ALA cyclopentyl	+/–	Cells in culture	(127)
ALA cyclo-pentyl-methyl	++	Cells in culture	(127)
ALA polyethyleneglycol ester	+	Cells in culture	(342)
ALA methyl ester peptide	–	Cells in culture	(342)
R,S-ALA-2-(hydroxymethyl)tetrahydropyranyl ester	++	Cells in culture	(103)
N-acetyl-ALA ester	–	Cells in culture	(103)

($ > $C3–4) [4,102,103,127,132,250,251]. The benefit of conjugating long-chain alcohols to 5-ALA is due to the improved rate of plasma membrane penetration (see below). Once located inside the cells esterases may deesterify the derivatives and 5-ALA can enter the heme pathway (Fig.2). Based on an assumed need for de-esterification of 5-ALA esters in order to form PpIX it may be concluded that cellular esterases deesterify 5-ALA esters with different rates [252]. It is not clear whether intracellular deesterification may be a rate-limiting step in the rate of PpIX accumulation.

5-ALA-methyl and -hexyl ester have been utilized clinically and shown promising for treatment of BCC, solar keratosis and bladder diagnosis [2–4]. However, most of the 5-ALA derivatives have only been evaluated in cells in culture. The possible improvements in tissue penetration by the 5-ALA derivatives therefore need further evaluation. That there may be differences between the improvements in the rate of PpIX formation in cells in culture and in tissues has been clearly documented in the case of 5-ALA methyl ester. 5-ALA methyl ester is less efficient in PpIX formation that ALA, while this seems to be opposite after topical application of these compounds [253]. In addition, the specificity of the compounds for the target tissues may differ as shown by comparing 5-ALA and 5-ALA esters [254].

8.6.3 Modulation of porphyrin synthesis induced by ALA

8.6.3.1 Modulation of porphyrin synthesis by iron availability
Combination of ALA-treatment with the iron chelators EDTA, CP94, CP20 and DEF has been shown to increase the rate of accumulation of photosensitizing porphyrins or to increase the sensitivity of the cells to photoinactivation [206,207,213,215, 231,255–259]. At similar concentrations DEF seems to be somewhat more efficient in accumulating PpIX than EDTA [213]. The increased sensitization of gastric cancer cells to photoinactivation induced by DEF in ALA-treated cells is reversed by extracellular concentrations exceeding 40 μM [260].

The effect of iron chelators on normal compared to tumor cells has not yet been considered, but a generally higher ferrochelatase activity in normal cells and a relatively low iron level in tumor tissues may indicate a reduced specificity of ALA-PDT for neoplastic tissues by treatment with iron chelators [258]. Systemic administration of 5-ALA is limited by adverse effects such as nausea, vomiting, headache, circulatory failure [261]. The transient elevation of liver enzymes may be due to the high levels of 5-ALA reaching the liver and not accumulation of PpIX. Coadministration of an iron-chelator may overcome these limitations [258]. In addition to the cellular need for iron in cytochromes and other enzymes like ribonucleotide reductase, it should be noted that iron depletion inhibits entry into S phase of the cell cycle [262]. The DEF has been included in treatment of advanced neuroblastoma [263,264] as well as in reducing iron stores in thalassemia [265] and is a relatively safe drug [266].

8.6.3.2 Other methods to modulate porphyrin synthesis and improve PDT efficacy
In addition to iron deprivation by chelators, incorpororation of iron into PpIX may be inhibited by photochemical inhibition of the ferrochelatase activity. Because PpIX may accumulate in mitochondria upon ALA-treatment, ferrochelatase may be photochemically inactivated by ALA-PDT and thereby stimulate PpIX accumulation. In

accordance with this hypothesis the ferrochelatase activity was reduced by about 50% in A431 human epidermoid carcinoma cells and HMEC-1 microvascular endothelial cells by similar PDT doses (<10% cell killing) [207,209]. This is similar to our finding in WiDr cells using an ALA-PDT dose inactivating approximately 90% of the cells (unpublished observations). Moreover, in the A431 cells this pretreatment with low PDT doses led to a significant increase in porphyrin accumulation, while only a minor and insignificant increase was observed in the HMEC-1 cells. The rate of PpIX accumulation was substantially reduced in R3230AC mammary adenocarcinoma cells after ALA-PDT when challenged with a 2nd treatment with ALA, suggested to be due to PDT-induced reduction in PBGD activity [234]. Fractionation of the light dose did not increase the rate of PpIX accumulation in normal rat oesophagus [237]. The sensitivity of the ferrochelatase to photoinactivation seems therefore to be cell type dependent. Photoinactivation and light-independent inactivation of the enzymes involved in the heme synthesis pathway by uro-, copro- and protoporphyrin may influence on the synthetic capacity [267–270].

The porphyrinogenic drugs 2-allyl-2-isopropylacetamide (AIA) and veronal has been shown to increase ALAS activity in rat liver homogenates, but not in mice heart [271,272]. The AIA, in the absence of exogenously added ALA, induces an increased accumulation of porphyrins in explants from liver, skin and heart [273], and increased the PpIX accumulation in B16 melanoma cells treated with dimethyl sulfoxide (DMSO) and suboptimal concentrations of 5-ALA [214]. The DMSO, known for its stimulation of differentiation [273], enhances induction of several enzymes in the porphyrin synthesis pathway [274–276]. This is at least partly due to transcriptional stimulation [277,278]. It should be noted that the chemotherapeutic DNA crosslinking agent mitomycin C, which has been shown promising in combination with Photofrin-based PDT [279], may have influence on the expression of ALAS activity [280]. Instead of overruling the heme pathway regulation by treatment with 5-ALA or derivatives thereof transduction of ALAS lacking regulatory motifs has been shown to induce accumulation of large quantities of fluorescing porphyrins [281]. This gene therapeutic approach to ALA-PDT does probably not require transgene activation in all the target cells since PpIX can diffuse out of the transduced or transfected cells and induce a bystander effect. PpIX has been shown to stimulate ALAS activity in rat liver, but it is not clear how this may influence the overall PpIX accumulation [176]. Another porphyrin biosynthesis modulator, the herb- and insecticide 1,10-phenanthroline, enhanced significantly the PpIX accumulation and sensitivity to photoinactivation in ALA-treated transformed cells both in vitro and in vivo [194]. Hemin and butyric acid, known stimulators of all enzymes in the heme synthesis pathway [276,278,282], were shown to stimulate PpIX accumulation in leukemic cells treated with zero or low concentrations of 5-ALA [200,283]. The progesterone receptor antagonist mifepristone was found to induce accumulation of PpIX when combined with desferrioxamine (in the absence of ALA] in primary chicken liver cells [284]. The androgen 5α-dihydrotestosterone increases the PpIX accumulation in an androgen-responsive human prostate cancer cell line, caused by stimulation of the cellular 5-ALA uptake [68].

PpIX and ALA-PDT are influenced by microenvironmental factors such as hypoxia and pH [239,285]. PpIX accumulation is attenuated, but not abolished, under anoxic conditions. This may be due to the oxygen-requirements of protoporphyrinogen oxidase and coproporphyrinogen oxidase [286,287]. The attenuation of PpIX accumulation under hypoxic conditions is highly cell density dependent in EMT6 mouse fibrosarcoma cells, i.e. there is a relatively small effect (1.4-fold) of hypoxia in exponentially growing cells while PpIX accumulation was attenuated by up to 98% in cells in the plateau phase [239]. The pH is usually lower in neoplastic tissues and the effect of pH on PpIX formation and PDT sensitivity is therefore of importance. The optimal pH for ALA-induced PpIX generation is 7.0 or 7.5 depending on the cell lines studies [248,285]. The PpIX formation correlates well with pH-dependent rate of 5-ALA uptake in WiDr adenocarcinoma cells [106]. The cause and the influence of extracellular pH on PpIX formation may depend on the extracellular concentration of 5-ALA since PpIX formation may not be dependent on the rate of 5-ALA uptake at high extracellular 5-ALA concentrations (>1 mM) [107]. The pH-dependent photocytotoxicity does not seem to follow the amounts of PpIX generated in the low pH range, possibly due to changes in intracellular pH (which occurs at extracellular pH below 6.5 [288]) and thereby changes in cellular defense mechanisms.

The human genetic disease *variegate porphyria* is caused by a deficiency in protoporphyrinogen oxidase activity, leading to an accumulation of protoporphyrinogen which may be spontaneously oxidized to PpIX. Thus, there has been some interest in inhibitors of this enzyme, such as sulfentrazone (FP846) and its analogues. There have been some promising reports on the use of FP846 in mammals [289,290], although acifluorfen, another protoporphyrinogen oxidase inhibitor, has been shown to inhibit PpIX accumulation in ALA-treated cancer cell lines [291,292]. The reason for this discrepancy is not clear, but may be due to a very slow oxidation of protoporphyrinogen intracellularly. ALA-PDT has also been combined with cyclophosphamide with a 30% potentiation of the PDT treatment [293].

The PDT effect on cells in culture may be enhanced by combining a sensitizing prodrug (ALA) with HpD under certain circumstances, depending on the sequence of sensitizer administration and on the incubation time [294]. However, this approach seems to be even more beneficial in vivo by topical 5-ALA treatment combined with a low systemically administrated dose of HpD or other synthetic photosensitizers given shortly before exposure to light in order to damage both the parenchyme cells and the vasculature [295]. The prolonged skin photosensitivity of systemic administration of synthetic sensitizers is avoided by the low concentrations needed. Folic acid has been used as adjuvant in chemotherapy and shown to improve antineoplastic efficacy. Similarly, folic acid has been shown to improve the therapeutic outcome of ALA-PDT of superficial skin malignancies [296].

8.7 Mechanisms and efficiency of ALA-based photosensitization

Because PpIX is formed in the mitochondria, PpIX may be located in the mitochondria after 5-ALA treatment. This has been confirmed by comparative double staining with rhodamine 123 and electron microscopical studies of photochemically treated bladder carcinoma cell lines [215,297], double staining with dihydrorhodamine [215], altered fluorescence decay rates of rhodamine 123 in cells treated with 5-ALA and exposed to 436 nm light [298], reduction in the mitochondrial dehydrogenase activity as measured by the MTT assay immediately after light exposure [186], and reduced oxygen consumption immediately after light exposure [299]. However, mitochondrial damage may not necessarily be equivalent with a reduced energy charge and ATP level since substantially higher doses of light were needed for a reduction of the energy charge than for inhibition of oxygen consumption [299]. ALA-PDT may exert its effect through formation of singlet oxygen ($^1O_2$) [300–302] although other reactive oxygen species may also be involved [209,285,303,304]. The range of action of $^1O_2$ in cells has been estimated to be approximately 0.01–0.02 μm [305], but may be even shorter if the photoactivated dye producing $^1O_2$ is closely associated with a target rich in molecules that can react with or quench $^1O_2$. Therefore, the primary site of action of ALA-PDT may be highly restricted to its close vicinity, and studies of the cellular localization of photosensitizers may yield valuable information about the identity of the main primary targets in ALA-PDT, although the primary damage may only be the first step in a cascade of reactions leading to cell death. The intracellular localization of ALA-induced fluorescing porphyrins based on fluorescence-microscopic studies includes mitochondria, plasma membrane, lysosomes, endoplasmic reticulum as well as diffuse extranuclear localization [101,199,215,298,306,307]. Although the localization pattern in no cases is contradictory to a mitochondrial localization of a fraction of the PpIX formed PpIX is also located in other subcellular compartments. The intracellular distribution seems to by dependent on time of incubation, i.e. immediately after initiation of ALA-treatment PpIX is located mainly in mitochondria while after continued incubation cytoplasmic areas, including endoplasmic reticulum, as well as the plasma membrane are stained. This is in accordance with the more rapid uptake of PpIX in isolated rat liver microsomes than in mitochondria [308]. Sustained 5-ALA incubation or treatment with high concentrations of 5-ALA may reduce the fluorescence quantum yield, i.e. increase the aggregation of the formed PpIX [307]. Long term incubation may also lead to lysosomal localization perhaps due to autophagic engulfment of mitochondria in autophage-competent cells [298]. The intracellular localization of PpIX formed endogenously differs from that of exogenously added PpIX [297,307,309,310]. Exogenously added PpIX accumulate mainly in the plasma membrane [310], and is less efficient in sensitizing Photofrin-mediated PDT resistant cells to photoinactivation than the non-resistant counterpart [297]. In contrast, intracellular localization, rate of PpIX formation and sensitivity to photoinactivation induced by 5-ALA was equal in the RIF radiation-induced fibrosarcomas resistant to Photofrin-mediated PDT and the parental cells [297].

Owing to the intracellular localization pattern, mitochondria have been shown to be damaged after ALA-PDT. Thus, ALA-based PDT leads to inhibition of mitochondrial dehydrogenases and inhibition of respiration as described above as well as failure of dihydrorhodamine to stain ALA-PDT treated cells [311], swelling of mitochondria [215,312], reduced sensitivity to photoinactivation by co-treatment with compounds with high affinity for mitochondrial (peripheral) benzodiazepine receptors [216,292], and reduced binding of the benzodiazepine PK 11195 to platelets after PpIX treatment [313]. Laser-based microirradiation of specific subcellular regions indicated that the nucleus and the perinuclear cytoplasma containing large amounts of mitochondria were more sensitive to light that the peripheral cytoplasma [314]. Additionally, other compartments, which not clearly have been found to contain fluorescing porphyrins, are damaged by ALA-PDT. This includes damage to endoplasmic reticulum [308,311], increased level of intracellular free calcium [315], and cellular loss of K^+ [214]. PpIX and light more rapidly damage the Ca^{2+} influx function of microsomes, i.e. endoplasmic reticulum particles, than that of mitochondria [308].

Lipid peroxidation has been shown to be induced by ALA-PDT and is most likely involved in the cytotoxic effect of such treatments [145,302]. Vitamine E and glutathione protects against ALA-PDT induced lipid peroxidation and glutathione reduces the cells sensitivity to photoinactivation while this is not so clear with respect to vitamin E [145,316]. Some anticancerdrugs such as cyclophosphamide, mitomycin C and adriamycin or radiation have been found to potentiate the effects of ALA-PDT possibly by action mechanisms other than those enhancing porphyrin biosynthesis [317–321]. Although it is not clear how cells are inactivated by ALA-PDT it has been shown that cells may undergo apoptotic cell death, i.e. V79 Chinese hamster lung fibroblasts undergo apoptotic cell death after ALA-PDT, while WiDr human adenocarcinoma cells are killed only through necrotic cell death [322]. ALA-PDT induces apoptosis also in HL60 leukemic cells [323]. The treatment was found to enhance the expression of 7 proteins and suppress or photo-oxidize 17 proteins of which 3 proteins were identified as endoplasmic reticulum associated chaperones involved in calcium homeostasis. It was proposed that the reduced amounts of endoplasmic reticulum chaperones triggered a release of Ca^{2+} into the cytosol where it triggers mitochondiral permeability transition pores leading at the end to activation of caspases and subsequently to apoptosis, or a direct activation of Ca^{2+}-dependent endonucleases also leading to apoptosis [323,324]. The mRNA expression of the (proto)oncogenes c-myc and bcl-2 was found to be transiently increased in transformed, but not in normal fibrablasts, after ALA-PDT [325]. In vitro and in vivo action spectrum studies indicate that the maximum cytotoxic effect is obtained by exposure to 635 nm light [218,326].

Monocytes (CD14 +) and dendritic cells (CD83 +) accumulate PpIX upon treatment with 5-ALA [327]. ALA-PDT of such antigen-presenting cells attenuates the T-cell response to antigens, which may be due to alterations in the MHC molecules or reduced phagocytic activity. Similar immunomodulatory responses have also been seen after photodynamic therapy with other photosensitizers [328,329].

Protoporphyrin IX is bleached upon exposure to light and 70–95% of PpIX surface fluorescence may be photobleached by clinically relevant light doses [330,331]. The degradation process is oxygen dependent and does not follow simple first order kinetics since it is dependent on the PpIX concentration [132,330,332]. The photobleaching of PpIX induces several products of which photoprotoporphyrin is the most abundant [333]. Photoprotoporphyrin is a chlorin type photosensitizer with a red-shifted Q-band (668 nm) with a high extinction coefficient. The photophysical properties of photo-protoporphyrin have previously been described [334,335].

The cell cycle dependency on sensitivity to PDT may be cell line dependent. In a bladder cancer cell line S- and G2-phase cells were more sensitive to PDT than cells in G1-phase [336]. This was caused by a different ability of the cells to accumulate PpIX in the different cell cycle phases. In a mammalian epithelial cell line a similar difference in PpIX accumulation was not observed [240]. Neoplastic cells tend to accumulate in G2 phase after treatment with ionizing radiation and this has been tried to be utilized to potentiate the sensitivity of cells to ALA-PDT. However, such a combination treatment has either been additive or only slightly synergistic, depending on the cell line and treatment regimen [317,319]. Similar results have been obtained when ionizing radiation has been combined with PDT using other photosensitizers [337–340].

References

1. Z. Malik, H. Lugaci (1987). Destruction of erythroleukaemic cells by photoactivation of endogenous porphyrins. *Br. J. Cancer*, **56**, 589–595.
2. A.M. Soler, T. Warloe, J. Tausjoand, K.E. Giercksky (2000). Photodynamic therapy of residual or recurrent basal cell carcinoma after radiotherapy using topical 5-aminolevulinic acid or methylester aminolevulinic acid [In Process Citation]. *Acta Oncol.*, **39**, 605–609.
3. C. Fritsch, B, Homey, W. Stahl, P. Lehmann, T. Ruzicka, H. Sies (1998). Preferential relative porphyrin enrichment in solar keratoses upon topical application of δ-aminolevulinic acid methylester. *Photochem. Photobiol.*, **68**, 218–221.
4. N. Lange, P. Jichlinski, M. Zellweger, M. Forrer, A. Marti, L. Guillou, P. Kucera, G. Wagnières, H. Van den Bergh (1999). Photodetection of early human bladder cancer based on the fluorescence of 5-aminolaevulinic acid hexylester-induced protoporphyrin IX: a pilot study. *Br. J. Cancer*, **80**, 185–193.
5. L. Gossner, M. Stolte, R. Sroka, K. Rick, A. May, E.G. Hahn, C. Ell (1998). Photodynamic ablation of high-grade dysplasia and early cancer in Barrett's esophagus by means of 5-aminolevulinic acid. *Gastroenterology*, **114**, 448–455.
6. P. Hillemanns, M. Untch, C. Dannecker, R. Baumgartner, H. Stepp, J. Diebold, H. Weingandt, F. Pröve, M. Korell (2000). Photodynamic therapy of vulvar intraepithelial neoplasia using 5-aminolevulinic acid. *Int. J. Cancer*, **85**, 649–653.
7. J.D. Zollo, N.C. Zeitouni (2000). The Roswell Park Cancer Institute experience with extramammary Paget's disease. *Br. J. Dermatol.*, **142**, 59–65.
8. S. Karrer, C. Abels, M. Landthaler, R.M. Szeimies (2000). Topical photodynamic therapy for localized scleroderma. *Acta Derm. Venereol.*, **80**, 26–27.
9. I.M. Stender, R.H. Na, H. Fogh, C. Gluud, H.C. Wulf (2000). Photodynamic therapy with 5-aminolaevulinic acid or placebo for recalcitrant foot and hand warts: randomised double-blind trial. *Lancet*, **355**, 963–966.
10. C.P.M. Fritsch, H. Menke, T. Ruzicka, G. Goerz, R.R. Olbrisch (1997). Successful surgery of multiple recurrent basal cell carcinomas guided by photodynamic diagnosis. *Aesthetic. Plast. Surg.*, **21**, 437–439.
11. C. Fritsch, G. Goerz, T. Ruzicka (1998). Photodynamic therapy in dermatology. *Arch. Dermatol.*, **134**, 207–214.
12. P.D. Cotter, H.A. Drabkin, T. Varkony, D.L. Smith, D.F. Bishop (1995). Assignment of the human housekeeping delta-aminolevulinate synthase gene (ALAS1) to chromosome band 3p21.1 by PCR analysis of somatic cell hybrids. *Cytogenet. Cell Genet.*, **69**, 207–208.
13. D. Richardson, E. Baker (1991). The uptake of inorganic iron complexes by human melanoma cells. *Biochimica et Biophysica Acta.*, **1093**, 20–28.
14. J.G. Conboy, T.C. Cox, S.S. Bottomley, M.J. Bawden, B.K. May (1992). Human erythroid 5-aminolevulinate synthase. Gene structure and species-specific differences in alternative RNA splicing. *J. Biol. Chem.*, **267**, 18753–18758.
15. G.C. Ferreira, H.A. Dailey (1993). Expression of mammalian 5-aminolevulinate synthase in Escherichia coli. Overproduction, purification, and characterization. *J. Biol. Chem.*, **268**, 584–590.
16. B.K. May, M.J. Bawden (1989). Control of heme biosynthesis in animals. *Semin. Hematol*, **26**, 150–156.

17. A.W. Scotto, L.F. Chang, D.S. Beattie. (1983) The characterization and submitochondrial localization of delta-aminolevulinic acid synthase and an associated amidase in rat liver mitochondria using an improved assay for both enzymes. *J. Biol. Chem.*, **258**, 81–90.

18. T. Emanuelli, J.B. Rocha, M.E. Pereira, P.C. Nascimento, D.O. Souza, F.A. Beber (1998). delta-Aminolevulinate dehydratase inhibition by 2,3-dimercaptopropanol is mediated by chelation of zinc from a site involved in maintaining cysteinyl residues in a reduced state. *Pharmacol. Toxicol.*, **83**, 95–103.

19. F.A. Beber, J. Wollmeister, M.J. Brigo, M.C. Silva, C.N. Pereira, J.B. Rocha (1998) delta-Aminolevulinate dehydratase inhibition by ascorbic acid is mediated by an oxidation system existing in the hepatic supernatant. *Int. J. Vitam. Nutr. Res.*, **68**, 181–188.

20. G.G. Guo, M. Gu, J.D. Etlinger (1994). 240-kDa proteasome inhibitor (CF-2) is identical to delta-aminolevulinic acid dehydratase. *J. Biol. Chem.*, **269**, 12399–12402.

21. M. Gross, S. Hessefort, A. Olin (1999). Purification of a 38-kDa protein from rabbit reticulocyte lysate which promotes protein renaturation by heat shock protein 70 and its identification as delta-aminolevulinic acid dehydratase and as a putative DnaJ protein. *J. Biol. Chem.*, **274**, 3125–3134.

22. M. Guolo, C. Machalinski, M. Biscoglio, A.M. Stella, C. Franco, L. Pataro, R.E. De Salamanca, A. Batlle (1999). Inhibition of erythrocyte aminolevulinate dehydratase by a 56.2-kD peptide from uremic plasma. *Exp. Nephrol.*, **7**, 236–241.

23. B. Norton, W.G. Lanyon, M.R. Moore, M. Porteous, G.R. Youngs, J.M. Connor (1993). Evidence for involvement of a second genetic locus on chromosome 11q in porphyrin metabolism. *Hum. Genet.*, **91**, 576–578.

24. W. Xu, C.A. Kozak, R.J. Desnick (1995). Uroporphyrinogen-III synthase: molecular cloning, nucleotide sequence, expression of a mouse full-length cDNA, and its localization on mouse chromosome 7. *Genomics*, **26**, 556–562.

25. M.R. Moore, K.E.L. McColl, C. Rimington, A. Goldberg (1987) In: *Disorders of Porphyrin Metabolism*. Plenum Press, New York.

26. A.G. Smith, J.E. Francis (1979) Decarboxylation of porphyrinogens by rat liver uroporphyrinogen decarboxylase. *Biochem. J.*, **183**, 455–458.

27. H. De Verneuil, B. Grandchamp, C. Foubert, D. Weil, V.C. N'Guyen, M.S. Gross, S. Sassa, Y. Nordmann (1984). Assignment of the gene for uroporphyrinogen decarboxylase to human chromosome 1 by somatic cell hybridization and specific enzyme immunoassay. *Hum. Genet.*, **66**, 202–205.

28. P.H. Romeo, N. Raich, A. Dubart, D. Beaupain, M. Pryor, J. Kushner, M. Cohen-Solal, M. Goossens (1986). Molecular cloning and nucleotide sequence of a complete human uroporphyrinogen decarboxylase cDNA. *J. Biol. Chem.*, **261**, 9825–9831.

29. F. De Matteis, C. Harvey, C. Reed, R. Hempenius (1988). Increased oxidation of uroporphyrinogen by an inducible liver microsomal system. Possible relevance to drug-induced uroporphyria. *Biochem. J.*, **250**, 161–169.

30. P.R. Sinclair, N. Gorman, I.B. Tsyrlov, U. Fuhr, H.S. Walton, J.F. Sinclair (1998). Uroporphyrinogen oxidation catalyzed by human cytochromes P450. *Drug Metab. Dispos.*, **26**, 1019–1025.

31. P.R. Sinclair, N. Gorman, T. Dalton, H.S. Walton, W.J. Bement, J.F. Sinclair, A.G. Smith, D.W. Nebert (1998) Uroporphyria produced in mice by iron and 5-aminolaevulinic acid does not occur in *Cyp1a2*(–/–) null mutant mice. *Biochem. J.*, **330**, 149–153.

32. S. Deam, G.H. Elder (1991). Uroporphyria produced in mice by iron and 5-aminolevulinic acid. *Biochem. Pharmacol.*, **41**, 2019–2022.

33. G.H. Elder, J.O. Evans (1978). Evidence that the coproporphyrinogen oxidase activity of rat liver is situated in the intermembrane space of mitochondria. *Biochem. J.*, **172**, 345–347.

34. B. Grandchamp, N. Phung, Y. Nordmann (1978). The mitochondrial localization of coproporphyrinogen III oxidase. *Biochem. J.*, **176**, 97–102.

35. M. Akthar, (1991). In: P.M. Jordan (Ed.), *Biosynthesis of Tetrapyrroles* (pp. 67–100). Elsevier, Amsterdam.

36. B. Grandchamp, D. Weil, Y. Nordmann, N. Van Cong, H. De Verneuil, C. Foubert, M.S. Gross (1983). Assignment of the human coproporphyrinogen oxidase to chromosome. *Hum. Genet*, **64**, 180–183.

37. A.M.d.C. Battle, A. Benson, C. Rimington (1965). Purification and properties of coproporphyrinogenase. *Biochem. J.*, **97**, 731–740.

38. G.C. Ferreira, T.L. Andrew, S.W. Karr, H.A. Dailey (1988). Organization of the terminal two enzymes of the heme biosynthetic pathway. Orientation of protoporphyrinogen oxidase and evidence for a membrane complex. *J. Biol. Chem.*, **263**, 3835–3839.

39. E. Rossi, K.A. Costin, P. Garcia-Webb (1988). Ferrochelatase activity in human lymphocytes, as quantified by a new high-performance liquid-chromatographic method. *Clin. Chem.*, **34**, 2481–2485.

40. J.M. Jacobs, P.R. Sinclair, J.F. Sinclair, N. Gorman, H.S. Walton, S.G. Wood, C. Nichols (1998). Formation of zinc protoporphyrin in cultured hepatocytes: effects of ferrochelatase inhibition, iron chelation or lead. *Toxicology*, **125**, 95–105.

41. R.F. Labbe, R.L. Rettmer (1989). Zinc protoporphyrin: a product of iron-deficient erythropoiesis. *Semin. Hematol.*, **26**, 40–46.

42. G.C. Ferreira, R. Franco, S.G. Lloyd, A.S. Pereira, I. Moura, J.J. Moura, B.H. Huynh (1994). Mammalian ferrochelatase, a new addition to the metalloenzyme family. *J. Biol. Chem.*, **269**, 7062–7065.

43. N.S. Punekar, R.S. Gokhale (1991). Factors influencing the stability of heme and ferrochelatase: role of oxygen. *Biotechnol. Appl. Biochem.*, **14**, 21–29.

44. S.S. Bottomley, U. Muller-Eberhard (1988). Pathophysiology of heme synthesis. *Semin. Hematol.*, **25**, 282–302.

45. E. Rossi, P.V. Attwood, P. Garcia-Webb, K.A. Costin (1990). Inhibition of human lymphocyte ferrochelatase activity by hemin. *Biochimica et Biophysica Acta*, **1038**, 375–381.

46. S.J. Wolfson, A. Bartczak, J.R. Bloomer (1979). Effect of endogenous heme generation on delta-aminolevulinic acid synthase activity in rat liver mitochondria. *J. Biol. Chem.*, **254**, 3543–3546.

47. S. Sassa, A. Kappas (1981). In: H. Harris, K. Hirshhorn (Eds), *Advances in Human Genetics* (pp. 121–231). Plenum Press, New York.

48. M. Williams, J, Van der Zee, J. Van Steveninck (1992). Toxic dark effects of protoporphyrin on the cytochrome P-450 system in rat liver microsomes. *Biochem. J.*, **288**, 155–159.

49. C.A. Louis, S.G. Wood, H.S. Walton, P.R. Sinclair, J.F. Sinclair (1998). Mechanism of the synergistic induction of CYP2H by isopentanol plus ethanol: comparison to glutethimide and relation to induction of 5-aminolevulinate synthase. *Arch. Biochem. Biophys.*, **360**, 239–247.

50. M.M. Iba, J. Alam, C. Touchard, P.E. Thomas, A. Ghosal, J. Fung (1999). Coordinate up-regulation of CYP1A1 and heme oxygenase-1 (HO-1) expression and modulation of δ-aminolevulinic acid synthase and tryptophan pyrrolase activities in pyridine-treated rats. *Biochem. Pharmacol.*, **58**, 723–734.

51. G. Braidotti, I.A. Borthwick, B.K. May (1993). Identification of regulatory sequences in the gene for 5-aminolevulinate synthase from rat. *J. Biol. Chem.*, **268**, 1109–1117.

52. B. Li, J.O. Holloszy, C.F. Semenkovich (1999). Respiratory uncoupling induces delta-aminolevulinate synthase expression through a nuclear respiratory factor-1-dependent mechanism in HeLa cells. *J. Biol. Chem.*, **274**, 17534–17540.

53. J.W. Hamilton, W.J. Bement, P.R. Sinclair, J.F. Sinclair, J.A Alcedo, K.E. Wetterhahn (1991). Heme regulates hepatic 5-aminolevulinate synthase mRNA expression by decreasing mRNA half-life and not by altering its rate of transcription. *Arch. Biochem. Biophys.*, **289**, 387–392.

54. P.D. Drew, I.Z. Ades (1989) Regulation of the stability of chicken embryo liver delta-aminolevulinate synthase mRNA by hemin. *Biochem. Biophys. Res. Commun.*, **162**, 102–107.

55. E.E Cable, J.A Pepe, N.C. Karamitsios, R.W. Lambrecht, H.L. Bonkovsky (1994). Differential effects of metalloporphyrins on messenger RNA levels of delta-aminolevulinate synthase and heme oxygenase. Studies in cultured chick embryo liver cells. *J. Clin. Invest.*, **94**, 649–654.

56. M. Yamamoto, S. Kure, J.D. Engel, K. Hiraga (1988). Structure, turnover, and heme-mediated suppression of the level of mRNA encoding rat liver delta-aminolevulinate synthase. *J. Biol. Chem.*, **263**, 15973–15979.

57. B.K. May, S.C. Dogra, T.J. Sadlon, C.R. Bhasker, T.C. Cox, S.S. Bottomley (1995). Molecular regulation of heme biosynthesis in higher vertebrates. *Prog. Nucleic. Acid. Res. Mol. Biol.*, **51**, 1–51.

58. I.A. Borthwick, G. Srivastava, A.R. Day, B.A. Pirola, M.A. Snoswell, B.K. May, W.H. Elliott (1985). Complete nucleotide sequence of hepatic 5-aminolaevulinate synthase precursor. *Eur. J. Biochem.*, **150**, 481–484.

59. K. Yamauchi, N. Hayashi, G. Kikuchi (1980). Translocation of delta-aminolevulinate synthase from the cytosol to the mitochondria and its regulation by hemin in the rat liver. *J. Biol. Chem.*, **255**, 1746–1751.

60. N. Hayashi, N. Watanabe, G. Kikuchi (1983). Inhibition by hemin of in vitro translocation of chicken liver delta-aminolevulinate synthase into mitochondria. *Biochem. Biophys. Res. Commun.*, **115**, 700–706.

61. J.T. Lathrop, M.P. Timko (1993). Regulation by heme of mitochondrial protein transport through a conserved amino acid motif. *Science*, **259**, 522–525.

62. J. Igarashi, N. Hayashi, G. Kikuchi (1976). delta-Aminolevulinate synthetases in the liver cytosol fraction and mitochondria of mice treated with allylisopropylacetamide and 3,5-dicarbethoxyl-1,4-dihydrocollidine. *J. Biochem. (Tokyo)*, **80**, 1091–1099.

63. K. Meguro, K. Igarashi, M. Yamamoto, H. Fujita, S. Sassa (1995). The role of the erythroid-specific delta-aminolevulinate synthase gene expression in erythroid heme synthesis. *Blood*, **86**, 940–948.

64. T. Houston, M.R. Moore, K.E. McColl, E. Fitzsimons (1991). Erythroid 5-aminolaevulinate synthase activity during normal and iron deficient erythropoiesis. *Br. J. Haematol.*, **78**, 561–564.

65. O. Melefors, B. Goossen, H.E. Johansson, R. Stripecke, N.K. Gray, M.W. Hentze (1993). Translational control of 5-aminolevulinate synthase mRNA by iron-responsive elements in erythroid cells. *J. Biol. Chem.*, **268**, 5974–5978.

66. G.C. Ferreira (1995). Ferrochelatase binds the iron-responsive element present in the erythroid 5-aminolevulinate synthase mRNA. *Biochem. Biophys. Res. Commun.*, **214**, 875–878.

67. W. Shoji, Y. Ohmori, M. Obinata (1993). c-Myc selectively regulates the latent period and erythroid-specific genes in murine erythroleukemia cell differentiation. *Jpn. J. Cancer. Res.*, **84**, 885–892.

68. T. Momma, M.R. Hamblin, T. Hasan (1997). Hormonal modulation of the accumulation of 5-aminolevulinic acid-induced protoporphyrin and phototoxicity in prostate cancer cells. *Int. J. Cancer.*, **72**, 1062–1069.

69. S.L. Gibson, L.T. Anderson, J.J. Havens, R. Hilf (1999). Effect of estrogenic perturbations

on δ-aminolevulinic acid-induced porphobilinogen deaminase and protoporphyrin IX levels in rat Harderian glands, liver, and R3230AC tumors. *Biochem. Pharmacol.*, **58**, 1821–1829.

70. S.L. Gibson, D.J. Cupriks, J.J. Havens, M.L. Nguyen, R. Hilf (1998). A regulatory role for porphobilinogen deaminase (PBGD) in δ-aminolaevulinic acid (δ-ALA)-induced photosensitization? *Br. J. Cancer*, **77**, 235–242.

71. E.C. Theil (1994). Iron regulatory elements (IREs): a family of mRNA non-coding sequences. *Biochem. J.*, **304**, 1–11.

72. B. Guo, Y. Yu, E.A. Leibold (1994). Iron regulates cytoplasmic levels of a novel iron-responsive element-binding protein without aconitase activity. *J. Biol. Chem.*, **269**, 24252–24260.

73. B. Guo, J.D. Phillips, Y. Yu, E.A. Leibold (1995). Iron regulates the intracellular degradation of iron regulatory protein 2 by the proteasome. *J. Biol. Chem.*, **270**, 21645–21651.

74. D.P. Mascotti, D. Rup, R.E. Thach (1995). Regulation of iron metabolism: translational effects mediated by iron, heme, and cytokines. *Annu. Rev. Nutr.*, **15**, 239–261.

75. L.C. Kuhn, M.W. Hentze (1992). Coordination of cellular iron metabolism by post-transcriptional gene regulation. *J. Inorg. Biochem.*, **47**, 183–195.

76. T.A. Rouault, R.D. Klausner (1996). Iron-sulfur clusters as biosensors of oxidants and iron [published erratum appears in Trends Biochem Sci 1996 Jul;21(7):246] [see comments]. *Trends Biochem. Sci.*, **21**, 174–177.

77. L.S. Goessling, S. Daniels-McQueen, M. Bhattacharyya-Pakrasi, J.J. Lin, R.E. Thach (1992). Enhanced degradation of the ferritin repressor protein during induction of ferritin messenger RNA translation. *Science*, **256**, 670–673.

78. J.H. Ward, I. Jordan, J.P. Kushner, J. Kaplan (1984). Heme regulation of HeLa cell transferrin receptor number. *J. Biol. Chem.*, **259**, 13235–13240.

79. E.W. Mullner, S. Rothenberger, A.M. Muller, L.C. Kuhn (1992). In vivo and in vitro modulation of the mRNA-binding activity of iron-regulatory factor. Tissue distribution and effects of cell proliferation, iron levels and redox state. *Eur. J. Biochem.*, **208**, 597–605.

80. H. Carvalho, E.J.H. Bechara, R. Meneghini, M. Demasi (1997). Haem precursor δ-aminolaevulinic acid induces activation of the cytosolic iron regulatory protein 1. *Biochem. J.*, **328**, 827–832.

81. E.C. Theil, R.S. Eisenstein (2000). Combinatorial mRNA regulation: Iron regulatory proteins and Iso-iron responsive elements (iso-IREs). *J. Biol. Chem.*

82. M. Muckenthaler, M.W. Hentze (1997). Mechanisms for posttranscriptional regulation by iron-responsive elements and iron regulatory proteins. *Prog. Mol. Subcell. Biol.*, **18**, 93–115.

83. D.J. Haile (1999). Regulation of genes of iron metabolism by the iron-response proteins. *Am. J. Med. Sci.*, **318**, 230–240.

84. P. Ponka (1999). Cell biology of heme. *Am. J. Med. Sci.*, **318**, 241–256.

85. D.R. Richardson, P. Ponka (1997). The molecular mechanisms of the metabolism and transport of iron in normal and neoplastic cells. *Biochim. Biophys. Acta Rev. Biomembr.*, **1331**, 1–40.

86. K. Thorstensen, I. Romslo (1990). The role of transferrin in the mechanism of cellular iron uptake. *Biochem. J.*, **271**, 1–9.

87. L.A. Applegate, P. Luscher, R.M. Tyrrell (1991). Induction of heme oxygenase: a general response to oxidant stress in cultured mammalian cells. *Cancer Res.*, **51**, 974–978.

88. I. Rublevskaya, M.D. Maines (1994). Interaction of Fe-protoporphyrin IX and heme analogues with purified recombinant heme oxygenase-2, the constitutive isozyme of the brain and testes. *J. Biol. Chem.*, **269**, 26390–26395.

89. G.F. Vile, R.M. Tyrrell (1993). Oxidative stress resulting from ultraviolet A irradiation of human skin fibroblasts leads to a heme oxygenase-dependent increase in ferritin. *J. Biol. Chem.*, **268**, 14678–14681.

90. A. Smith, J. Alam, P.V. Escriba, W.T. Morgan (1993). Regulation of heme oxygenase and metallothionein gene expression by the heme analogs, cobalt-, and tin-protoporphyrin. *J. Biol. Chem.*, **268**, 7365–7371.

91. S. Basu-Modak, R.M.Tyrrell (1993). Singlet oxygen: a primary effector in the ultraviolet A/near-visible light induction of the human heme oxygenase gene. *Cancer Res.*, **53**, 4505–4510.

92. D. Bressoud, V. Jomini, R.M. Tyrrell (1992). Dark induction of haem oxygenase messenger RNA by haematoporphyrin derivative and zinc phthalocyanine; agents for photodynamic therapy. *J. Photochem. Photobiol. B.*, **14**, 311–318.

93. M. Novo, G. Huttmann, H. Diddens (1996). Chemical instability of 5-aminolevulinic acid used in the fluorescence diagnosis of bladder tumours. *J. Photochem. Photobiol. B.*, **34**, 143–148.

94. A.R. Butler, S. George (1992). The nonenzymatic cyclic dimerisation of 5-aminolevulinic acid. *Tetrahedron.*, **48**, 7879–7886.

95. B. Franck, H. Stratmann (1981). Condensation products of the porphyrin precursor 5-aminolevulinic acid. *Heterocycles.*, **15**, 919–923.

96. A. Bunke, O. Zerbe, H. Schmid, G. Burmeister, H.P. Merkle, B.Gander (2000). Degradation mechanism and stability of 5-aminolevulinic acid. *J. Pharm. Sci.*, **89**, 1335–1341.

97. B. Elfsson, I. Wallin, S. Eksborg, K. Rudaeus, A.M. Ros, H. Ehrsson (1999). Stability of 5-aminolevulinic acid in aqueous solution. *Eur. J. Pharm. Sci.*, **7**, 87–91.

98. S.C. Chang, A.J. MacRobert, S.G. Bown (1996). Photodynamic therapy on rat urinary bladder with intravesical instillation of 5-aminolevulinic acid: light diffusion and histological changes. *J. Urol.*, **155**, 1749–1753.

99. B.N. Roy, D.A. van Vugt, G.E. Weagle, R.H. Pottier, R.L. Reid (1997). Effect of continuous and multiple doses of 5-aminolevulinic acid on protoporphyrin IX concentrations in the rat uterus. *J. Photochem. Photobiol. B Biology*, **41**, 122–127.

100. S.I. Shedlofsky, P.R. Sinclair, H.L. Bonkovsky, J.F. Healey, A.T. Swim, J.M. Robinson (1987). Haem synthesis from exogenous 5-aminolaevulinate in cultured chick-embryo hepatocytes. Effects of inducers of cytochromes P-450. *Biochem. J.*, 248, 229–236.

101. J.M. Gaullier, K. Berg, Q. Peng, H. Anholt, P.K. Selbo, L.W. Ma, J. Moan (1997). Use of 5-aminolevulinic acid esters to improve photodynamic therapy on cells in culture. *Cancer Res.*, **57**, 1481–1486.

102. P. Uehlinger, M. Zellweger, G. Wagnieres, L. Juillerat-Jeanneret, H. van den Bergh, N. Lange (2000). 5-Aminolevulinic acid and its derivatives: physical chemical properties and protoporphyrin IX formation in cultured cells. *J. Photochem. Photobiol. B: Biol.*, **54**, 72–80.

103. J. Kloek, G.M.J.B. Van Henegouwen (1996). Prodrugs of 5-aminolevulinic acid for photodynamic therapy. *Photochem. Photobiol.*, **64**, 994–1000.

104. R. Washbrook, H. Fukuda, A. Battle, P. Riley (1997). Stimulation of tetrapyrrole synthesis in mammalian epithelial cells in culture by exposure to aminolaevulinic acid. *Br. J. Cancer.*, **75**, 381–387.

105. D.M. Becker, S. Kramer, J.D. Viljoen (1974). Delta-aminolevulinic acid uptake by rabbit brain cerebral cortex. *J. Neurochem.*, **23**, 1019–1023.

106. E. Rud, K. Berg (1998). In: B. Ehrenberg, K. Berg (Eds), *Photochemotherapy:Photodynamic Therapy and Other Modalities IV* (pp. 28–37).

107. E. Rud, O. Gederaas, A. Hogset, K. Berg (2000 May). 5-aminolevulinic acid, but not

5-aminolevulinic acid esters, is transported into adenocarcinoma cells by system BETA transporters. *Photochem. Photobiol.*, **71**(5), 640–647.

108. A. Novotny, J. Xiang, W. Stummer, N.S. Teuscher, D.E. Smith, R.F. Keep (2000 July). Mechanisms of 5-aminolevulinic acid uptake at the choroid plexus. *J. Neurochem.*, **75**(1), 321–328.

109. V.A. Percy, M.C. Lamm, J.J. Taljaard (1981). delta-Aminolaevulinic acid uptake, toxicity, and effect on [14C]gamma-aminobutyric acid uptake into neurons and glia in culture. *J. Neurochem.*, **36**, 69–76.

110. M.J. Brennan, R.C. Cantrill (1979). The effect of delta-aminolaevulinic acid on the uptake and efflux of [3H]GABA in rat brain synaptosomes. *J. Neurochem.*, **32**, 1781–1786.

111. H. Kostron, A. Obwegeser, R. Jakober (1996). Photodynamic therapy in neurosurgery: a review. *J. Photochem. Photobiol B.*, **36**, 157–168.

112. S.C. García, M.B. Moretti, E. Ramos, A. Batlle (1997). Carbon and nitrogen sources regulate δ-aminolevulinic acid and gamma-aminobutyric acid transport in. *Saccharomyces cerevisiae. Int. J. Biochem. Cell. Biol.*, **29**, 1097–1101.

113. S. Bermudez Moretti, S.R. Correa Garcia, M.S. Chianelli, E.H. Ramos, J.R. Mattoon, A. Batlle (1995). Evidence that 4-aminobutyric acid and 5-aminolevulinic acid share a common transport system into Saccharomyces cerevisiae. *Int. J. Biochem. Cell. Biol.*, **27**, 169–173.

114. M. Bermudez Moretti, S. Correa Garcia, E. Ramos, A. Batlle (1996). delta-Aminolevulinic acid uptake is mediated by the gamma-aminobutyric acid-specific permease UGA4. *Cellular & Molecular Biology*, **42**, 519–523.

115. M. Palacin, R. Estevez, J. Bertran, A. Zorzano (1998). Molecular biology of mammalian plasma membrane amino acid transporters. *Physiol. Rev.*, **78**, 969–1054.

116. F. Conti, L.V. Zuccarello, P. Barbaresi, A. Minelli, N.C. Brecha, M. Melone (1999). Neuronal, glial, and epithelial localization of gamma-aminobutyric acid transporter 2, a high-affinity gamma-aminobutyric acid plasma membrane transporter, in the cerebral cortex and neighboring structures. *J. Comp. Neurol.*, **409**, 482–494.

117. V.K. Ramanathan, C.M. Brett, K.M. Giacomini (1997). Na$^+$-dependent gamma-aminobutyric acid (GABA) transport in the choroid plexus of rabbit. *Biochimica et Biophysica Acta.*, **1330**, 94–102.

118. V.K. Ramanathan, S.J. Chung, K.M. Giacomini, C.M. Brett (1997). Taurine transport in cultured choroid plexus. *Pharm. Res.*, **14**, 406–409.

119. F. Jursky, N. Nelson (1999). Developmental expression of the neurotransmitter transporter GAT3. *J. Neurosci. Res.*, **55**, 394–399.

120. Z.N. Canellakis, L.M. Milstone, L.L. Marsh, P.R. Young, P.K. Bondy (1983). GABA from putrescine is bound in macromolecular form in keratinocytes. *Life Sci.*, **33**, 599–603.

121. S.L. Erdo, J.R. Wolff (1990). gamma-Aminobutyric acid outside the mammalian brain. *J. Neurochem.*, **54**, 363–372.

122. Y. Yoshida, A. Sadata, W.P. Zhang, K. Saito, N. Shinoura, H. Hamada (1998). Generation of fiber-mutant recombinant adenoviruses for gene therapy of malignant glioma. *Hum. Gene Ther.*, **9**, 2503–2515.

123. M.J. Brennan R.C. Cantrill (1979) Delta-aminolaevulinic acid is a potent agonist for GABA autoreceptors. *Nature*, **280**, 514–515.

124. M. Uusi-Oukari, S. Soini, J. Heikkila, A. Koivisto, K. Neuvonen, P. Pasanen, S.T. Sinkkonen, J.K. Laihia, C.T. Jansen, E.R. Korpi (2000). Stereospecific modulation of GABA(A) receptor function by urocanic acid isomers. *Eur. J. Pharmacol.*, **400**, 11–17.

125. F. Döring, J. Walter, J. Will, M. Föcking, M. Boll, S. Amasheh, W. Clauss, H, Daniel (1998). Delta-aminolevulinic acid transport by intestinal and renal peptide transporters and its physiological and clinical implications. *J. Clin. Invest.*, **101**, 2761–2767.

126. W. Stummer, S. Stocker, A. Novotny, A. Heimann, O. Sauer, O. Kempski, N. Plesnila, J. Wietzorrek, H.J. Reulen (1998). In vitro and in vivo porphyrin accumulation by C6 glioma cells after exposure to 5-aminolevulinic acid. *J. Photochem. Photobiol. B Biol.*, **45**, 160–169.

127. C.J. Whitaker, S.H. Battah, M.J. Forsyth, C. Edwards, R.W. Boyle, E.K. Matthews (2000). Photosensitization of pancreatic tumour cells by delta-aminolaevulinic acid esters. *Anticancer Drug Des.*, **15**, 161–170.

128. S. Langer, C. Abels, A. Botzlar, S. Pahernik, K. Rick, R.M. Szeimies, A.E. Goetz (1999). Active and higher intracellular uptake of 5-aminolevulinic acid in tumors may be inhibited by glycine. *J. Invest. Dermatol.*, **112**, 723–728.

129. K. Kalka, C. Fritsch, K. Bolsen, B. Verwohlt, G. Goerz (1997). Influence of indoles (melatonin, serotonin and tryptophan) on the porphyrin metabolism in vitro. *Skin Pharmacol.*, **10**, 221–224.

130. C. Cheeks, R.P. Wedeen (1986). Renal tubular transport of delta-aminolevulinic acid in rat. *Proc. Soc. Exp. Biol. Med.*, **181**, 596–601.

131. O.A. Gederaas, A. Holroyd, S.B. Brown, D. Vernon, J. Moan, K. Berg (2001). 5-Aminolevulinic acid methyl ester transport on amino acid carriers in a human colon adenocarcinoma cell line. *Photochem. Photobiol.*, **73**.

132. A. Marti, N. Lange, H. Van den Bergh, D. Sedmera, P. Jichlinski, P. Kucera (1999). Optimisation of the formation and distribution of protoporphyrin IX in the urothelium: An in vitro approach. *Journal of Urology*, **162**, 546–552.

133. M.J. Pine, U. Kim, C. Ip (1982). Free amino acid pools of rodent mammary tumors. *J. Natl. Cancer Inst.*, **69**, 729–735.

134. H.P. Monteiro, D.S. Abdalla, A. Faljoni-Alario, E.J. Bechara (1986). Generation of active oxygen species during coupled autoxidation of oxyhemoglobin and delta-aminolevulinic acid. *Biochim. Biophys. Acta.*, **881**, 100–106.

135. H.P. Monteiro, D.S. Abdalla, O. Augusto, E.J. Bechara (1989). Free radical generation during delta-aminolevulinic acid autoxidation: induction by hemoglobin and connections with porphyrinpathies. *Arch. Biochem. Biophys.*, **271**, 206–216.

136. T. Douki, J. Onuki, M.H.G. Medeiros, E.J.H. Bechara, J. Cadet, P. Di Mascio (1998). Hydroxyl radicals are involved in the oxidation of isolated and cellular DNA bases by 4-aminolevulinic acid. *FEBS. Lett.*, **428**, 93–96.

137. F.G. Princ, A.A. Juknat, A.A. Amitrano, A. Batlle (1998). Effect of reactive oxygen species promoted by δ-Aminolevulinic acid on porphyrin biosynthesis and glucose uptake in rat cerebellum. *Gen. Pharmacol.*, **31**, 143–148.

138. M. Hermes-Lima, R.F. Castilho, V.G. Valle, E.J. Bechara, A.E. Vercesi (1992). Calcium-dependent mitochondrial oxidative damage promoted by 5-aminolevulinic acid. *Biochim. Biophys. Acta.*, **1180**, 201–206.

139. M. Hermes-Lima, V.G. Valle, A.E. Vercesi, E.J. Bechara (1991). Damage to rat liver mitochondria promoted by delta-aminolevulinic acid-generated reactive oxygen species: connections with acute intermittent porphyria and lead-poisoning. *Biochim. Biophys. Acta.*, **1056**, 57–63.

140. M. Hermes-Lima (1995). How do Ca^{2+} and 5-aminolevulinic acid-derived oxyradicals promote injury to isolated mitochondria? *Free Radic. Biol. Med.*, **19**, 381–390.

141. P.I. Oteiza, E.J. Bechara (1993). 5-aminolevulinic acid induces lipid peroxidation in cardiolipin-rich liposomes. *Arch. Biochem. Biophys.*, **305**, 282–287.

142. C.G. Fraga, J. Onuki, F. Lucesoli, E.J. Bechara, P. Di Mascio (1994). 5-Aminolevulinic acid mediates the in vivo and in vitro formation of 8-hydroxy-2′-deoxyguanosine in DNA. *Carcinogenesis*, **15**, 2241–2244.

143. J. Onuki, M.H. Medeiros, E.J. Bechara, P. Di Mascio (1994). 5-Aminolevulinic acid

induces single-strand breaks in plasmid pBR322 DNA in the presence of Fe^{2+} ions. *Biochim. Biophys. Acta.*, **1225**, 259–263.

144. P. Di Mascio, P.C. Teixeira, J. Onuki, M.H.G. Medeiros, D. Dörnemann, T. Douki, J. Cadet (2000). DNA damage by 5-aminolevulinic and 4,5-dioxovaleric acids in the presence of ferritin. *Arch. Biochem. Biophys.*, **373**, 368–374.

145. O.A. Gederaas, J.W. Lagerberg, O. Brekke, K. Berg, T.M. Dubbelman (2000). 5-Aminolevulinic acid induced lipid peroxidation after light exposure on human colon carcinoma cells and effects of alpha-tocopherol treatment. *Cancer Lett.*, **159**, 23–32.

146. T. Douki, J. Onuki, M.H.G. Medeiros, E.J.H. Bechara, J. Cadet, P. Di Mascio (1998). DNA alkylation by 4,5-dioxovaleric acid, the final oxidation product of 5-aminolevulinic acid. *Chem. Res. Toxicol.*, **11**, 150–157.

147. P.I. Oteiza, C.G. Kleinman, M. Demasi, E.J. Bechara (1995). 5-Aminolevulinic acid induces iron release from ferritin. *Arch. Biochem. Biophys.*, **316**, 607–611.

148. R. Neal, P. Yang, J. Fiechtl, D.Yildiz, H. Gurer, N. Ercal (1997). Pro-oxidant effects of δ-aminolevulinic acid (δ-ALA) on Chinese hamster ovary (CHO) cells. *Toxicol. Lett.*, **91**, 169–178.

149. M. Yusof, D. Yildiz, N. Ercal (1999). *N*-acetyl-L-cysteine protects against δ-aminolevulinic acid-induced 8-hydroxydeoxyguanosine formation. *Toxicol. Lett.*, **106**, 41–47.

150. F.G. Princ, A.G. Maxit, C. Cardalda, A. Batlle, A.A. Juknat (1998). In vivo protection by melatonin against δ-aminolevulinic acid-induced oxidative damage and its antioxidant effect on the activity of haem enzymes. *J. Pineal Res.*, **24**, 1–8.

151. R.C. Carneiro, R.J. Reiter (1998). Delta-aminolevulinic acid-induced lipid peroxidation in rat kidney and liver is attenuated by melatonin: an in vitro and in vivo study. *J. Pineal Res.*, **24**, 131–136.

152. R.C. Carneiro, R.J. Reiter (1998). Melatonin protects against lipid peroxidation induced by delta-aminolevulinic acid in rat cerebellum, cortex and hippocampus. *Neuroscience*, **82**, 293–299.

153. M. Karbownik, D.X. Tan, R.J. Reiter (2000). Melatonin reduces the oxidation of nuclear DNA and membrane lipids induced by the carcinogen delta-aminolevulinic acid. *Int. J. Cancer*, **88**, 7–11.

154. R.J. Reiter (1998). Oxidative damage in the central nervous system: protection by melatonin. *Prog. Neurobiol.*, **56**, 359–384.

155. R.J. Reiter, J.M. Guerrero, J.J. Garcia, D. Acuna-Castroviejo (1998). Reactive oxygen intermediates, molecular damage, and aging. Relation to melatonin. *Ann. N.Y. Acad. Sci.*, **854**, 410–424.

156. J. Cadet, V.M. Carvalho, J. Onuki, T. Douki, M.H. Medeiros, P.D. Di Mascio (1999). Purine DNA adducts of 4,5-dioxovaleric acid and 2,4-decadienal. *IARC. Sci. Publ.*, 103–113.

157. L.J. Marnett, H.K. Hurd, M.C. Hollstein, D.E. Levin, H. Esterbauer, B.N. Ames (1985). Naturally occurring carbonyl compounds are mutagens in Salmonella tester strain TA104. *Mutat. Res.*, **148**, 25–34.

158. V.L. Dellarco (1988). A mutagenicity assessment of acetaldehyde. *Mutat. Res.*, **195**, 1–20.

159. D.M. Fiedler, P.M. Eckl, B. Krammer (1996). Does delta-aminolaevulinic acid induce genotoxic effects? *J. Photochem. Photobiol B.*, **33**, 39–44.

160. A. Gorchein, R. Webber (1987). delta-Aminolaevulinic acid in plasma, cerebrospinal fluid, saliva and erythrocytes: studies in normal, uraemic and porphyric subjects. *Clin. Sci.*, **72**, 103–112.

161. J.G. Gubler, M.J. Bargetzi, U.A. Meyer (1990). Primary liver carcinoma in two sisters with acute intermittent porphyria [letter] [see comments]. *Am. J. Med.*, **89**, 540–541.

162. F. Lithner, L. Wetterberg (1984). Hepatocellular carcinoma in patients with acute

intermittent porphyria. *Acta Med. Scand.*, **215**, 271–274.

163. P.L. Thunnissen, J. Meyer, R.W. de Koning (1991). Acute intermittent porphyria and primary liver-cell carcinoma. *Neth. J. Med.*, **38**, 171–174.

164. K. Rick, R. Sroka, H. Stepp, M. Kriegmair, R.M. Huber, K. Jacob, R. Baumgartner (1997). Pharmacokinetics of 5-aminolevulinic acid-induced protoporphyrin IX in skin and blood. *J. Photochem. Photobiol. B Biol.*, **40**, 313–319.

165. J. Van den Boogert, R. van Hillegersberg, F.W. de Rooij, R.W. de Bruin, A. Edixhoven-Bosdijk, A.B. Houtsmuller, P.D. Siersema, J.H. Wilson, H.W. Tilanus (1998). 5-Aminolaevulinic acid-induced protoporphyrin IX accumulation in tissues: pharmacokinetics after oral or intravenous administration. *J. Photochem. Photobiol. B Biol.*, **44**, 29–38.

166. M. Demasi, C.A. Costa, C. Pascual, S. Llesuy, E.J.H. Bechara (1997). Oxidative tissue response promoted by 5-aminolevulinic acid promptly induces the increase of plasma antioxidant capacity. *Free Radic. Res.*, **26**, 235–243.

167. F.B. McGillion, C.G. Thompson, A. Goldberg (1975). Tissue uptake of o-aminolaevulinic acid. *Biochem. Pharmacol.*, **24**, 99–301.

168. C. Fritsch, C. Abels, A.E. Goetz, W. Stahl, K. Bolsen, T. Ruzicka, G. Goerz, H. Sies (1997). Porphyrins preferentially accumulate in a melanoma following intravenous injection of 5-aminolevulinic acid. *Biol. Chem. Hoppe Seyler.*, **378**, 51–57.

169. A. Wennberg, O. Larko, P. Lonnroth, G. Larson, A. Krogstad (2000). Delta-aminolevulinic acid in superficial basal cell carcinomas and normal skin-a microdialysis and perfusion study. *Clin. Exp. Dermatol.*, **25**, 317–322.

170. C. Fritsch, B. Verwohlt, K. Bolsen, T. Ruzicka, G. Goerz (1996). Influence of topical photodynamic therapy with 5-aminolevulinic acid on porphyrin metabolism. *Arch. Dermatol. Res.*, **288**, 517–521.

171. W.E. Muller, S.H. Snyder (1977). delta-Aminolevulinic acid: influences on synaptic GABA receptor binding may explain CNS symptoms of porphyria. *Ann. Neurol.*, **2**, 340–342.

172. M.G. Cutler, M.R. Moore, F.G. Ewart (1979). Effects of delta-aminolaevulinic acid administration on social behaviour in the laboratory mouse. *Psychopharmacology (Berl.)*, **61**, 131–135.

173. B. Pereira, R. Curi, E. Kokubun, E.J. Bechara (1992). 5-aminolevulinic acid-induced alterations of oxidative metabolism in sedentary and exercise-trained rats. *J. Appl. Physiol.*, **72**, 226–230.

174. M. Demasi, C.A. Penatti, R. DeLucia, E.J. Bechara (1996). The prooxidant effect of 5-aminolevulinic acid in the brain tissue of rats: implications in neuropsychiatric manifestations in porphyrias. *Free Radic. Biol. Med.*, **20**, 291–299.

175. J. Van Steveninck, J.P. Boegheim, T.M. Dubbelman, J. Van der Zee (1988). The influence of porphyrins on iron-catalysed generation of hydroxyl radicals. *Biochem. J.*, **250**, 197–201.

176. S. Afonso, G. Vanore, A. Batlle (1999). Protoporphyrin IX and oxidative stress. *Free Radic. Res.*, **31**, 161–170.

177. R. Stocker (1990). Induction of haem oxygenase as a defence against oxidative stress. *Free Radic. Res. Commun.*, **9**, 101–112.

178. M.A. Herman, J. Webber, D. Fromm, D. Kessel (1998). Hemodynamic effects of 5-aminolevulinic acid in humans. *J. Photochem. Photobiol. B Biol.*, **43**, 61–65.

179. W. Chung, J. Lee, W. Lee, Y. Surh, K. Park (2000). Protective effects of hemin and tetrakis(4-benzoic acid)porphyrin on bacterial mutagenesis and mouse skin carcinogenesis induced by 7,12-dimethylbenz[a]anthracene. *Mutat. Res.*, **472**, 139–145.

180. I.M. Stender, N. Bech-Thomsen, T. Poulsen, H.C. Wulf (1997). Photodynamic therapy with topical δ-aminolevulinic acid delays UV photocarcinogenesis in hairless mice. *Photochem.*

Photobiol., **66**, 493–496.

181. C. Fritsch, J. Batz, K.Bolsen, K.W. Schulte, M. Zumdick, T. Ruzicka, G. Goerz (1997). Ex vivo application of δ-aminolevulinic acid induces high and specific porphyrin levels in human skin tumors: Possible basis for selective photodynamic therapy. *Photochem. Photobiol.*, **66**, 114–118.

182. R.T. Walters, T.G. Gribble, H.C. Schartz (1963). Synthesis of haem in normal and leukemic leukocytes. *Nature*, **197**, 1213–1214.

183. S.N. Datta, C.S. Loh, A.J. MacRobert, S.D. Whatley, P.N. Matthews (1998). Quantitative studies of the kinetics of 5-aminolaevulinic acid induced fluorescence in bladder transitional cell carcinoma. *Br. J. Cancer*, **78**, 1113–1118.

184. C. Fritsch, K. Lang, W. Neuse, T. Ruzicka, P. Lehmann (1998). Photodynamic diagnosis and therapy in dermatology. *Skin Pharmacol. Appl. Skin Physiol.*, **11**, 358–373.

185. N.M. Navone, C.F. Polo, A.L. Frisardi, N.E. Andrade, A.M. Battle (1990). Heme biosynthesis in human breast cancer-mimetic in vitro studies and some heme enzymic activity levels. *Int. J. Biochem.*, **22**, 1407–1411.

186. R. Riesenberg, C. Fuchs, M. Kriegmair (1996). Photodynamic effects of 5-aminolevulinic acid-induced porphyrin on human bladder carcinoma cells in vitro. *Eur. J. Cancer [A]*, **32A**, 328–334.

187. R.C. Krieg, S. Fickweiler, O.S. Wolfbeis, R. Knuechel (2000). Cell-type specific protoporphyrin IX metabolism in human bladder cancer in vitro. *Photochem. Photobiol.*, **72**, 226–233.

188. E.R. Gallegos, I. DeLeon Rodriguez, L.A. Martinez Guzman, A.J. Perez Zapata (1999). In vitro study of biosynthesis of protoporphyrin IX induced by delta-aminolevulinic acid in normal and cancerous cells of the human cervix. *Arch. Med. Res.*, **30**, 163–170.

189. G. Li, M.R. Szewczuk, L. Raptis, J.G. Johnson, G.E. Weagle, R.H. Pottier, J.C. Kennedy (1999). Rodent fibroblast model for studies of response of malignant cells to exogenous 5-aminolevulinic acid. *Br. J. Cancer.*, **80**, 676–684.

190. L. Wyld, J.L. Burn, M.W.R. Reed, N.J. Brown (1997). Factors affecting amninolaevulinic acid-induced generation of protoporphyrin IX. *Br. J. Cancer.*, **76**, 705–712.

191. P. Chakrabarti, E. Orihuela, N. Egger, D.E. Neal Jr., R. Gangula, A. Adesokun, M. Motamedi (1998). Delta-aminolevulinic acid-mediated photosensitization of prostate cell lines: Implication for photodynamic therapy of prostate cancer. *Prostate*, **36**, 211–218.

192. N.G. Egger, J.A.Schoenecker Jr., W.K. Gourley, M. Motamedi, K.E. Anderson, S.A. Weinman (1997). Photosensitization of experimental hepatocellular carcinoma with protoporphyrin synthesized from administered δ-aminolevulinic acid: Studies with cultured cells and implanted tumors. *J. Hepatol.*, **26**, 913–920.

193. D. Grebenova, H. Cajthamlova, J. Bartosova, J. Marinov, H. Klamova, O. Fuchs, Z. Hrkal (1998). Selective destruction of leukaemic cells by photo-activation of 5-aminolaevulinic acid-induced protoporphyrin-IX. *J. Photochem. Photobiol. B Biol.*, **47**, 74–81.

194. N. Rebeiz, C.C. Rebeiz, S. Arkins, K.W. Kelley, C.A. Rebeiz (1992). Photodestruction of tumor cells by induction of endogenous accumulation of protoporphyrin IX: enhancement by 1,10-phenanthroline. *Photochem. Photobiol.*, **55**, 431–435.

195. E.A. Hryhorenko, K. Rittenhouse-Diakun, N.S. Harvey, J. Morgan, C.C. Stewart, A.R. Oseroff (1998). Characterization of endogenous protoporphyrin IX induced by δ-aminolevulinic acid in resting and activated peripheral blood lymphocytes by four-color flow cytometry. *Photochem. Photobiol.*, **67**, 565–572.

196. S. Eleouet, J. Carre, V. Vonarx, D. Heyman, Y. Lajat, T. Patrice (1997). Delta-aminolevulinic acid-induced fluorescence in normal human lymphocytes. *J. Photochem. Photobiol. B Biol..*, **41**, 22–29.

197. E. Schick, R. Kaufmann, A. Ruck, A. Hainzl, W.H. Boehncke (1995). Influence of

activation and differentiation of cells on the effectiveness of photodynamic therapy. *Acta Derm. Venereol.*, **75**, 276–279.

198. P. Collas, P. Alestrom (1997). Nuclear localization signals: a driving force for nuclear transport of plasmid DNA in zebrafish. [Review] [37 refs]. *Biochemistry & Cell Biology*, **75**, 633–640.

199. P. Steinbach, H. Weingandt, R. Baumgartner, M. Kriegmair, F. Hofstadter, R. Knuchel (1995). Cellular fluorescence of the endogenous photosensitizer protoporphyrin IX following exposure to 5-aminolevulinic acid. *Photochem. Photobiol.*, **62**, 887–895.

200. Z. Malik, B. Ehrenberg, A. Faraggi (1989). Inactivation of erythrocytic, lymphocytic and myelocytic leukemic cells by photoexcitation of endogenous porphyrins. *J. Photochem. Photobiol B.*, **4**, 195–205.

201. J.C. Tsai, Y.Y. Hsiao, L.J. Teng, C.T. Chen, M.C. Kao (1999). Comparative study on the ALA photodynamic effects of human glioma and meningioma cells. *Lasers Surg. Med.*, **24**, 296–305.

202. G. Li, M.R. Szewczuk, R.H. Pottier, J.C. Kennedy (1999). Effect of mammalian cell differentiation on response to exogenous 5-aminolevulinic acid. *Photochem. Photobiol.*, **69**, 231–235.

203. S.N. Datta, R. Allman, C.S. Loh, M. Mason, P.N. Matthews (1997). Photodynamic therapy of bladder cancer cell lines. *Br. J. Urol.*, **80**, 421–426.

204. S.L. Gibson, M.L. Nguyen, J.J. Havens, A. Barbarin, R. Hilf (1999). Relationship of δ-aminolevulinic acid-induced protoporphyrin IX levels to mitochondrial content in neoplastic cells in vitro. *Biochem. Biophys. Res. Commun.*, **265**, 315–321.

205. M. Kondo, N. Hirota, T. Takaoka, M. Kajiwara (1993). Heme-biosynthetic enzyme activities and porphyrin accumulation in normal liver and hepatoma cell lines of rat. *Cell Biol. Toxicol.*, **9**, 95–105.

206. H.W. Lim, S. Behar, D. He (1994). Effect of porphyrin and irradiation on heme biosynthetic pathway in endothelial cells. *Photodermatol. Photoimmunol. Photomed.*, **10**, 17–21.

207. D. He, S. Sassa, H.W. Lim (1993). Effect of UVA and blue light on porphyrin biosynthesis in epidermal cells. *Photochem. Photobiol.*, **57**, 825–829.

208. D. He, E. Karas, S.Sassa, H.W. Lim (1993). Porphyrin synthesis by murine epidermal cells. *Skin Pharmacol.*, **6**, 20–25.

209. D. He, S. Behar, N. Nomura, S.Sassa, H.W. Lim (1995). The effect of ALA and radiation on porphyrin/heme biosynthesis in endothelial cells. *Photochem. Photobiol.*, **61**, 656–661.

210. S.L. Gibson, J.J. Havens, T.H. Foster, R. Hilf (1997). Time-dependent intracellular accumulation of δ-aminolevulinic acid, induction of porphyrin synthesis and subsequent phototoxicity. *Photochem. Photobiol.*, **65**, 416–421.

211. Princ, F.G., Juknat, A.A., and Batlle, A.M. (1994). Porphyrinogenesis in rat cerebellum. Effect of high delta-aminolevulinic acid concentration. *Gen. Pharmacol.*, **25**, 761–766.

212. A.A. Juknat, M.L. Kotler, A.M. Batlle (1995). High delta-aminolevulinic acid uptake in rat cerebral cortex: effect on porphyrin biosynthesis. *Comparative Biochemistry & Physiology, Part C Pharmacology*, 143–150.

213. K. Berg, H. Anholt, O. Bech, J. Moan (1996). The influence of iron chelators on the accumulation of protoporphyrin IX in 5-aminolaevulinic acid-treated cells. *Br. J. Cancer*, **74**, 688–697.

214. N. Schoenfeld, R. Mamet, Y. Nordenberg, M. Shafran, T. Babushkin, Z. Malik (1994). Protoporphyrin biosynthesis in melanoma B16 cells stimulated by 5-aminolevulinic acid and chemical inducers: characterization of photodynamic inactivation. *Int. J. Cancer*, **56**, 106–112.

215. S. Iinuma, S.S. Farshi, B. Ortel, T. Hasan (1994). A mechanistic study of cellular

photodestruction with 5-aminolaevulinic acid-induced porphyrin [published erratum appears in *Br. J. Cancer* 1994 Dec;70(6):1283]. *Br. J. Cancer*, **70**, 21–28.

216. S.L. Ratcliffe, E.K. Matthews (1995). Modification of the photodynamic action of delta-aminolaevulinic acid (ALA) on rat pancreatoma cells by mitochondrial benzodiazepine receptor ligands. *Br. J. Cancer*, **71**, 300–305.

217. Ø Bech, Q. Peng, K. Berg, J. Moan (1992). In: P. Spinelli, M. Dal Fante, R. Marchsini (Eds), *Photodynamic Therapy and Biomedical Lasers* (pp. 521–525), Elsevier Science Publishers B.V., Amsterdam.

218. R.M. Szeimies, C. Abels, C. Fritsch, S. Karrer, P. Steinbach, W. Baumler, G. Goerz, A.E. Goetz, M. Landthaler (1995). Wavelength dependency of photodynamic effects after sensitization with 5-aminolevulinic acid in vitro and in vivo. *J. Invest. Dermatol.*, **105**, 672–677.

219. W. Dietel, K. Bolsen, E. Dickson, C. Fritsch, R. Pottier, R. Wendenburg (1996). Formation of water-soluble porphyrins and protoporphyrin IX in 5-aminolevulinic-acid-incubated carcinoma cells. *J. Photochem. Photobiol B.*, **33**, 225–231.

220. N.M. Navone, A.L. Frisardi, E.R. Resnik, A.M.d.C. Battle, C.F. Polo (1988). Porphyrin biosynthesis in human breast cancer. Preliminary mimetic in vitro studies. *Med. Sci. Res.*, **16**, 61–62.

221. I. Durko, A. Juhasz (1986). Porphyrin synthesis in primary nervous tissue cultures from 10(−3) M delta-aminolaevulinic acid in the presence of melatonin and neuropeptides. *Neurochem. Res.*, **11**, 607–615.

222. O.A. Gederaas, K. Berg, I. Romslo (2000). A comparative study of normal and reverse phase high pressure liquid chromatography for analysis of porphyrins accumulated after 5-aminolaevulinic acid treatment of colon adenocarcinoma cells. *Cancer Lett.*, **150**, 205–213.

223. P. Hinnen, F.W.M. De Rooij, M.L.F. Van Velthuysen, A. Edixhoven, R. van Hillegersberg, H.W. Tilanus, J.H.P. Wilson, P.D. Siersema (1998). Biochemical basis of 5-aminolaevulinic acid induced protoporphyrin IX accumulation: a study in patients with (pre)malignant lesions of the oesophagus. *Br. J. Cancer*, **78**, 679–682.

224. P. Hinnen, F.W. de Rooij, E.M. Terlouw, A. Edixhoven, H. Van Dekken, R. van Hillegersberg, H.W. Tilanus, J.H. Wilson, P.D. Siersema (2000). Porphyrin biosynthesis in human Barrett's oesophagus and adenocarcinoma after ingestion of 5-aminolaevulinic acid. *Br. J. Cancer*, **83**, 539–543.

225. N. Schoenfeld, O. Epstein, M. Lahav, R. Mamet, M. Shaklai, A. Atsmon (1988). The heme biosynthetic pathway in lymphocytes of patients with malignant lymphoproliferative disorders. *Cancer Lett.*, **43**, 43–48.

226. H.A. Dailey, A. Smith (1984). Differential interaction of porphyrins used in photoradiation therapy with ferrochelatase. *Biochem. J.*, **223**, 441–445.

227. J.D. Laskey, P. Ponka, H.M. Schulman (1986). Control of heme synthesis during Friend cell differentiation: role of iron and transferrin. *J. Cell Physiol.*, **129**, 185–192.

228. K. Rittenhouse-Diakun, H. Van Leengoed, J. Morgan, E. Hryhorenko, G. Paszkiewicz, J.E. Whitaker, A.R. Oseroff (1995). The role of transferrin receptor (CD71) in photodynamic therapy of activated and malignant lymphocytes using the heme precursor delta-aminolevulinic acid (ALA). *Photochem. Photobiol.*, **61**, 523–528.

229. L.M. Neckers (1991). Regulation of transferrin receptor expression and control of cell growth. *Pathobiology*, **59**, 11–18.

230. C. Pourzand, O. Reelfs, E. Kvam, R.M. Tyrrell (1999). The iron regulatory protein can determine the effectiveness of 5-aminolevulinic acid in inducing protoporphyrin IX in human primary skin fibroblasts. *J. Invest. Dermatol.*, **112**, 419–425.

231. D. He, E. Karas, S. Sassa, H.W. Lim (1993). Porphyrin synthesis by murine epidermal cells. *Skin Pharmacol.*, **6**, 20–25.

232. B. Ortel, A. Tanew, H. Honigsmann (1993). Lethal photosensitization by endogenous porphyrins of PAM cells – modification by desferrioxamine. *J. Photochem. Photobiol. B.*, **17**, 273–278.

233. G.H. Elder (1982). Enzymatic defects in porphyria: an overview. *Semin. Liver Dis.*, **2**, 87–99.

234. S.L. Gibson, J.J. Havens, M.L. Nguyen, R. Hilf (1999). δ-Aminolaevulinic acid-induced photodynamic therapy inhibits protoporphyrin IX biosynthesis and reduces subsequent treatment efficacy in vitro. *Br. J. Cancer*, **80**, 998–1004.

235. R. Hilf, J.J. Havens, S.L. Gibson (1999). Effect of δ-aminolevulinic acid on protoporphyrin IX accumulation in tumor cells transfected with plasmids containing porphobilinogen deaminase DNA. *Photochem. Photobiol.*, **70**, 334–340.

236. S. Afonso, R.E. De Salamanca, A. Batlle (1998). Porphyrin-induced protein structural alterations of heme enzymes II: Protection of 5-aminolevulinic acid dehydratase and porphobilinogen deaminase from the photodynamic and non-photodynamic effects of URO and PROTO. *Int. J. Biochem. Cell Biol.*, **30**, 535–543.

237. J. Van den Boogert, H.J. van Staveren, R.W.F. De Bruin, F.W.M. De Rooij, A. Edixhoven-Bosdijk, P.D. Siersema, R. van Hillegersberg (2000). Fractionated illumination in oesophageal ALA-PDT: effect on ferrochelatase activity. *J. Photochem. Photobiol. B*, **56**, 53–60.

238. J. Moan, O. Bech, J.M. Gaullier, T. Stokke, H.B. Steen, L.W. Ma, K. Berg (1998). Protoporphyrin IX accumulation in cells treated with 5-aminolevulinic acid: dependence on cell density, cell size and cell cycle. *Int. J. Cancer*, **75**, 134–139.

239. I. Georgakoudi, P.C. Keng, T.H. Foster (1999). Hypoxia significantly reduces aminolaevulinic acid-induced protoporphyrin IX synthesis in EMT6 cells. *Br. J. Cancer*, **79**, 1372–1377.

240. H. Fukuda, A.M. Batlle, P.A. Riley (1993). Kinetics of porphyrin accumulation in cultured epithelial cells exposed to ALA. *Int. J. Biochem.*, **25**, 1407–1410.

241. R.O. Hynes (1992). Integrins: versatility, modulation, and signaling in cell adhesion. *Cell*, **69**, 11–25.

242. R. Knuechel, P. Keng, F. Hofstaedter, V. Langmuir, R.M. Sutherland, D.P. Penney (1990). Differentiation patterns in two- and three-dimensional culture systems of human squamous carcinoma cell lines. *Am. J. Pathol.*, **137**, 725–736.

243. C. Rosales, V. O'Brien, L. Kornberg, R. Juliano (1995). Signal transduction by cell adhesion receptors. *Biochim. Biophys. Acta.*, **242**, 77–98.

244. R. Raghow (1994). The role of extracellular matrix in postinflammatory wound healing and fibrosis. *FASEB. J.*, **8**, 823–831.

245. J. Hanania, Z. Malik (1992). The effect of EDTA and serum on endogenous porphyrin accumulation and photodynamic sensitization of human K562 leukemic cells. *Cancer Lett.*, **65**, 127–131.

246. J. Bartosova, Z. Hrkal (2000). Accumulation of protoporphyrin-IX (PpIX) in leukemic cell lines following induction by 5-aminolevulinic acid (ALA). *Comp. Biochem. Physiol. C. Toxicol. Pharmacol.*, **126**, 245–252.

247. C. Fuchs, R. Riesenberg, J. Siegert, R. Baumgartner (1997). H-dependent formation of 5-aminolaevulinic acid-induced protoporphyrin IX in fibrosarcoma cells. *J. Photochem. Photobiol. B Biol.*, **40**, 49–54.

248. Ø. Bech, K. Berg, J. Moan (1997). The pH dependency of protoporphyria IX formation in cells incubated with 5-aminolevulinic acid. *Cancer Lett.*, **113**, 25–29.

249. K.A. Kennedy, J.D. McGurl, L. Leondaridis, O. Alabaster (1985). pH dependence of mitomycin C-induced cross-linking activity in EMT6 tumor cells. *Cancer Res.*, **45**, 3541–3547.

250. A. Casas, A.M.D. Batlle, A.R. Butler, D. Robertson, E.H. Brown, A. MacRobert, P.A. Riley (1999). Comparative effect of ALA derivatives on protoporphyrin IX production in human and rat skin organ cultures. *Br. J. Cancer.*, **80**, 1525–1532.

251. R. Washbrook, P.A. Riley (1997). Comparison of δ-aminolaevulinic acid and its methyl ester as an inducer of porphyrin synthesis in cultured cells. *Br. J. Cancer*, **75**, 1417–1420.

252. J. Kloek, W. Akkermans, G.M.J.B. Van Henegouwen (1998). Derivatives of 5-aminolevulinic acid for photodynamic therapy: Enzymatic conversion into protoporphyrin. *Photochem. Photobiol.*, **67**, 150–154.

253. Q. Peng, J. Moan, T. Warloe, V. Iani, H.B. Steen, A. Bjørseth, J.M. Nesland (1996). Build-up of esterified aminolevulinic-acid-derivative-induced porphyrin fluorescence in normal mouse skin. *J. Photochem. Photobiol. B.*, **34**, 95–96.

254. R. Sørensen, P. Juzenas, V. Iani, J. Moan (2000). In: B. Ehrenberg, K. Berg, (Eds), *Photochemotherapy of Cancer and Other Diseases* (pp. 77–81).

255. P. Tompa, E. Schad, A. Baki, A. Alexa, J. Batke, P. Friedrich (1995). An ultrasensitive, continuous fluorometric assay for calpain activity. *Anal. Biochem.*, **228**, 287–293.

256. B. Ortel, A. Tanew, H. Honigsmann (1993). Lethal photosensitization by endogenous porphyrins of PAM cells – modification by desferrioxamine. *J. Photochem. Photobiol. B Biol.*, **17**, 273–278.

257. S.C. Chang, A.J. MacRobert, J.B. Porter, S.G. Bown (1997). The efficacy of an iron chelator (CP94) in increasing cellular protoporphyrin IX following intravesical 5-aminolaevulinic acid administration: an in vivo study. *J. Photochem. Photobiol. B Biol.*, **38**, 114–122.

258. A. Curnow, B.W. McIlroy, M.J. Postle-Hacon, J.B. Porter, A.J. MacRobert, S.G. Bown (1998). Enhancement of 5-aminolaevulinic acid-induced photodynamic therapy in normal rat colon using hydroxypyridinone iron-chelating agents. *Br. J. Cancer*, **78**, 1278–1282.

259. A. Maffei, K. Papadopoulos, P.E. Harris (1997). MHC class I antigen processing pathways. *Hum. Immunol.*, **54**, 91–103.

260. W.C. Tan, N. Krasner, P. O'Toole, M. Lombard (1997). Enhancement of photodynamic therapy in gastric cancer cells by removal of iron. *Gut*, **41**, 14–18.

261. J. Regula, A.J. MacRobert, A. Gorchein, G.A. Buonaccorsi, S.M. Thorpe, S.M., Spencer, A.R. Hatfield, S.G. Bown (1995). Photosensitisation and photodynamic therapy of oesophageal, duodenal, and colorectal tumours using 5 aminolaevulinic acid induced protoporphyrin IX-a pilot study. *Gut*, **36**, 67–75.

262. C. Brodie, G. Siriwardana, J. Lucas, R, Schleicher, N. Terada, A. Szepesi, E. Gelfand, P. Seligman (1993). Neuroblastoma sensitivity to growth inhibition by deferrioxamine: evidence for a block in G1 phase of the cell cycle. *Cancer Res.*, **53**, 3968–3975.

263. A. Donfrancesco, G. Deb, L. De Sio, R. Cozza, A. Castellano (1996). Role of deferoxamine in tumor therapy. *Acta Haematol*, **95**, 66–69.

264. A. Donfrancesco, G. Deb, C. Dominici, D. Pileggi, M.A. Castello, L. Helson (1990). Effects of a single course of deferoxamine in neuroblastoma patients. *Cancer Res.*, **50**, 4929–4930.

265. N.F. Olivieri, G. Koren, D. Matsui, P.P. Liu, L. Blendis, R. Cameron, R.A. McClelland, D.M. Templeton (1992). Reduction of tissue iron stores and normalization of serum ferritin during treatment with the oral iron chelator L1 in thalassemia intermedia. *Blood*, **79**, 2741–2748.

266. Y. Bentur, M. McGuigan, G. Koren (1991). Deferoxamine (desferrioxamine). New toxicities for an old drug. *Drug Saf.*, **6**, 37–46.

267. S.G. Afonso, S. Chinarro, J.J. Munoz, R.E. De Salamanca, A.M. Batlle (1990). Photodynamic and non-photodynamic action of several porphyrins on the activity of some heme-enzymes. *J. Enzym. Inhib.*, **3**, 303–310.

268. A.M. Batlle, R.E. De Salamanca, S. Chinarro, S.G. Afonso, A.M. Stella (1986). Photodynamic inactivation of red cell uroporphyrinogen decarboxylase by porphyrins. *Int. J. Biochem.*, **18**, 143–147.

269. S.G. Afonso, S. Chinarro, R.E. De Salamanca, A.M. Batlle (1991). Further evidence on the photodynamic and the novel non-photodynamic inactivation of uroporphyrinogen decarboxylase by uroporphyrin I. *J. Enzym. Inhib.*, **5**, 225–233.

270. S.G. Afonso, C.F. Polo, R.E. De Salamanca, A. Batlle (1996). Mechanistic studies on uroporphyrin I-induced photoinactivation of some heme-enzymes. *Int. J. Biochem. Cell Biol.*, **28**, 415–420.

271. D.W. Briggs, L.W. Condie, R.M. Sedman, T.R. Tephly (1976). Delta-Aminolevulinic acid synthetase in the heart. *J. Biol. Chem.*, **251**, 4996–5001.

272. F. De Matteis, A. Gibbs (1972). Stimulation of liver 5-aminolaevulinate synthetase by drugs and its relevance to drug-induced accumulation of cytochrome P-450. Studies with phenylbutazone and 3,5-diethoxycarbonyl-1,4-dihydrocollidine. *Biochem. J.*, **126**, 1149–1160.

273. S.G. Afonso, C.F. Polo, N.M. Navone, E.S. Vazquez, A.M. Buzaleh, E. Schoua, A.M. Batlle (1990). In vivo prevention and reversal by the antimitotic colchicine and other related compounds of some porphyrinogenic drug action on heme pathway. *Gen. Pharmacol.*, **21**, 423–426.

274. P.S. Ebert, I. Wars, D.N. Buell (1976). Erythroid differentiation in cultured Friend leukemia cells treated with metabolic inhibitors. *Cancer Res.*, **36**, 1809–1813.

275. F. Iwasa, R.A. Galbraith, S. Sassa (1988). Effects of dimethyl sulphoxide on the synthesis of plasma proteins in the human hepatoma HepG2. Induction of an acute-phase-like reaction. *Biochem. J.*, **253**, 927–930.

276. T. Rutherford, G.G. Thompson, M.R. Moore (1979). Heme biosynthesis in Friend erythroleukemia cells: control by ferrochelatase. *Proc. Natl. Acad. Sci. USA*, **76**, 833–836.

277. H. Fujita, M. Yamamoto, T. Yamagami, N. Hayashi, T.R. Bishop, H. De Verneuil, T. Yoshinaga, S. Shibahara, R. Morimoto, S. Sassa (1991). Sequential activation of genes for heme pathway enzymes during erythroid differentiation of mouse Friend virus-transformed erythroleukemia cells. *Biochim. Biophys. Acta.*, **1090**, 311–316.

278. Y. Fukuda, H. Fujita, S. Taketani, S. Sassa (1993). Dimethyl sulphoxide and haemin induce ferrochelatase mRNA by different mechanisms in murine erythroleukaemia cells. *Br. J. Haematol.*, **83**, 480–484.

279. L.W. Ma, J. Moan, K. Berg, Q. Peng, H.B. Steen (1993). Potentiation of photodynamic therapy by mitomycin C in cultured human colon adenocarcinoma cells. *Radiat. Res.*, **134**, 22–28.

280. R.M. Caron, J.W. Hamilton (1995). Preferential effects of the chemotherapeutic DNA crosslinking agent mitomycin C on inducible gene expression in vivo. *Environ. Mol. Mutagen.*, **25**, 4–11.

281. J. Gagnebin, M. Brunori, M. Otter, L. Juillerat-Jeanneret, P. Monnier, R. Iggo (1999). A photosensitising adenovirus for photodynamic therapy. *Gene Ther.*, **6**, 1742–1750.

282. N. Kawasaki, K. Morimoto, T. Tanimoto, T. Hayakawa (1996). Control of hemoglobin synthesis in erythroid differentiating K562 cells. 1. Role of iron in erythroid cell heme synthesis. *Arch. Biochem. Biophys.*, **328**, 289–294.

283. Z. Malik, S.D. Chitayat, Y. Langzam (1988). Hemin dependent morphological maturation and endogenous porphyrin synthesis by K562 leukemic cells. *Cancer Lett.*, **41**, 203–209.

284. E.E. Cable, J.A. Pepe, S.E. Donohue, R.W. Lambrecht, H.L. Bonkovsky (1994). Effects of mifepristone (RU-486) on heme metabolism and cytochromes P-450 in cultured chick embryo liver cells, possible implications for acute porphyria. *Eur. J. Biochem.*, **225**, 651–657.

285. L. Wyld, M.W.R. Reed, N.J. Brown (1998). The influence of hypoxia and pH on aminolaevulinic acid-induced photodynamic therapy in bladder cancer cells in vitro. *Br. J. Cancer*, **77**, 1621–1627.

286. S. Sano, S. Granick (1961). Mitochondrial coproporphyrinogen oxidase and protoporhyrin formation. *J. Biol. Chem.*, **236**, 1173–1180.

287. R. Poulson, W.J. Polglase (1975). The enzymic conversion of protoporphyrinogen IX to protoporphyrin IX. Protoporphyrinogen oxidase activity in mitochondrial extracts of Saccharomyces cerevisiae. *J. Biol. Chem.*, **250**, 1269–1274.

288. E. Musgrove, M. Seaman, D. Hedley (1987). Relationship between cytoplasmic pH and proliferation during exponential growth and cellular quiescence. *Exp. Cell Res.*, **172**, 65–75.

289. V.H. Fingar, T.J. Wieman, K.S. McMahon, P.S. Haydon, B.P. Halling, D.A. Yuhas, J.W. Winkelman (1997). Photodynamic therapy using a protoporphyrinogen oxidase inhibitor. *Cancer Res.*, **57**, 4551–4556.

290. B.P. Halling, D.A. Yuhas, V.H. Fingar, J.W. Winkelman (1994). Protoporphyrinogen oxidase inhibitors for tumor therapy. *Am. Chem. Soc. Symp. Ser.*, **559**, 280–290.

291. J. Carre, S. Eleouet, N. Rousset, V. Vonarx, D. Heyman, Y. Lajat, T. Patrice (1999). Protoporphyrin IX fluorescence kinetics in C6 glioblastoma cells after delta-aminolevulinic acid incubation: Effect of a protoporphyrinogen oxidase inhibitor. *Cell Mol. Biol.*, **45**, 433–444.

292. M. Mesenholler, E.K. Matthews (2000). A key role for the mitochondrial benzodiazepine receptor in cellular photosensitisation with delta-aminolaevulinic acid. *Eur. J. Pharmacol.*, **406**, 171–180.

293. A. Casas, H. Fukuda, A.M. Batlle (1998). Potentiation of the 5-aminolevulinic acid-based photodynamic therapy with cyclophosphamide. *Cancer Biochem. Biophys.*, **16**, 183–196.

294. H. Messmann, M. Geisler, U. Gross, C. Abels, R.M. Szeimies, P. Steinbach, R. Knuchel, M. Doss, J. Scholmerich, A. Holstege (1997). Influence of a haematoporphyrin derivative on the protoporphyrin IX synthesis and photodynamic effect after 5-aminolaevulinic acid sensitization in human colon carcinoma cells. *Br. J. Cancer*, **76**, 878–883.

295. Q. Peng, K. Berg, J. Moan, M. Kongshaug, J.M. Nesland (1997). 5-aminolevulinic acid-based photodynamic therapy: Principles and experimental research. *Photochem. Photobiol.*, **65**, 235–251.

296. R.H. Jindra, A. Kubin, H. Kolbabek, G. Alth, W. Dobrowsky (1999). Ambulant photodynamic therapy of superficial malignomas with 5-ALA in combination with folic acid and use of noncoherent light. *Drugs Exp. Clin. Res.*, **25**, 37–41.

297. B.C. Wilson, M. Olivo, G. Singh (1997). Subcellular localization of Photofrin(R) and aminolevulinic acid and photodynamic cross-resistance in vitro in radiation-induced fibrosarcoma cells sensitive or resistant to photofrin-mediated photodynamic therapy. *Photochem. Photobiol.*, **65**, 166–176.

298. J.M. Gaullier, M. Geze, R. Santus, M.T. Sa, E.J.C. Maziere, M. Bazin, P. Morliere, L. Dubertret (1995). Subcellular localization of and photosensitization by protoporphyrin IXhuman keratinocytes and fibroblasts cultivated with 5-aminolevulinic acid. *Photochem. Photobiol.*, **62**, 114–122.

299. I. Shevchuk, V. Chekulayev, J. Moan, K. Berg (1996). Effects of the inhibitors of energy metabolism, lonidamine and levamisole, on 5-aminolevulinic-acid-induced photochemotherapy. *Int. J. Cancer*, **67**, 791–799.

300. J. Mosinger, K. Losinská, T. Abrhámová, S. Veiserová, Z. Micka, I. Nemcová, B. Mosinger (2000). Determination of singlet oxygen production and antibacterial effect of nonpolar porphyrins in heterogeneous systems. *Anal. Lett.*, **33**, 1091–1104.

301. I. Georgakoudi, T.H. Foster (1998). Singlet oxygen-versus nonsinglet oxygen-mediated mechanisms of sensitizer photobleaching and their effects on photodynamic dosimetry. *Photochem. Photobiol.*, **67**, 612–625.

302. S. Karrer, W. Baumler, C. Abels, U. Hohenleutner, M. Landthaler, R.M. Szeimies (1999). Long-pulse dye laser for photodynamic therapy: investigations in vitro and in vivo. *Lasers Surg. Med.*, **25**, 51–59.

303. Y. Gilaberte, D. Pereboom, F.J. Carapeto, J.O. Alda (1997). Flow cytometry study of the role of superoxide anion and hydrogen peroxide in cellular photodestruction with 5-aminolevulinic acid-induced protoporphyrin IX. *Photodermatol. Photoimmunol. Photomed.*, **13**, 43–49.

304. A.K. Haylett, F.I. McNair, D. McGarvey, N.J. Dodd, E. Forbes, T.G. Truscott, J.V. Moore (1997). Singlet oxygen and superoxide characteristics of a series of novel asymmetric photosensitizers. *Cancer Lett.*, **112**, 233–238.

305. J. Moan, K. Berg (1991). The photodegradation of porphyrins in cells can be used to estimate the lifetime of singlet oxygen. *Photochem. Photobiol.*, **53**, 549–553.

306. K.P. Uberriegler, E. Banieghbal, B. Krammer (1995). Subcellular damage kinetics within co-cultivated WI38 and VA13-transformed WI38 human fibroblasts following 5-aminolevulinic acid-induced protoporphyrin IX formation. *Photochem. Photobiol.*, **62**, 1052–1057.

307. Z. Malik, M. Dishi, Y. Garini (1996). Fourier transform multipixel spectroscopy and spectral imaging of protoporphyrin in single melanoma cells. *Photochem. Photobiol.*, **63**, 608–614.

308. F. Ricchelli, P. Barbato, M. Milani, S. Gobbo, C. Salet, G. Moreno (1999). Photodynamic action of porphyrin on Ca^{2+} influx in endoplasmic reticulum: a comparison with mitochondria. *Biochem. J.*, **338** (Pt 1), 221–227.

309. A.A. Schothorst, D. Suurmond, J.S. Ploem (1977). In vitro studies on the protoporphyrin uptake and photosensitivity of normal skin fibroblasts and fibroblasts from patients with erythropoietic protoporphyria. *J. Invest. Dermatol.*, **69**, 551–557.

310. K. Tabata, S. Ogura, I. Okura (1997). Photodynamic efficiency of protoporphyrin IX: Comparison of endogenous protoporphyrin IX induced by 5-aminolevulinic acid and exogenous porphyrin IX. *Photochem. Photobiol.*, **66**, 842–846.

311. K.P. Uberriegler, E. Banieghbal, B. Krammer (1995). Subcellular damage kinetics within co-cultivated WI38 and VA13-transformed WI38 human fibroblasts following 5-aminolevulinic acid-induced protoporphyrin IX formation. *Photochem. Photobiol.*, **62**, 1052–1057.

312. S. Radakovic-Fijan, K. Rappersberger, A. Tanew, H. Honigsmann, B. Ortel (1999). Ultrastructural changes in PAM cells after photodynamic treatment with delta-aminolevulinic acid-induced porphyrins or photosan. *J. Invest. Dermatol.*, **112**, 264–270.

313. J. Odber, M. Cutler, S. Dover, M.R. Moore (1994). Haem precursor effects on [3H]-PK 11195 binding to platelets. *Neuroreport*, **5**, 1093–1096.

314. H. Liang, D.S. Shin, Y.E. Lee, D.C. Nguyen, T.C. Trang, A.H. Pan, S.L. Huang, D.H. Chong, M.W. Berns (1998). Subcellular phototoxicity of 5-aminolaevulinic acid (ALA). *Lasers Surg. Med.*, **22**, 14–24.

315. O.A. Gederaas, K. Thorstensen, I. Romslo (1996). The effect of brief illumination on intracellular free calcium concentration in cells with 5-aminolevulinic acid-induced protoporphyrin IX synthesis. *Scand. J. Clin. Lab. Invest.*, **56**, 583–589.

316. J.M Gaullier, A. Valla, M. Bazin, M. Giraud, L. Dubertret, R. Santus (1997). N-conjugates of 2,5-disubstituted pyrrole and glutathione. Evaluation of their potency as antioxidants against photosensitization of NCTC 2544 keratinocytes by excess endogenous proto-porphyrin IX. *J. Photochem. Photobiol. B.*, **39**, 24–29.

317. K.Berg, Z. Luksiene, J. Moan, L. Ma (1995). Combined treatment of ionizing radiation and photosensitization by 5-aminolevulinic acid-induced protoporphyrin IX. *Radiat. Res.*, **142**, 340–346.

318. A. Casas, H. Fukuda, A.M.D. Batlle (1997). Metabolic changes in the heme pathway driven by cyclophosphamide treatment in mice. *Cell. Mol. Biol.*, **43**, 95–101.

319. S.N. Datta, R. Allman, C. Loh, M. Mason, P.N. Matthews (1997). Effect of photodynamic therapy in combination with mitomycin C on a mitomycin-resistant bladder cancer cell line. *Br. J. Cancer*, **76**, 312–317.

320. P. Baas, C. Michielsen, H. Oppelaar, N. van Zandwijk, F.A. Stewart (1994). Enhancement of interstitial photodynamic therapy by mitomycin C and EO9 in a mouse tumour model. *Int. J. Cancer*, **56**, 880–885.

321. P. Baas, I.P. van Geel, H. Oppelaar, M. Meyer, J.H. Beynen, N. van Zandwijk, F.A. Stewart (1996). Enhancement of photodynamic therapy by mitomycin C: a preclinical and clinical study. *Br. J. Cancer*, **73**, 945–951.

322. B.B. Noodt, K. Berg, T. Stokke, Q. Peng, J.M. Nesland (1996). Apoptosis and necrosis induced with light and 5-aminolaevulinic acid-derived protoporphyrin IX. *Br. J. Cancer*, **74**, 22–29.

323. D. Grebenova, P. Halada, J. Stulik, V. Havlicek, Z. Hrkal (2000 July). Protein changes in HL60 leukemia cells associated with 5-aminolevulinic acid-based photodynamic therapy. Early effects on endoplasmic reticulum chaperones. *Photochem. Photobiol.*, **72**, 16–22.

324. E. Ben-Hur, T.M. Dubbelman (1993). Cytoplasmic free calcium changes as a trigger mechanism in the response of cells to photosensitization. *Photochem. Photobiol.*, **58**, 890–894.

325. T. Verwanger, G. Schnitzhofer, B. Krammer (1998). Expression kinetics of the (proto) oncogenes c-myc and bcl-2 following photodynamic treatment of normal and transformed human fibroblasts with 5-aminolaevulinic acid-stimulated endogenous protoporphyrin IX. *J. Photochem. Photobiol. B Biol.*, **45**, 131–135.

326. S. Stocker, R. Knüchel, R. Sroka, M. Kriegmair, P. Steinbach, R. Baumgartner (1997). Wavelength dependent photodynamic effects on chemically induced rat bladder tumors following intravesical instillation of 5-aminolevulinic acid. *Journal of Urology*, **157**, 357–361.

327. E.A. Hryhorenko, A.R. Oseroff, J. Morgan, K. Rittenhouse-Diakun (1998). Antigen specific and nonspecific modulation of the immune response by aminolevulinic acid based photodynamic therapy. *Immunopharmacology*, **40**, 231–240.

328. S. Gruner, H.D. Volk, F. Noack, H. Meffert, R. von Baehr (1986). Inhibition of HLA-DR antigen expression and of the allogeneic mixed leukocyte reaction by photochemical treatment. *Tissue Antigens.*, **27**, 147–154.

329. M.O.K. Obochi, L.G. Ratkay, J.G. Levy (1997). Prolonged skin allograft survival after photodynamic therapy associated with modification of donor skin antigenicity. *Transplantation*, **63**, 810–817.

330. D.J. Robinson, H.S. De Bruijn, N. van der Veen, M.R. Stringer, S.B. Brown, W.M. Star (1998). Fluorescence photobleaching of ALA-induced protoporphyrin IX during photo-dynamic therapy of normal hairless mouse skin: The effect of light dose and irradiance and the resulting biological effect. *Photochem. Photobiol.*, **67**, 140–149.

331. H.S. De Bruijn, N. van der Veen, D.J. Robinson, W.M. Star (1999). Improvement of systemic 5-aminolevulinic acid-based photodynamic therapy in vivo using light fractionation with a 75-minute interval. *Cancer Res.*, **59**, 901–904.

332. J. Moan, G. Streckyte, S. Bagdonas, O. Bech, K. Berg (1997). Photobleaching of protoporphyrin IX in cells incubated with 5-aminolevulinic acid. *Int. J. Cancer*, **70**, 90–97.

333. S. Bagdonas, L.W. Ma, V. Iani, R. Rotomskis, P. Juzenas, J. Moan (2000). Photo-transformations of 5-aminolevulinic acid-induced protoporphyrin IX in vitro: a spectroscopic study. *Photochem. Photobiol.*, **72**, 186–192.

334. P. Charlesworth, T.G. Truscott (1993). The use of 5-aminolevulinic acid (ALA) in photodynamic therapy (PDT). *J. Photochem. Photobiol. B.*, **18**, 99–100.

335. P. Valat, G.D. Reinhart, D.M. Jameson (1988). Application of time-resolved fluorometry to the resolution of porphyrin-photoproduct mixtures. *Photochem. Photobiol.*, **47**, 787–790.

336. L. Wyld, O. Smith, J. Lawry, M.W.R. Reed, N.J. Brown (1998). Cell cycle phase influences tumour cell sensitivity to aminolaevulinic acid-induced photodynamic therapy in vitro. *Br. J. Cancer*, **78**, 50–55.

337. G.M. Schnitzhofer, B. Krammer (1996). Photodynamic treatment and radiotherapy: Combined effect on the colony-forming ability of V79 Chinese hamster fibroblasts. *Cancer Lett.*, **108**, 93–99.

338. J.P. Boegheim, T.M. Dubbelman, L.H. Mullenders, J. Van Steveninck (1987). Photo-dynamic effects of haematoporphyrin derivative on DNA repair in murine L929 fibroblasts. *Biochem. J.*, **244**, 711–715.

339. H. Kostron, M.R. Swartz, D.C. Miller, R.L. Martuza (1986). The interaction of hematoporphyrin derivative, light, and ionizing radiation in a rat glioma model. *Cancer*, **57**, 964–970.

340. D.A. Bellnier, T.J. Dougherty (1986). Haematoporphyrin derivative photosensitization and gamma-radiation damage interaction in Chinese hamster ovary fibroblasts. *Int. J. Radiat. Biol. Relat. Stud. Phys. Chem. Med.*, **50**, 659–664.

341. S. Eleouet, N. Rousset, J. Carre, L. Bourre, V. Vonarx, Y. Lajat, G.M. van T. Patrice (2000). In vitro fluorescence, toxicity and phototoxicity induced by delta-aminolevulinic acid (ALA) or ALA-esters. *Photochem.Photobiol.*, **71**, 447–454.

342. Y. Berger, A. Greppi, O. Siri, L. Juillerat, R. Neier (1999). New 5-aminolevulinic acid derivatives for selective protoporphyrin IX production in cells. 2nd Internet Conference on Photochemistry and Photobiology(http://www.photobiology.com/photobiology99/contrib/neier/index.htm).

343. R. van Hillegersberg, J.W. Van den Berg, W.J. Kort, O.T. Terpstra, J.H. Wilson (1992). Selective accumulation of endogenously produced porphyrins in a liver metastasis model in rats. *Gastroenterology*, **103**, 647–651.

Part III:

Fluorescence Diagnosis

Photodynamic Therapy and Fluorescence Diagnosis in Dermatology
P.-G. Calzavara-Pinton, R.-M. Szeimies and B. Ortel, editors.

Chapter 9

Fluorescence diagnosis

Christoph Abels and Günther Ackermann

Table of contents

Abstract

To determine the localization and extension of malignant tumors intraoperatively is still a major problem in surgical oncology. 5-aminolevulinic acid (ALA) is a metabolite of the heme biosynthesis which is taken up selectively in neoplastic tissue following either topical or systemic application. Thus, bypassing the rate limiting step of heme biosynthesis ALA is metabolized and fluorescent porphyrins accumulate primarily in the tumor cells. Upon irradiation with light the tumor becomes visible and can be delineated from the surrounding tissue. This procedure called fluorescence diagnosis (FD) will enable the dermatologist to perform either a directed biopsy or very likely a controlled and complete resection of the tumor sparing vital tissue.

9.1 Introduction

Fluorescence diagnosis (*FD*) represents a promising, however, still not routinely used procedure for the in vivo diagnosis of dysplastic or neoplastic tissue. In regard to photodynamic therapy (PDT) the term "photodynamic diagnosis" (PDD) is often used. However, the "dynamic effect" of PDT, which means the excitation of a photosensitizer into the triplet state and the subsequent generation of reactive oxygen species (ROS) is unwanted. Therefore, the intensity of the excitation light used for FD is 100 fold lower as compared to PDT and thus *fluorescence diagnosis* a distinct and better term.

For FD, similar as for PDT, a fluorescent chromophore is applied either topically or systemically, which accumulates thereafter rather selectively in the target tissue. Due to the irradiation with light matching the absorption spectrum of the chromophore the emitted fluorescence allows the detection of superficial tumors in different hollow organs, e.g. bladder, gastrointestinal tract or lung [1–3]. The used chromophores differ regarding their pharmacological (mode of application, accumulation, plasma half life, clearance and metabolism, side effects) and physical properties (absorption- and emission spectra, fluorescence or triplet quantum yield) [4] and none of the currently used substances is clinically approved for this indication. The procedure described in this article has thus to be considered experimental.

9.2 Development of fluorescence diagnosis

The first to report on the detection of tumors by the red fluorescence of porphyrins was Figge in 1948 [5]. Then to detect neoplasia of the cervix purified hematoporphyrin derivative (HpD) was used [6]. However, the selectivity, which is given by the ratio of fluorescence in tumor vs. surrounding tissue, of the HPD is limited. Therefore detection systems with expensive and complicated technical equipment had to be used to subtract autofluorescence or calculate spectral ratio's [7]. Moreover, HpD or Photofrin had to be injected intravenously, resulting in a generalized photosensitization of the skin lasting for weeks. The observed phototoxic reactions range from slight erythema to severe edema with blistering [8]. In addition a topical application of HpD or Photofrin was not successful due to the high molecular weight [9].

9.3 Fluorescence diagnosis with 5-aminolevulinic acid

The substance of interest for FD during the last years due to its high selectivity for neoplastic tissue is 5-aminolevulinic acid (ALA). ALA is a metabolite of the heme biosynthesis which is taken up actively and higher into tumor cells [10]. Under physiologic conditions ALA-synthase regulates the formation of ALA in the cell (Fig. 1). Due to the exogenous application of ALA this bottle neck is bypassed and the exogenous ALA is metabolized to porphyrins, in particular protoporphyrin IX (PPIX) [11,12]. ALA can be applied topically and even after systemic application the photosensitization lasts only up to 48 h.

The clinical application and benefit of ALA-induced fluorescence was first demonstrated for bladder tumors [1]. In the meantime several medical specialties perform successfully, though not routinely, FD. It has been already shown that there is a higher sensitivity regarding tumor positivity of the taken biopsies as compared with conventional cystoscopy [13] or gastrointestinal endoscopy [2].

In dermatology FD may provide a useful tool to highlight superficial skin tumors or even delineate the tumor margins for surgery as shown recently [14]. However, the described setup using Wood's light for excitation of the porphyrin fluorescence is hardly specific, since also the surrounding or even inflammatory skin may emit red fluorescence following application of ALA. Therefore a detection system is necessary to enable the quantification of the fluorescence. Thus, using digital image analysis, an user independent decision can be made regarding the investigated skin lesions.

9.3.1 Excitation and Fluorescence detection

For the excitation of the induced porphyrins in the skin lesion of interest following topical application of ALA an incoherent light source is sufficient. In the presented system a light source (D-Light, Karl Storz GmbH, Tutlingen) was used equipped with a xenon high pressure lamp and a blue bandpass filter ($\lambda_{ex} = 380$–430 nm). The area of interest was irradiated by using a light guide and a lens system to focus the emitted light (Fig. 2). In the tissue the penetrating photons are absorbed by the induced porphyrin molecules in the tumor cells. The excited molecules relax to the ground state under emission of red fluorescence ($\lambda_{em} = 610$–705 nm). Here, the emitted fluorescence was separated by a long pass filter from the excitation light and detected by a b/w CCD camera. A frame grabber digitized the analog video signal of the camera and it was displayed on a PC screen with a resolution of 752×582 Pixel in **real-time** (50 frames/s) to enable the online marking of the biopsy site. In addition to the fluorescence image a congruent, clinical image was captured using a RGB-camera for documentation and follow up.

To illustrate the application of FD in dermatology a case is presented, where the use of this new diagnostic procedure leads to the early detection and excision of a basal cell carcinoma in a high risk patient.

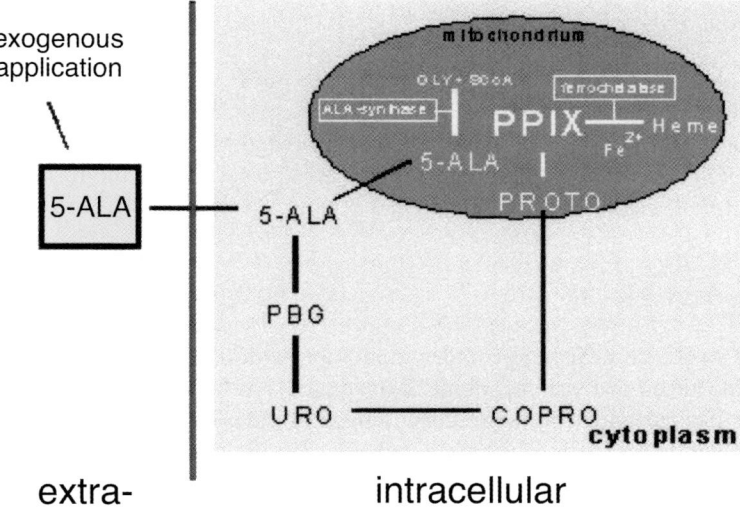

Figure 1. Heme biosynthesis.

9.3.2 Case report

The 75 year old, male patient suffered since his forties from recurrent, superficial basal cell carcinomas only on the right lateral thigh. He has been treated by cryotherapy and twice by extensive CO_2 laser treatment. After the recurrence of basal cell carcinomas PDT with ALA was performed with good result. Due to a follow up visit the patient presented again in our department 6 months later. On the ventral and lateral right thigh there was a 15×15 cm^2 area with multiple, disseminated and confluent erythematous lesions and infiltrated plaques in a scar tissue with hyper- and hypopigmentation. Clinically there were no signs of recurrence, however, structure of the skin hardly allowed a diagnosis. Due to the patients history and to exclude a recurrence a FD was performed following topical application of ALA.

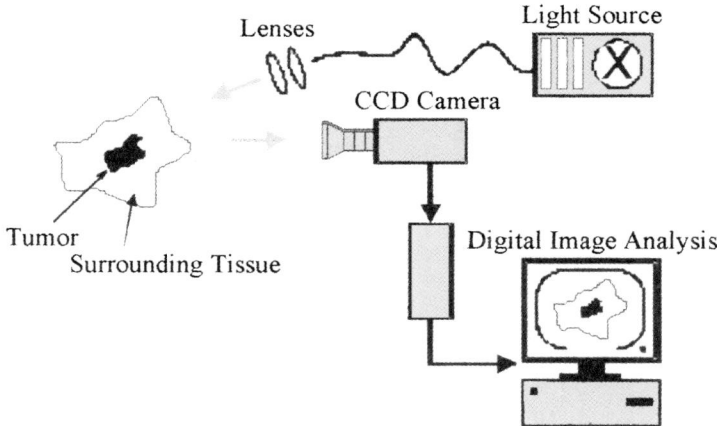

Figure 2. Components of a fluorescence diagnosis system.

9.3.3 Topical application of ALA

ALA was obtained as hydrochloride from Medac GmbH, Hamburg, Germany. A cream (W/O emulsion) containing 20% ALA was freshly prepared by the hospital pharmacy. For 4 h the cream was applied under occlusion (Tegaderm, 3M, Borken, Germany) to the right thigh covered with aluminium foil to avoid bleaching of the formed porphyrin molecules (*photobleaching*).

9.3.4 Performance of FD

In contrast to the clinical image under irradiation with blue light there were multiple, also with the naked eye visible, sharply demarcated, red fluorescent areas in the weakly fluorescent background. The fluorescence image of the b/w-CCD-camera showed these areas with high intensity as white spots, whereas the dark area represents scarred tissue (for comparison see fig. 4b). To correct for intra- and interindividual differences the measured fluorescence intensities were normalized to the surrounding skin as shown by the scale. Interestingly, there was a single lesion which exhibits a significantly higher fluorescence intensity as compared with the surrounding tissue, non-fluorescent skin appeared blue. Consequently, this lesion was marked and a biopsy taken. Histology revealed again a superficial basal cell carcinoma, which was excised *in toto*. Thus, due to the sensitivity of FD this BCC with only a size of ca. $600 \times 400 \ \mu m$ could be excised by a punch biopsy prior to clinical detection.

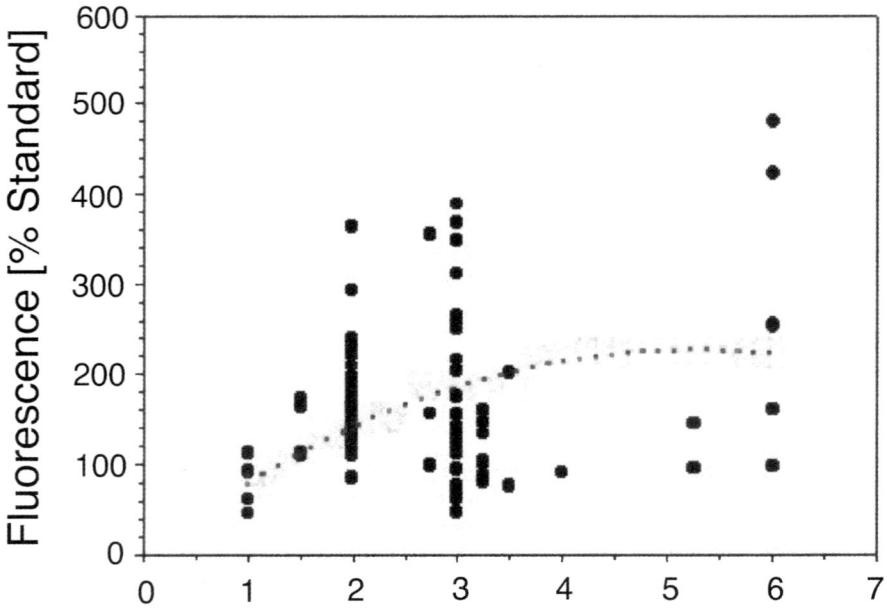

Figure 3. Fluorescence intensity of basal cell carcinomas following topical application of 20% ALA in a W/O-emulsion (n = 83).

9.4 Discussion

The detection of the fluorescence emitted by the ALA-induced porphyrins may be visible without any additional technical equipment, but exhibiting some disadvantages (Table 1). The naked eye can serve as detector for the red fluorescence using a Wood's

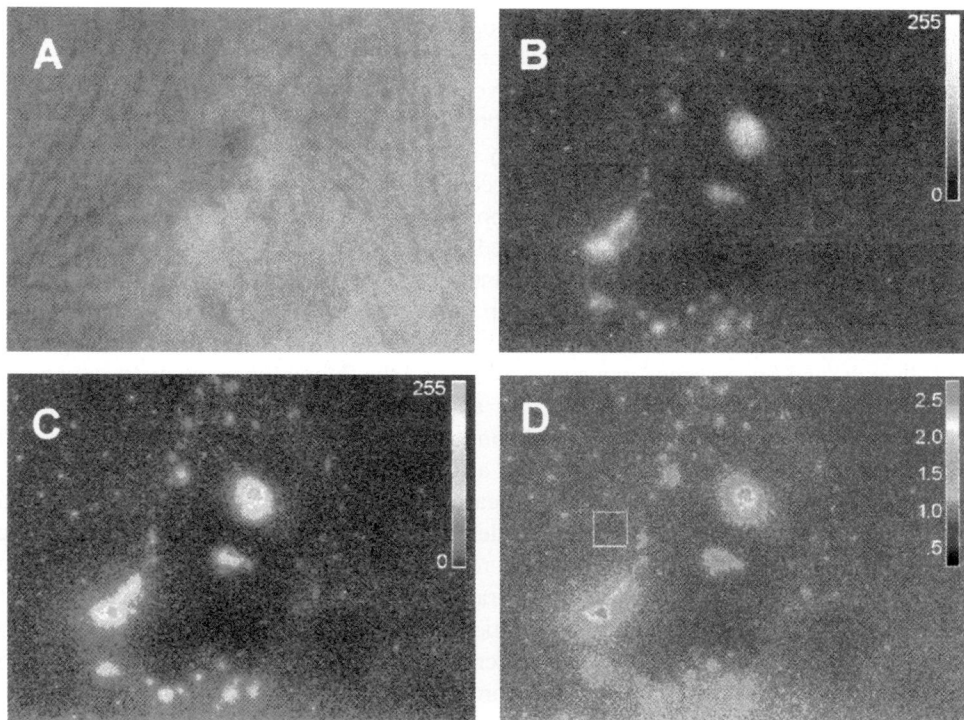

Figure 4. Clinical image of suspicious area on temple (A). b/w image without (B) and false-colour-coded image with shading correction (C). Processed image normalized to the surrounding skin (D). The two lesions (ratio > 2) after biopsy proved to be basal cell carcinomas.

Table 1. Methods of fluorescence detection with/without image storage

Detector	Disadvantage	Advantage
eye	bad contrast	no technical equipment
eye + band pass filter	no quantitative analysis	border detection
ratiofluorometer	time consuming, no image	border detection, different wavelengths
CCD camera + band pass filter	costs, only fluorescence image	quantitative image analysis
2 CCD cameras + dichroic mirror	costs	quantitative image analysis without loss of information

light [14], but this is, if anyone wants, also possible with the presented light source. However, inhomogeneity of the irradiation and unspecific background fluorescence lead to a false impression of the distribution of the fluorescence intensity. A quantitative read-out and an objective evaluation is not possible using this method. To enhance the contrast employing this simple method of FD a long pass filter can be used to avoid interference with the bright excitation light.

A ratiofluorometer giving an acustic or optical signal will provide a higher sensitivity as compared to the previously mentioned methods due to the rationing of the intensity measured at different wavelengths. However, the measurement at certain points is time consuming for larger skin areas as in the presented case [15].

Because of these reasons in dermatology only an image providing procedure covering a large area as described here will be successfully used. Moreover, a quantification of the measured fluorescence intensities is possible using either a reference signal or by giving the ratio in regard to a determined reference intensity, e.g. surrounding skin. These calibration methods allow the interindividual comparison of the different fluorescent images, which is a prerequisite for clinical trials or scientific questions. On the other hand calibration to the surrounding tissue enhances the contrast of the specific vs. background fluorescence in a single image and the fluorescence intensity is given as a ratio of the reference signal. Another advantage of digital image analysis is the exclusion of tissue autofluorescence. The inhomogeneous irradiation of the investigated skin by the light source (*shading*) should also be corrected for a quantitative read-out. Additional parameters influencing the measured fluorescence intensity, e.g. age, skin type or inflammation, can thus be considered and corrected by employing respective algorithms. The multiplication of the resulting fluorescence image with certain parameters will make use of the dynamic range of the CCD-camera to yield the highest possible contrast. A false-colour presentation will make it easier for the investigator to evaluate the distribution of the fluorescence intensity, because the human eye can differentiate colours markedly better as compared to grey shades.

The above described simple image processing makes it possible to detect suspicious areas which showed a fluorescence ratio of >2.0 as compared to normal tissue indicating a neoplastic transformation. The biopsy taken confirmed the presence of a superficial basal cell carcinoma. Due to the exact marking of the biopsy site during FD the basal cell carcinoma was excised *in toto* just by a punch biopsy. The slightly hyperkeratotic stratum corneum without any ulceration and the intact epidermis show, that it is feasible to detect malignant cell clusters in the dermis using FD.

The depth of detection of a FD-system is limited by the penetration of the used chromophore and the penetration depth of the excitation light. ALA is a small hydrophilic molecule (MW 170) and topically applied ALA penetrates very well into the skin [16]. As shown, blue light was used for irradiation because of the absorption maximum of the induced porphyrin, the so called Soret band. However, blue light penetrates only 1–2 mm into the skin [17]. Since porphyrins exhibit in addition to the Soret band other absorption maxima, the Q bands, excitation using deeper penetrating light, e.g. green or red, is principally possible, but the Q bands are 10–20 fold smaller as compared to the Soret band.

As described the formulation for the topical application of ALA was a W/O emulsion (cream). In previous experiments it was shown, that there is a fluorescence ratio of >2, tumor vs. surrounding skin, 3 h following application using this FD system [18]. Thus the ratio determined in the presented case is representative for basal cell carcinomas. In contrast to FD a gel formulation containing 40% DMSO is used for PDT to achieve better penetration of ALA into the tissue. The result is a higher fluorescence intensity in tumor, but also in the surrounding tissue reducing the contrast in-between tumor and surrounding tissue. Comparing different incubation times after topical application of ALA in a W/O-emulsion after 4–6 h the maximal fluorescence intensity is reached and longer incubation did not result in a further fluorescence increase (Fig. 3). Moreover, with increasing incubation time the contrast between tumor and surrounding tissue is decreasing [18]. Therefore an incubation time of approx. 3 h is recommended for FD when using a W/O-emulsion.

The FD of dysplastic or neoplastic tissue following the application of ALA either systemically or topically is already used for some times in other medical specialties. In urology a recent study with 104 patients showed employing fluorescence cystoscopy after intravesical instillation of 3% ALA a sensitivity of 96.9% in the detection of bladder tumors as compared to 72.7% detected by conventional cystoscopy [13]. Better results, which means higher sensitivity regarding the positivity of the taken biopsies, using fluorescence endoscopy as compared with conventional endoscopy were obtained in the gastrointestinal tract [2]. In addition, the early detection of tumors in the bronchial tract [3] and longer survival after fluorescence diagnosis based resection of malignant gliomas were reported [19].

The presented case demonstrates the benefit of a fluorescence directed biopsy of a superficial basal cell carcinoma after topical application of ALA using an image processing system. The fluorescence diagnosis (FD) of skin tumors is not yet a routine procedure, but in the case of clinically difficult cases this method will provide a useful additional device for the dermatologist, similar as dermatoscopy in the detection of malignant melanoma. Using digital image processing allows a quantitative and user independent analysis of the detected fluorescence intensities and enhances significantly the contrast in-between tumor and surrounding tissue.

Further investigations will determine a threshold of the fluorescence intensity for different lesions, e.g. basal cell carcinoma or Bowen's disease, which enable the objective discrimination into suspicious or non-suspicious tissue. Having established such an algorithm FD might provide additional information regarding the extension of the tumor margins prior to excision. Moreover, the acquired images, RGB and fluorescent image, will be subjected to different image analysis algorithms, e.g. neuronal networks, for pattern recognition to reveal additional typical characteristics of tumors, which might contribute to make a correct diagnosis.

References

1. M. Kriegmair, R. Baumgartner, R. Knüchel, H. Stepp, F. Hofstädter, A. Hofstetter (1996). Detection of early bladder cancer by 5-aminolevulinic acid induced porphyrin fluorescence. *J. Urol.*, **155**, 105–110.
2. H. Messmann, R. Knüchel, W. Bäumler, A. Holstege, J. Schölmerich (1999). Endoscopic fluorescence detection of dysplasia in patients with Barrett's esophagus, ulcerative colitis or

adenomatous polyps after 5-aminolevulinic acid-induced protoporphyrin IX sensitization. *Gastrointest. Endosc.*, **49**, 97–101.

3. R. Baumgartner, R.M. Huber, H. Schulz, H. Stepp, K. Rick, F. Gamarra, A. Leberig, C. Roth (1996). Inhalation of 5-aminolevulinic acid: a new technique for fluorescence detection of early stage lung cancer. *J. Photochem. Photobiol. B.*, **36**, 169–174.

4. G. Jori (1996). Tumour photosensitizers: approaches to enhance the selectivity and efficiency of photodynamic therapy. *J. Photochem. Photobiol. B.*, **36**, 87–93.

5. F.H.J. Figge, G.S. Weiland, C.J. Manganiello (1948). Cancer detection and therapy, affinity of neoplastic, embryonic and traumatized tissue for porphyrins and metalloporphyrins. *Proc. Soc. Exp. Biol. Med.*, **68**, 640–641.

6. R.L. Lipson, J.H. Pratt, E.J. Baldes (1964). Hematoporphyrin derivative for detection of cervical cancer. *Obstet. Gynecol.*, **24**, 78.

7. D.W. Rogers, R.J. Lanzafame, J. Blackman, J.O. Naim, H.R. Herrera, J.R. Hinshaw (1990). Methods for the endoscopic photographic and visual detection of helium cadmium laser-induced fluorescence of Photofrin II. *Lasers Surg. Med.*, **10**, 45–51.

8. H. Lui (1994). Photodynamic therapy in dermatology with porfimer sodium and benzoporphyrin derivative: an update. *Semin. Oncol.*, **21**, 11–14.

9. E. Bretschko, R.M. Szeimies, M. Landthaler, G. Lee (1996). Topical 5-Aminolevulinic acid for photodynamic therapy of basal cell carcinoma. Evaluation of stratum corneum permeability in vitro. *J. Contr. Release*, **42**, 203–208.

10. S. Langer, C. Abels, A. Botzlar, S. Pahernik, K. Rick, R.M. Szeimies, A.E. Goetz (1999). Active and higher intracellular uptake of 5-aminolevulinic acid in tumours may be inhibited by glycine. *J. Invest. Dermatol.*, **112**, 723–728.

11. H. Heyerdahl, I. Wang, D.L. Liu, R. Berg, S. Anderson-Engels, Q. Peng, J. Moan, S. Svanberg, K. Svanberg (1997). Pharmacokinetic studies on 5-aminolevulinic acid-induced protoporphyrin IX accumulation in tumour and normal tissue. *Cancer Lett.*, **112**, 225–231.

12. J.C. Kennedy, R.H. Pottier (1992). Endogenous protoporphyrin IX, a clinically useful photosensitizer for photodynamic therapy. *J. Photochem. Photobiol. B.*, **14**, 275–292.

13. M. Kriegmair, D. Zaak, R. Knuechel, R. Baumgartner, A. Hofstetter (1999). 5-Aminolevulinic acid-induced fluorescence endoscopy for the detection of lower urinary tract tumors. *Urol. Int.*, **63**, 27–31.

14. C. Fritsch, P.M. Becker-Wegerich, H. Menke, T. Ruzicka, G. Goerz, R.R. Olbrisch (1997). Successful surgery of multiple recurrent basal cell carcinoma guided by photodynamic diagnosis. *Aesth. Plast. Surg.*, **21**, 437–439.

15. S. Lam, J.Y. Hung, S.M. Kennedy, J.C. Leriche, S. Vedal, B. Nelems, C.E. Macaulay, B. Palcic (1992). Detection of dysplasia and carcinoma in situ by ratio fluorometry. *Am. Rev. Respir. Dis.*, **146**, 1458–1461.

16. R.M. Szeimies, T. Sassy, M. Landthaler (1994). Penetration potency of topical applied delta-aminolevulinic acid for photodynamic therapy of basal cell carcinoma. *Photochem. Photobiol.*, **59**, 73–76.

17. J. Moan, V. Iani, L. Ma (1996). Choice of proper wavelength for photochemotherapy proceedings of photochemotherapy and other modalities. *Proc. SPIE*, **2625**, 544–549.

18. G. Ackermann, C. Abels, W. Bäumler, S. Langer, M. Landthaler, E.W. Lang, R.M. Szeimies (1998). Simulations on the selectivity of 5-aminolevulinic acid-induced fluorescence in vivo. *J. Photochem. Photobiol. B.*, **47**, 121–128.

19. W. Stummer, S. Stocker, S. Wagner, H. Stepp, C. Fritsch, C. Goetz, A.E. Goetz, R. Kiefmann, H.J. Reulen (1998). Intraoperative detection of malignant gliomas by 5-aminole-vulinic acid-induced porphyrin fluorescence. *Neurosurgery*, **42**, 518–526.

Part IV:

Systemic Photodynamic Therapy

Photodynamic Therapy and Fluorescence Diagnosis in Dermatology
P.-G. Calzavara-Pinton, R.-M. Szeimies and B. Ortel, editors.

Chapter 10

Systemic sensitization – oncologic indications in dermatology

Christoph Abels and Rolf-Markus Szeimies

Table of contents

Abstract

Systemic PDT using first generation photosensitizers, e.g. porfimer sodium (Photofrin®), which induce prolonged generalized cutaneous photosensitivity, may provide still an option for multiple or large tumors, in particular basal cell carcinomas, in elderly patients unsuitable for surgery or any other therapeutic modality. From the literature a concentration of 2 mg/kg b.w. and irradiation with a fluence of approx. 100 J/cm² (100–150 mW/cm²) using red light should be used for successful treatment. However, new photosensitizers absorbing in the near infrared part of the visible spectrum and not inducing prolonged cutaneous photosensitization are under investigation. Indications for systemic PDT with these drugs, whose target is primarily the microvasculature, will be probably unpigmented solid tumors with increased microvascular density and a tumor thickness < 10 mm.

10.1 Introduction

In 1978 the first study on 25 patients with cutaneous or subcutaneous malignant tumors was performed using HpD (2.5–5 mg/kg b.w.) intravenously injected and subsequent irradiation with a xenon arc lamp ($\lambda_{ex} = 600$–700 nm). From the lesions treated with systemic PDT 111/113 showed either complete or at least partial remission [1]. Up to now the porphyrin mixture HpD or the partially purified and clinically approved – however, not for dermatological indications – form porfimer sodium (Photofrin®) are the systemic agents for which extensive clinical data are currently available. Since this first study on skin tumors systemic PDT has been used in particular for the treatment of Bowen's disease, basal cell carcinoma, squamous cell carcinoma, and recurrent metastatic breast cancer (Table 1). Although a large number of clinical studies on systemic PDT of non-melanoma skin cancer have been published, most of them were uncontrolled open trials with only a small number of patients and different parameters regarding tumor type, drug concentration, light dose, and follow-up. Another important draw back of systemic PDT using HpD or porfimer sodium is the generalized cutaneous photosensitization lasting for several weeks. This is due to the must of an i.v. injection of these drugs, because after topical application there is an insufficient penetration into the lesions [2], and the subsequent accumulation of the drug in the skin.

10.2 PDT with hematoporphyrin derivatives

10.2.1 Bowen's disease

Systemic PDT for Bowen's disease is very effective. Intravenous injection of porfimer sodium (2 mg/kg b.w.) and subsequent irradiation with light of a wavelength of $\lambda_{ex} = 630$ nm (20–50 J/cm²) induced complete remission in 98–100% of the treated lesions [3–5]. Complete remission of Bowen's disease (n = 8) was also achieved using higher fluences of light (630 nm, 185–250 J/cm²) [6]. However, irradiation with low

Table 1. Systemic PDT for oncological indications using HpD or porfimer sodium

Author	Number of Lesions	Sensitizer Concentration	Wavelength & Fluence	Complete Remission
Bowen's disease				
Waldow et al. 1987	3	Photofrin 2.0 mg/kg	630 nm 40–60 J/cm^2	100%
Robinson et al. 1988	> 500 90	Photofrin 2.0 vs. 1.0 mg/kg	628 nm 25 vs. 50 J/cm^2	100% 50%
Buchanan et al. 1989	50	Photofrin 2.0 mg/kg	630 nm 50 J/cm^2	98%
McCaughan et al. 1989	2	HpD/Photofrin 3.0/2.0 mg/kg	630 nm 20–30 J/cm^2	50%
Jones et al. 1992	8	Photofrin 1.0 mg/kg	630 nm 185–250 J/cm^2	100%
Squamous Cell Carcinoma				
Pennington et al. 1988	32	HpD 5.0 mg/kg	630 nm 30 J/cm^2	< 50%
McCaughan et al. 1989	5	HpD/Photofrin 3.0/2.0 mg/kg	630 nm 20–30 J/cm^2	40%
Gross et al. 1990	1	Photofrin 2.0 mg/kg	630 nm 150 J/cm^2	100%
Feyh et al. 1993	7	Photosan-3 2.0 mg/kg	630 nm 100 J/cm^2	86%
Basal Cell Carcinoma				
Tse et al. 1984	40	HpD 3.0 mg/kg	600–700 nm 38–180 J/cm^2	83%
Bandieramonte et al. 1984	42	HpD 3.0 mg/kg	480–515, 630 nm 60–120 J/cm^2	60%
Waldow et al. 1987	6	Photofrin 1.5–2.0 mg/kg	630 nm 40–60 J/cm^2	100%
Pennington et al. 1988	21	HpD 5.0 mg/kg	630 nm 30 J/cm^2	0%
Robinson et al. 1988	15	Photofrin 2.0 mg/kg	628 nm 50 J/cm^2	93%
Buchanan et al. 1989	13	Photofrin 1.5–2.0 mg/kg	630 nm 50–100 J/cm^2	39%
McCaughan et al. 1989	27	HpD/Photofrin 3.0/2.0 mg/kg	630 nm 20–30 J/cm^2	15%

Table 1. Continued

Author	Number of Lesions	Sensitizer Concentration	Wavelength & Fluence	Complete Remission
Basal Cell Carcinoma				
Feyh et al. 1993	67	Photosan-3 2.0 mg/kg	630 nm 100 J/cm^2	97%
Calzavara et al. 1991	17	HpD/Photofrin 3.0/2.5–3.0 mg/kg	600–700 nm 25–225 J/cm^2	59%
Wilson et al. 1992	151	Photofrin 1.0 mg/kg	630 nm 72–288 J/cm^2	89%
Hintschich et al. 1993	27	Photosan-3 2.0 mg/kg	630 nm 100 J/cm^2	52%
Metastases of Breast Cancer				
Dougherty 1981	35	HpD 2.5–5.0 mg/kg	? ?	97%
Schuh et al. 1987	30	Photofrin 1.0–2.0 mg/kg	630 nm 36–288 J/cm^2	80%
McCaughan et al. 1989	29	HpD/Photofrin 3.0/2.0 mg/kg	630 nm 20–30 J/cm^2	100%
Sperduto et al. 1991	20	Photofrin 1.5 mg/kg	630 nm 20–359 J/cm^2	65%

fluences of only 50–100 J/cm^2 and 1 mg/kg b.w. porfimer sodium to minimize cutaneous photosensitization resulted in complete remission of only 50% of Bowen's disease [3,4].

10.2.2 Squamous cell carcinoma

Squamous cell carcinomas (SCC) of the skin do not respond as well to systemic PDT as Bowen's disease. A 50% recurrence rate was reported within 6 months after PDT of 32 SCC treated with HpD (5.0 mg/kg b.w.) [7]. A very likely explanation for this poor result could be the low fluence (30 J/cm^2) used in this study. The poor response of SCC is supported by McCaughan and coworkers. They used HpD (3 mg/kg b.w.) or porfimer sodium (2 mg/kg b.w.) and also low fluences of 20–30 J/cm^2 [8]. One year after PDT there was only a 40% remission rate (2/5 SCC). Only one case report showed the complete remission of a large squamous cell carcinoma of the lower lip during a follow-up 6 months after systemic PDT (porfimer sodium, 2 mg/kg b.w., 630 nm, 150 J/cm^2) [9].

10.2.3 Basal cell carcinoma

For systemic PDT of basal cell carcinomas (BCC) there are more reports in the literature. Already in 1981 Dougherty [10] used HpD (5 mg/kg b.w.) for curative PDT of three BCC located in the face of a 72-year-old man. Irradiation was performed twice with a xenon-lamp (600–700 nm, 100 mW/cm^2, 120 J/cm^2) on day 4 and 5 after HpD injection. Feyh and coworkers treated 67 BCC with HpD (2 mg/kg b.w.) and an argon-ion pumped dye laser (630 nm, 100 J/cm^2). Only three recurrences were reported after 4.5 year follow up [11]. Another study published by the same group showed poor response rates for BCC located on the eyelid using the same treatment protocol, probably due to inhomogenous light dosimetry [12]. Tse and co-workers treated 40 BCC in 3 patients with nevoid BCC syndrome using HpD and irradiation with either a dye laser or a xenon-lamp [13]. All tumors resolved clinically within 4–6 weeks after PDT. However, the histological examination of the treated sites revealed nests of tumor cells in 17.5% (7/40). During the follow up period of 12–14 months the recurrence rate was 10.8%. Recurrences of large, ulcerative, or crusted tumors, probably disturbing the light delivery to the tumor cells, accounted for much of the poor response rate [13].

10.2.4 Metastases of breast cancer

Another indication for systemic PDT, early followed on, is the palliative treatment of cutaneous metastases of breast cancer, first reported by [14]. However, in all reported cases, PDT was performed after conventional therapies had failed (radiation, chemotherapy, hormone-therapy, conventional surgery). Thus, the goal of PDT in these cases was primarily the reduction of tumor mass in order to avoid ulceration or excessive bleeding. Dougherty achieved partial remissions ($>50\%$ reduction of tumor bulk) in 34 of 35 patients with cutaneous or subcutaneous metastases of breast cancer 4–6 weeks after PDT using HpD (2.5–5.0 mg/kg b.w.) [10]. McCaughan showed partial or even complete remission in all patients with cutaneous metastases of breast cancer after treatment with HpD (3 mg/kg b.w.) or porfimer sodium (2 mg/kg b.w.) and subsequent irradiation with a fluence of 60–120 J/cm^2 [8]. Summarizing all patients ($n = 118$, total number of lesions = 846) with cutaneous metastases of breast cancer treated with systemic PDT, Schlag et al. reported a complete remission rate of 63% of all treated lesions, a partial remission of 20%, and no response in 16% [15]. They noted that the best results were achieved when tumors were <2 cm.

10.2.5 Other indications

In 1980, Forbes treated successfully a patient with cutaneous T-cell lymphoma (plaque stage) using HpD (5 mg/kg b.w.) and successive irradiation with an incoherent light source ($\lambda_{ex} = 620$–640 nm, 40 mW/cm^2, 48–96 J/cm^2, irradiation 72 h and 96 h after injection) [16]. He also obtained partial remission in a patient with metastatic Kaposi's sarcoma using the same treatment protocol [16]. Dougherty [10] and Calzavara [17] applied 2.5–3.0 mg/kg b.w. HpD and a light dose of either 120 J/cm^2 or 50–200 J/cm^2

in patients suffering from classic Kaposi's sarcoma. All tumors (100%) in three patients and 85% of the tumors in the other four patients treated with PDT showed complete remission.

In another study Schweitzer and Visscher treated 5 patients with AIDS-associated Kaposi's sarcoma using 2 mg/kg b.w. porfimer sodium and irradiation with red light either as surface or interstitial irradiation (50–200 J/cm^2) [18]. Complete or partial remission of cutaneous or mucosal nodular lesions was observed 8 weeks after PDT only in 54/92 lesions (58.7%). Similar results were reported by Hebeda et al. for 8 HIV-positive patients with a total of 83 Kaposi's sarcomas irradiated with a dye laser (630 nm, 70–120 J/cm^2) after injection of 2 mg/kg b.w. porfimer sodium. Although the remission rate was sufficient 60–70%, the cosmetic results were unsatisfactory since long lasting hyperpigmentation and scar formation occurred [19].

10.3 Second generation photosensitizers

Benzoporphyrin derivative-monoacid ring A (BPD-MA, Visudyne®), already approved for age-related macular degeneration, represents one of a number of so called second generation photosensitizers absorbing in the near infrared part of the visible spectrum without induction of long lasting cutaneous photosensitivity. BPD-MA is a semi-synthetic porphyrin synthesized from protoporphyrin IX (PpIX). A major advantage of BPD-MA is that both drug and the irradiation procedure can be administered on the same day. In contrast to HpD, BPD-MA is metabolized and excreted in an inactive form [20]. The duration of potential susceptibility to cutaneous photosensitivity is therefore less than 72 h [21]. Phase-I/II studies with BPD-MA have confirmed the efficacy of this photosensitizer in the treatment of some epithelial skin tumors after intravenous administration of 0.375–0.50 mg/kg b.w. in a liposomal formulation followed by irradiation at 690 nm (50–150 J/cm^2) 2–6 h later. Remission rates of up to 100% were achieved [22].

Phase-I studies with lutetium texaphyrin, a compound with an expanded porphyrin macrocycle, have been completed in 19 patients with BCC, Kaposi's sarcoma, or metastatic melanoma. This agent exhibits a peak absorption at 732 nm, which allows also for a deeper light penetration in skin. Lutetium texaphyrin is believed to be highly selective for tumors as compared to normal skin and shows no significant skin phototoxicity when given systemically [23].

SnET$_2$, a tin etiopurpurin, is currently in phase-II trials for the treatment of cutaneous metastatic breast cancer or Kaposi's sarcoma [24]. An overall 75% complete response rate in 121 Kaposi's sarcoma lesions with an excellent cosmetic result was reported [25]. A complete response rate of 92% was achieved in 8 patients with advanced breast cancer metastases (number of lesions = 86) after a single PDT with SnET$_2$ (1.2 mg/kg b.w.) and laser light (~ 660 nm, 150 mW/cm^2, 200 J/cm^2) [26].

Another chlorin photosensitizer tetra(m-hydroxyphenyl)chlorin (mTHPC), which is undergoing clinical trials for head and neck cancer in Europe and the United States, appears to be the most active of all photosensitizers requiring only very low drug concentrations and fluences, but exhibiting again rather long cutaneous photosensitivity [27].

The activity of another new photosensitizer, mono-L-aspartyl chlorin e_6 (Npe6) was evaluated in a phase I-clinical study including 11 patients with a variety of solid skin tumors (BCC, SCC, recurrent breast cancer) [28]. Npe6 seems to offer the advantage of a rapid clearance from tissues resulting in limited cutaneous photosensitivity. Using concentrations of Npe6 of 2.5–3.5 mg/kg b.w. and irradiation with a fluence of 100 J/cm^2 resulted in 66% of sites remaining tumor free 12 weeks after PDT.

A promising new photosensitizer approved already for diagnostic indications, e.g. measurement of liver function, cardiac output and plasma volume, is indocyanine green (ICG). In a pilot study ICG was used as photosensitizer in combination with a diode laser to treat AIDS-associated Kaposi's sarcoma (KS) in three patients [29]. Directly and up to 50 min after intravenous administration of ICG (2–4 mg/kg b.w.), KS (n = 57), mainly plaque-type, were irradiated continously using a diode laser ($\lambda_{em} = 805$ nm, 100 J/cm^2, 0.5–5 W/cm^2) matching the absorption maximum. Complete remission of KS (n = 16) was achieved when the KS were irradiated 1–30 min after injection of the second bolus of ICG (2×2 mg/kg b.w., 30 min apart) with 3–5 W/cm^2 and 100 J/cm^2. Biopsies (n = 3) revealed necrosis of the tumour 24 h and complete remission 4 weeks following PDT. There were no systemic side effects observed and the cosmetic results were very good. However, as described by others, hyperpigmentation occurred in lesions located on the lower extremities, but only temporarily. These results showed that photosensitization with ICG and subsequent irradiation with an appropriate diode laser cannot only be applied in ophthalmology but also for highly vascularized solid tumors of the skin. Nevertheless, additional clinical studies have to show the efficacy for the treatment of neoplasms of the skin.

10.4 Conclusion

Systemic PDT in dermatology using first generation photosensitizers is restricted to elderly patients suffering from large or multiple tumors, in particular basal cell carcinomas, unsuitable for surgery or any other therapeutic modality due to the induction of prolonged generalized cutaneous photosensitivity. However, new photo-sensitizers absorbing in the near infrared part of the visible spectrum and not inducing prolonged cutaneous photosensitization will definetely provide additional treatment options despite the widespread use of topical PDT with ALA. Indications for systemic PDT with these drugs, which targets primarily the microvasculature, will be unpigmented solid tumors with increased microvascular density and a tumor thickness < 10 mm using surface irradiation or even larger volumes using interstitial fibers. Therefore, systemic PDT shows still a great potential once second generation photosensitizers are approved.

References

1. T.J. Dougherty, J.E. Kaufman, A. Goldfarb (1978). Photoradiation therapy for the treatment of malignant tumors. *Cancer Res.*, **38**, 2628–2635.
2. E. Bretschko, R.M. Szeimies, M. Landthaler, G. Lee (1996). Topical 5-aminolevulinic acid for photodynamic therapy of basal cell carcinoma. Evaluation of stratum corneum permeability in vitro. *J. Contr. Release*, **42**, 203–208.

3. R.B. Buchanan, J.A.S. Carruth, A.L. McKenzieL, S.R. Williams (1989). Photodynamic therapy in the treatment of malignant tumours of the skin and head and neck. *Eur. J. Surg. Oncol.*, **15**, 400–406.

4. P.J. Robinson, J.A.S. Carruth, G.M. Fairris (1988). Photodynamic therapy: a better treatment for widespread Bowen's disease. *Br. J. Dermatol.*, **119**, 59–61.

5. S.M. Waldow, R.V. Lobraico, I.K. Kohler, S. Wallk, H.T. Fritts (1987). Photodynamic therapy for treatment of malignant cutaneous lesions. *Lasers Surg. Med.*, **7**, 451–456.

6. C.M. Jones, T. Mang, M. Cooper, D.B. Wilson, H.L. Stoll (1992). Photodynamic therapy in the treatment of Bowen's disease. *J. Am. Acad. Dermatol.*, **27**, 979–982.

7. D.G. Pennington, M. Waner, A. Knox (1988). Photodynamic therapy for multiple skin cancers. *Plast. Reconstr. Surg.*, **82**, 1067–1071.

8. J.S. McCaughan Jr., J.T. Guy, W. Hicks, L. Laufman, T.A. Nims, J. Walker (1989). Photodynamic therapy for cutaneous and subcutaneous malignant neoplasms. *Arch. Surg.*, **124**, 211–216.

9. D.J. Gross, M. Waner, R.H. Schosser, S.M. Dinehart (1990). Squamous cell carcinoma of the lower lip involving a large cutaneous surface: Photodynamic therapy as an alternative therapy. *Arch. Dermatol.*, **126**, 1148–1150.

10 T.J. Dougherty (1981). Photoradiation therapy for cutaneous and subcutaneous malignancies. *J. Invest. Dermatol.*, **77**, 122–124.

11. J. Feyh, R. Gutmann, A. Leunig (1993) Die photodynamische Lasertherapie im Bereich der Hals-, Nasen-, Ohrenheilkunde. *Laryngo. Rhino. Otol.*, **72**(8), 273–278.

12. C. Hintschich, J. Feyh, C. Beyer-Machule, K. Riedel, K. Ludwig (1993). Photodynamic laser therapy of basal-cell carcinoma of the lid. *Ger. J. Ophthalmol.*, **2**, 212–217.

13. D.T. Tse, R.D. Kersten, R.L. Anderson (1984). Hematoporphyrin derivative photoradiation therapy in managing nevoid basal cell carcinoma syndrome. *Arch. Ophthalmol.*, **102**, 990–994.

14. R.L. Lipson, M.J. Gray, E.J. Baldes (1966). Hematoporphyrin derivativ for detectin and management of cancer. *Proc. 9th Int. Cancer Congr.*, 393.

15. P. Schlag, M. Hünerbein, J. Stern, J. Gahlen, G. Graschew (1992). Photodynamische Therapie – Alternative bei lokal rezidiviertem Mamma-Karzinom. *Dt. Ärztebl.*, **89**, 680–687.

16. I.J. Forbes, P.A. Cowled, A.S.Y. Leong, A.D. Ward, R.B. Black, A.J. Blake, F.J. Jacka (1980). Phototherapy of human tumours using hematoporphyrin derivative. *Med. J. Aust.*, **2**, 489–493.

17. F. Calzavara, L. Tomio (1991). Photodynamic therapy: clinical experience at the department of radiotherapy at Padova general hospital. *J. Photochem. Photobiol. B. Biol.*, **11**, 91–95.

18. V.G. Schweitzer, D. Visscher (1990). Photodynamic therapy for treatment of AIDS-related oral Kaposi's sarcoma. *Otolaryngol. Head Neck Surg.*, **102**, 639–649.

19. K.M. Hebeda, M.T. Huizing, P.A. Brouwer, F.W. van der Meulen, H.J. Hulsebosch, P. Reiss, J.H. Oosting, C.H.N. Veenhof, P.J.M. Bakker (1995). Photodynamic therapy in AIDS-related cutaneous Kaposi's sarcoma. *J. Acquir. Immune Defic. Syndr. Hum. Retrovirol.*, **10**, 61–70.

20. A.M. Richter, A.K. Jain, A.J. Canaan, E. Waterfield, E.D. Sternberg, J.G. Levy (1992). Photosensitizing efficiency of two regio-isomers of the benzoporphyrin derivative monoacid ring A. *Biochem. Pharmacol.*, **43**, 2349–2358.

21. S.T. Wolford, D.L. Novicki, B. Kelly (1995). Comparative skin phototoxicity in mice with two photosensitizing drugs: Benzoporphyrin derivative monoacid ring A and porfimer sodium (Photofrin). *Fundam. Appl. Toxicol.*, **24**, 52–56.

22. J.G. Levy, C.A. Jones, L.A. Pilson (1994). The preclinical and clinical development and potential application of benzoporphyrin derivative. *Int. Photodynam.*, **1**, 3–5.

23. M.F. Renschler, A. Yuen, T.J. Panella, T.J. Wieman, C. Julius, M. Panjehpour, S. Taber, V. Fingar, S. Horning, R.A. Miller, E. Lowe, J. Engel, K. Woodburn, S.W. Young (1997). Photodynamic therapy trials with lutetium texaphyrin PCI-0123 (Lu-Tex). *Photochem. Photobiol.*, **65**, 47S.
24. T.J. Dougherty, C.J. Gomer, B.W. Henderson, G. Jori, D. Kessel, M. Korbelik, J. Moan, Q. Peng (1998). Photodynamic therapy. *J. Natl. Cancer Inst.*, **90**, 889–905.
25. R.R. Allison, T.S. Mang (1997). A phase II/III clinical study for the treatment of HIV-associated cutaneous Kaposi's sarcoma with tin ethyl etiopurpurin (SnET2)-induced photodynamic therapy, presented at the European Cancer Conference (ECC09), Hamburg, Germany.
26. R.R. Allison, T.S. Mang, B.D. Wilson (1998). Photodynamic therapy for the treatment of nonmelanomatous cutaneous malignancies. *Semin. Cut. Med. Surg.*, **17**, 153–163.
27. M.L. De Jode, A. Rowntree-Taylor (1997). m-THPC photodynamic therapy for head and neck cancer. *Lasers Med. Sci.*, **11**, 23–30.
28. S.W. Taber, V.H. Fingar, C.T. Coots, T.J. Wieman (1998). Photodynamic therapy using mono-L-aspartyl chlorin e_6 (Npe6) for the treatment of cutaneous disease: a phase I clinical study. *Clin. Cancer Res.*, **4**, 2741–2746.
29. C. Abels, S. Karrer, W. Bäumler, A.E. Goetz, M. Landthaler, R.M. Szeimies (1998). Indocyanine green and laser light for the treatment of AIDS-associated cutaneous Kaposi's Sarcoma. *Br. J. Cancer*, **77**, 1021–1024.

Photodynamic Therapy and Fluorescence Diagnosis in Dermatology
P.-G. Calzavara-Pinton, R.-M. Szeimies and B. Ortel, editors.

Chapter 11

Systemic sensitization – non-oncologic indications in dermatology

Robert Bissonnette and Harvey Lui

Table of contents

11.1 Introduction

During the early development of photodynamic therapy (PDT) at the beginning of the twentieth century, topical and systemic photosensitizers were used for a number of non-oncologic skin diseases including cutaneous tuberculosis, condyloma acuminata, and psoriasis [1,2]. In 1937, Silver published a case report of a patient with extensive psoriasis who improved with two intramuscular injections of hematoporphyrin followed by ultraviolet light irradiation [2]. Since the Second World War, there have been only a limited number of published studies on the use of systemic PDT photosensitizers for non-malignant conditions in dermatology.

The bias towards the use of PDT for cancer is related to a number of factors. First, prior to the last decade the only photosensitizers that were readily available were parenteral porphryins such as hematoporphyrin, hematoporphyrin derivative, and porfimer sodium, all of which induced generalized and prolonged cutaneous photosensitivity following their administration. The strict light avoidance precautions that were necessary with these agents was perhaps justifiable only for patients with cancer, and indeed until recently PDT for all other organ systems including the lung, bladder, and upper gastrointestinal tract was essentially synonymous with using systemic porphyrins and lasers to destroy malignant tissue. Furthermore in the past, PDT dosimetry was focused on generating sufficient singlet oxygen to induce necrosis of target tissues rather than on altering pathophysiologic responses such a immunomodulation and neovascularization. With the understanding that "low-level PDT" using lower doses of drug and/or light could induce more subtle biologic effects, the spectrum of disorders amenable to PDT broadened significantly.

The recent approval of verteporfin for ocular macular degeneration, as well as preclinical and pilot clinical studies demonstrating the potential efficacy of systemic photosensitizers for other non-oncologic applications have increased the impetus to use PDT for inflammatory, infectious, and vascular disorders in dermatology. Skin diseases such as psoriasis that manifest more widespread involvement are particularly suitable for this type of approach, as it is relatively easy and efficient to administer whole body light exposure after systemic photosensitizer administration. This chapter will review the published studies on the use of systemic PDT photosensitizers for non-oncologic applications in dermatology in the context of specific advantages and disadvantages of systemic versus topical PDT.

11.2 Advantages and disadvantages of systemic photosensitizers

For the most part, pre-formed tetrapyrrole-based photosensitizers such as porphyrins, chlorins, and purpurins are relatively large molecules that exhibit very limited percutaneous and gastrointestinal absorption. In addition when these particular drugs are applied topically to the skin, most of the drug actually remains on the surface where it largely remains inactive. Since these agents are also deeply colored and strongly absorbing dyes, not only do they cause staining, but their presence on the skin surface following topical application also limits the amount of activating light that can reach the intended target sites within the epidermis and dermis. Parenteral systemic administration is therefore necessary when these types of photosensitizers are used. In dermatology this requirement imposes practical limitations since the overwhelming majority of clinical practice settings do not currently have the capacity for, or experience with administering intravenous drugs.

In contrast photosensitizer precursors such as aminolevulinic acid (ALA) and its esters are smaller and colorless compounds that are photo-inert unless absorbed and metabolized within living cells to protoporphyrin IX. Nevertheless, even though ALA is now available for topical use in treating focal skin lesions, it is somewhat impractical for treating extensive skin surfaces such as may occur in psoriasis or eczema. In these settings not only is drug cost a significant factor, but effective and reliable delivery methods for ensuring uniform and consistent skin photosensitization following topical ALA administration have also not yet been established. Unlike the pre-formed photosensitizers above, ALA and its esters can induce PDT photosensitivity when given orally.

As discussed previously systemic photosensitizers do have the potential disadvantage of inducing whole body light sensitivity. While this property is actually necessary for the therapeutic effect, the duration of photosensitivity should ideally be relatively short. At the present time certain agents such as verteporfin and ALA cause photosensitivity for less than a couple of days, which is a significant advance over using first generation photosensitizers where outdoor activities must be curtailed for several weeks at a time. Systemic administration may also theoretically increase the risk of extracutaneous systemic side effects. For example systemic ALA can induce elevations of serum liver transaminase levels [3].

In summary systemic photosensitizers are ideal for clinical settings where uniform photosensitization of large or multiple areas of the skin is necessary, and the development of safe and effective orally active PDT drugs would perhaps represent the most significant factor in advancing this therapy.

11.3 Hematoporphyrin derivative and porfimer sodium

Hematoporphyrin derivative is a complex mixture of different dimers and oligomers of porphyrin. Porfimer sodium is a commercial and more purified version of hematoporphyrin derivative that has been approved in several countries for oncologic indications such as esophageal and lung cancer. In rodent models PDT with hematoporphyrin derivative can prevent contact hypersensitivity and skin graft rejection [4,5]. Although the exact mechanisms involved in this immunosuppressive effect are unknown, altered cytokine expression has been reported following PDT with porfimer sodium [6]. Increased IL-6 and IL-10 expression was observed on exposed skin as compared to increased IL-6 and decreased IL-10 within implanted tumors [4]. IL-10 has been reported to impair the induction of contact hypersensitivity [7]. Keratinocyte IL-10 is increased following UV irradiation and this cytokine may play a central role in UV-induced immunosuppression [8]. Similar mechanisms may explain the immunosuppressive effects of PDT.

Improvement of a psoriatic plaque within the pubic region was reported following a single PDT session with hematoporphyrin derivative and 630 nm laser light [9]. Eschars developed after PDT but the area was reported to have healed normally. In another study, eight patients received a single injection of porfimer sodium at 0.5 mg/kg followed by 9–20 light treatments over the next 3–4 weeks to selected plaques [10]. Three light sources were used: an argon dye laser (630 nm), fluorescent UVA, and a krypton laser (405 nm). Improvements of up to 85–100% were seen in patients treated with the 630 nm laser light. The administration of multiple light exposures following a single drug injection is a potential advantage of PDT with porfimer sodium for the treatment of psoriasis. However the prolonged skin photosensitivity of up to 2 months may explain why this photosensitizer has not been studied more extensively for psoriasis [11].

The ability of systemic PDT drugs to photosensitize vascular tissue has been exploited in preclinical studies for treating vascular malformations such as port wine stains with porfimer sodium [12]. Topical photosensitizers would not be expected to selectively target cutaneous blood vessels very efficiently in comparison to intravenous drugs which deliver drug directly to the intended target.

11.4 Verteporfin

Verteporfin (benzoporphyrin derivative, BPD) is a semi-synthetic porphyrin derivative that has recently been approved in Europe and North America for the treatment of ocular macular degeneration. Verteporfin has been studied in a number of animal models for the treatment of inflammatory and autoimmune diseases. Whole-body PDT with verteporfin has been shown to prevent autoimmune encephalitis [13], arthritis [14], and contact hypersensitivity [15], as well as prolong skin graft survival in mice [16]. PDT with verteporfin also induces IL-10 expression in mice, and it has also been shown that this PDT-induced decrease in contact hypersensitivity could be prevented with injection of IL-12 or anti-IL10 antibodies [17]. Furthermore transgenic mice deficient in IL-10 did not demonstrate a reduction in contact hypersensitivity following PDT with verteporfin. This immune modulation probably involves other mechanisms as suggested by a decrease in the expression of co-stimulatory molecules such as CD80 and CD86 on antigen presenting cells isolated from mouse skin after PDT with verteporfin [16], and preferential accumulation of verteporfin in mitogen-activated murine splenocytes as compared to their resting counterparts [18].

Clinical improvement in psoriasis following PDT with intravenous verteporfin was first reported using 690 nm light from an argon dye laser [19]. What was noteworthy about this particular study was that complete clearing was achieved in some of the laser-exposed sites following only a single PDT session without concomitant eschar formation. This is in distinct contrast to conventional psoriasis phototherapy where at least 20–25 treatment sessions are typically required to achieve clearing. Because lasers are expensive and impractical for treating large surfaces, subsequent studies of psoriasis with verteporfin were performed with non-coherent light. Seventeen patients with moderate to severe psoriasis and psoriatic arthritis received weekly injections of verteporfin at 8 mg/m^2 followed by half or whole body exposure to UVA light [20]. The minimal phototoxic dose (MPD) to verteporfin was determined following the first injection, and patients subsequently received escalating doses of UVA light (20–80% of MPD). In five out of 17 patients the PASI improved by at least 35%. Both the drug and total body light exposures were well tolerated, and clinical photosensitivity reactions were not seen. Although patients subjectively reported a global improvement in arthritis, objective rheumatologic assessments did not reveal a significant improvement in arthritis. An open clinical study of multiple PDT sessions with verteporfin for rheumatoid arthritis is currently underway in Canada.

In a different study 15 patients with psoriasis received 5 weekly injections of verteporfin at 8 mg/m^2 followed by irradiation of plaques with 60 J/cm^2 of red light (600–700 nm) over 3 hours [21]. The clinical severity scores decreased from 4.0 to 2.5 after the 5 PDT sessions. There were no significant adverse events apart from a moderate sunburn that developed on the forearms of a patient who did not comply with photoprotection precautions. A phase II trial involving weekly verteporfin administration followed by whole body fluorescent blue light exposure has recently been completed and the results of this trial are pending.

11.5 Aminolevulinic acid

ALA is an endogenous precursor of protoporphyrin IX (PpIX) that has recently been approved by the US Food and Drug Administration as a topical treatment for actinic keratoses when combined with blue light. ALA can be administered via the topical, oral, or intravenous routes. Topical ALA has been shown to improve psoriasis [22], but is impractical for patients with widespread involvement. An additional rationale for using ALA to treat immunologic disorders is the observation that activated lymphocytes appear to be more efficient at generating PpIX from exogenous ALA [23]. Following oral administration of ALA there is appreciable accumulation of PpIX in the skin. [3]. Patients are light sensitive for only 1–2 days, which corresponds to the duration of time that PpIX can be detected in normal skin after ALA administration [24]. Oral ALA has been studied most in gastroenterology where it has been used to assist in the detection of dysplastic and/or malignant epithelium such as in Barrett's oesophagus [25–27]. Systemic ALA at doses of 30–60 mg/kg has been associated with nausea, vomiting, systolic hypotension, as well as elevation of hepatic transaminases [3,24]. These side effects are potentially problematic if the further development of systemic ALA-PDT requires either higher ALA doses or multiple PDT sessions. As these side effects are ALA dose-dependant the use of lower ALA doses may overcome this problem [24,28].

In a phase I study, ALA has been administered orally at either 10, 20 or 30 mg/kg to 12 patients with moderate to severe psoriasis [29]. In vivo fluorescence spectroscopy showed a preferential accumulation of PpIX in psoriatic plaques as compared to adjacent normal skin. Under Wood's lamp examination, bright homogenous fluorescence was observed in psoriatic plaques after ALA administration. This contrasts with the more heterogeneous fluorescence that is observed after topical ALA application on psoriatic plaques (unpublished observations). PpIX fluorescence peaked between 4–5 hours after administration of 20 and 30 mg/kg of ALA. Light exposure was not performed in this pharmacokinetic study. Systemic ALA may increase the efficacy of ALA-PDT for the treatment of psoriasis by providing a more homogenous distribution of the photosensitizer throughout the plaque. A dose ranging study looking at the effect of different oral ALA and light doses is also underway.

Endogenous PpIX is also present in psoriatic skin without ALA application [30], and the reservoir for this endogenous PpIX appears to be within the psoriatic scale. It remains to be seen whether photoactivation of this endogenous PpIX with visible or UVA radiation would have a clinically beneficial effect on psoriasis.

11.6 Other photosensitizers

Few other photosensitizers have been studied for non-oncologic applications in dermatology. PDT with tin-protoporphyrin was studied in 10 patients with psoriasis [31]. Following a single injection of tin-protoporphyrin, patients received whole body UVA exposure 4–5 times per week for 3 weeks. The mean PASI score decreased from 7.9 to 6.6. Skin photosensitivity may last up to 3 months following administration of tin-protoporphyrin [32]. Assuming that singlet oxygen is central to the immunomodulating and effects of PDT, it could be assumed that other systemic photosensitizers such as tin ethyl etiopurpurin and meso-tetrahydroxyphenylchlorin would also be potentially useful for treating skin disorders other than cancer.

11.7 Future directions

Like PUVA, PDT with systemic photosensitizers is especially well suited for the treatment of certain extensive dermatologic diseases such as psoriasis or atopic dermatitis. PUVA remains one of the most efficient treatments for psoriasis. As opposed to PDT where the main mechanism of action involves type II photochemical reactions generating reactive oxygen species, PUVA generates DNA adducts. PUVA has thus been associated with an increased risk of developing squamous cell carcinoma [33] and malignant melanoma [34]. Because PDT photosensitizers may react primarily with membranes rather than DNA, the carcinogenic potential is presumably much less than that of psoralens. Ongoing and future studies with verteporfin, ALA, and other systemic photosensitizers will define the place of these molecules in the treatment of psoriasis. PDT with these drugs or with newer molecules that have not yet entered the clinic may one day supplant PUVA in our therapeutic armamentarium for psoriasis. The immunomodulatory effects of PDT could also be applied to other PUVA-responsive disorders such as atopic dermatitis, alopecia areata, scleroderma/morphea, graft versus host disease, and vitiligo.

References

1. H. Von Tappeiner, A. Jesionek (1903). Therapeutische Versuche mit fluorescierenden Stoffen. *Munch. Med. Wochenschr.*, **47**, 2042–2044.
2. H. Silver (1937). Psoriasis vulgaris treated with hematoporphyrin. *Arch. Dermatol. Syph.*, **36**, 1118–1119.
3. J. Webber, D. Kessel, D. Fromm (1997). Side effects and photosensitization of human tissues after aminolevulinic acid. *J. Surg. Res.*, **68**(1), 31–37.
4. S.O. Gollnick, X. Liu, B. Owczarczak, D.A. Musser, B.W. Henderson (1997). Altered expression of interleukin 6 and interleukin 10 as a result of photodynamic therapy in vivo. *Cancer Res.*, **57**(18), 3904–3909.

5. B. Qin, S.H. Selman, K.M. Payne, R.W. Keck, D.W. Metzger (1993). Enhanced skin allograft survival after photodynamic therapy. Association with lymphocyte inactivation and macrophage stimulation. *Transplantation*, **56**(6), 1481–1486.

6. S. Herman, Y. Kalechman, U. Gafter, B. Sredni, Z. Malik (1996). Photofrin II induces cytokine secretion by mouse spleen cells and human peripheral mononuclear cells. *Immunopharmacology*, **31**(2–3), 195–204.

7. A.H. Enk, J. Saloga, D. Becker, M. Mohamadzadeh, J. Knop (1994). Induction of hapten-specific tolerance by interleukin 10 in vivo. *J. Exp. Med.*, **179**(4), 1397–1402.

8. S.E. Ullrich (1995). The role of epidermal cytokines in the generation of cutaneous immune reactions and ultraviolet radiation-induced immune suppression. *Photochem. Photobiol.*, **62**(3), 389–401.

9. M.W. Berns, M. Rettenmaier, J. McCullough, J. Coffey, A. Wile, M. Berman et al. (1984). Response of psoriasis to red laser light (630 nm) following systemic injection of hematoporphyrin derivative. *Lasers Surg. Med.*, **4**(1), 73–77.

10. G.D. Weinstein, J.L. McCullough, J.S. Nelson, M.W. Berns, A. McCormick (1991). Low dose Photofrin II photodynamic therapy of psoriasis. *Clin. Res.*, **39**, 509A.

11. R.S. Wooten, K.C. Smith, D.A. Ahlquist, S.A. Muller, R.K. Balm (1988). Prospective study of cutaneous phototoxicity after systemic hematoporphyrin derivative. *Lasers Surg. Med.*, **8**(3), 294–300.

12. A. Orenstein, J.S. Nelson, L.H. Liaw, R. Kaplan, S. Kimel, M.W. Berns (1990). Photochemotherapy of hypervascular dermal lesions: a possible alternative to photothermal therapy? *Lasers Surg. Med.*, **10**(4), 334–343.

13. S. Leong, A.H. Chan, J.G. Levy, D.W. Hunt (1996). Transcutaneous photodynamic therapy alters the development of an adoptively transferred form of murine experimental autoimmune encephalomyelitis. *Photochem. Photobiol.*, **64**(5), 751–757.

14. R.K. Chowdhary, L.G. Ratkay, H.C. Neyndorff, A. Richter, M. Obochi, J.D. Waterfield et al. (1994). The use of transcutaneous photodynamic therapy in the prevention of adjuvant-enhanced arthritis in MRL/lpr mice. *Clin. Immunol. Immunopathol.*, **72**(2), 255–263.

15. G.O. Simkin, D.E. King, J.G. Levy, A.H. Chan, D.W. Hunt (1997). Inhibition of contact hypersensitivity with different analogs of benzoporphyrin derivative. *Immunopharmacology*, **37**(2–3), 221–230.

16. M.O. Obochi, L.G. Ratkay, J.G. Levy (1997). Prolonged skin allograft survival after photodynamic therapy associated with modification of donor skin antigenicity. *Transplantation*, **63**(6), 810–817.

17. G.O. Simkin, J.S. Tao, J.G. Levy, D.W. Hunt (2000). IL-10 contributes to the inhibition of contact hypersensitivity in mice treated with photodynamic therapy. *J. Immunol.*, **164**(5), 2457–2462.

18. M.O. Obochi, A.J. Canaan, A.K. Jain, A.M. Richter, J.G. Levy (1995). Targeting activated lymphocytes with photodynamic therapy: susceptibility of mitogen-stimulated splenic lymphocytes to benzoporphyrin derivative (BPD) photosensitization. *Photochem. Photobiol.*, **62**(1), 169–175.

19. L. Hruza, H. Lui, G. Hruza, D. McLean, R. Anderson (1995). Response of psoriasis to photodynamic therapy using benzoporphyrin derivative monoacid ring A. *Lasers Surg. Med. Suppl.*, **4**, 43.

20. R. Bissonnette, D.I. McLean, G. Reid, J. Kelsall, H. Lui (1998). Photodynamic therapy of psoriasis and psoriatic arthritis with BPD verteporfin. In: *Proceedings of the 7th Biennial Congress, International Photodynamic Association* (p. 73). Nantes, France.

21. W.H. Boehncke, T. Elshorst-Schmidt, R. Kaufmann (2000). Systemic photodynamic therapy is a safe and effective treatment for psoriasis [letter]. *Arch. Dermatol.*, **136**(2), 271–272.

22. W.H. Boehncke, W. Sterry, R. Kaufmann (1994). Treatment of psoriasis by topical photodynamic therapy with polychromatic light [letter]. *Lancet*, **343**(8900), 801.
23. E.A. Hryhorenko, K. Rittenhouse-Diakun, N.S. Harvey, J. Morgan, C.C. Stewart, A.R. Oseroff (1998). Characterization of endogenous protoporphyrin IX induced by delta-aminolevulinic acid in resting and activated peripheral blood lymphocytes by four-color flow cytometry, *Photochem. Photobiol.*, **67**(5) 565–572.
24. K. Rick, R. Sroka, H. Stepp, M. Kriegmair, R.M. Huber, K. Jacob et al. (1997). Pharmacokinetics of 5-aminolevulinic acid-induced protoporphyrin IX in skin and blood. *J. Photochem. Photobiol. B*, **40**(3), 313–319.
25. J. Regula, A.J. MacRobert, A. Gorchein, G.A. Buonaccorsi, S.M. Thorpe, G.M. Spencer et al. (1995). Photosensitisation and photodynamic therapy of oesophageal, duodenal, and colorectal tumours using 5 aminolaevulinic acid induced protoporphyrin IX – a pilot study. *Gut*, **36**(1), 67–75.
26. H. Messmann, R. Knuchel, W. Baumler, A. Holstege, J. Scholmerich (1999). Endoscopic fluorescence detection of dysplasia in patients with Barrett's esophagus, ulcerative colitis, or adenomatous polyps after 5-aminolevulinic acid-induced protoporphyrin IX sensitization. *Gastrointest. Endosc.*, **49**(1), 97–101.
27. B. Mayinger, H. Reh, J. Hochberger, E.G. Hahn (1999). Endoscopic photodynamic diagnosis: oral aminolevulinic acid is a marker of GI cancer and dysplastic lesions. *Gastrointest. Endosc.*, **50**(2), 242–246.
28. I. Wang, L.P. Clemente, R.M. Pratas, E. Cardoso, M.P. Clemente, S. Montan et al. (1999). Fluorescence diagnostics and kinetic studies in the head and neck region utilizing low-dose delta-aminolevulinic acid sensitization. *Cancer Lett.*, **135**(1), 11–19.
29. R. Bissonnette, H. Zeng, M. Korbelik, D.I. McLean, H. Lui (1998). Protoporphyrin IX fluorescence in psoriatic plaques and circulating blood cells after oral 5-ALA administration. In: *Proceedings of the 7th Biennial Congress, International Photodynamic Association* (p. 75). Nantes, France.
30. R. Bissonnette, H. Zeng, M. Korbelik, D.I. McLean, H. Lui (1998). Psoriatic plaques exhibit red autofluorescence that is due to protoporphyrin IX. *J. Invest. Dermatol.*, **111**(4), 586–591.
31. L. Emtestam, L. Berglund, B. Angelin, G.S. Drummond, A. Kappas (1989). Tin-protoporphyrin and long wave length ultraviolet light in treatment of psoriasis. *Lancet*, **1**(8649) 1231–1233.
32. L. Emtestam, B. Angelin, L. Berglund, G.S. Drummond, A. Kappas (1993). Photodynamic properties of Sn-protoporphyrin: clinical investigations and phototesting in human subjects. *Acta Derm. Venereol.*, **73**(1), 26–30.
33. R. Stern (1994). Metastatic squamous cell cancer after psoralen photochemotherapy. *Lancet*, **344**(8937), 1644–1645.
34. R.S. Stern, K.T. Nichols, L.H. Vakeva (1997). Malignant melanoma in patients treated for psoriasis with methoxsalen (psoralen) and ultraviolet A radiation (PUVA). The PUVA Follow-Up Study. *N. Engl. J. Med.*, **336**(15), 1041–1045.

Part V:

Topical Photodynamic Therapy

Photodynamic Therapy and Fluorescence Diagnosis in Dermatology
P.-G. Calzavara-Pinton, R.-M. Szeimies and B. Ortel, editors.

Chapter 12

Topical sensitization – oncologic indications – actinic keratoses

Alexis Sidoroff

Table of contents

Abstract

Actinic keratoses are very common cutaneous neoplasms usually arising on sun exposed areas of elderly fair skinned people. Histological features and the same chromosomal mutations found in squamous cell carcinoma indicate that actinic keratoses are carcinoma in situ rather than *pre*malignant lesions. Although the probability of progressing into invasive and destructive, even metastatic disease is rather low, this risk cannot reliably be predicted for an individual lesion. Thus treatment of actinic keratoses is often necessary. Among the many modalities used to remove actinic keratoses the most common are cryotherapy and topical 5-fluorouracil treatment. As a rather new modality photodynamic therapy (PDT) has proven effective in many studies and has established itself as a valuable treatment alternative in many centers, but treatment protocols differ quite significantly in the preparation of the sensitizers, the application modalities, the type of light source used, as well as the illumination parameters. This chapter provides an overview of the various modalities of PDT using topical sensitizers in the treatment of actinic keratoses. Emphasis is put on procedures that use δ-aminolevulinic acid (5-ALA) – a precursor of the photosensitizer protoporphyrin IX.

12.1 Introduction

Actinic keratoses (AKs, syn. solar keratoses) have for a long time been designated as precursors of malignancy, namely squamous cell carcinoma. In different studies the risk that AK lesions will progress into invasive disease has been reported to be between 0.025% [1] and 16% [2] depending on the study design with an estimated average between 8 and 10% [3]. Current concepts – propagated especially by dermatohistopathologists – regard AKs and it's analogues, arsenic keratosis and radiation keratosis, as already being (non invasive) squamous cell carcinoma (SCC) defined by criteria like the lesions being an epithelial neoplasm with crowded, pleomorphic and large nuclei, eosinophilic cytoplasm, and signs of abnormal cornification (dys- and parakeratosis) [4,5]. Furthermore the genetic changes of SCC can also be found in the cells of AKs [6]. This makes AKs the most common malignant cutaneous neoplasms. Although spontaneous regression or persistence in the in situ state are possible [7], the clinical potential of AK lesions to turn into invasive malignancies often makes the treatment of such lesions necessary and should not be regarded as mere cosmetic procedure.

12.2 Pathogenesis and epidemiology

With the current state of knowledge the development of AKs can be explained by the mechanism of multi-step carcinogenesis [8]. Ultraviolet radiation (UVR) initiates the process by causing a permanent mutation in keratinocytes (e.g. p53 tumor suppressor gene and/or ras proto-oncogenes). It has been demonstrated that ultraviolet B light (UVB) causes a very specific mutation in the tumor suppressor gene p53 in which cytosine is changed into thymine when cytosine is adjacent to thymine or two cytosines are adjacent to each other [6,9]. In the case of non-melanoma skin cancer UVR is not only the initiator, but also the promoter [10] and cells sufficiently damaged by repeated exposure to UVR do not undergo apoptosis (which can histologically be seen as "sunburn cells") but proliferate and clonally expand.

AKs and SCCs preferentially arise on sun exposed skin of elderly pale skinned individuals after chronic UVR exposure [11,12] and prevalence of AKs increases with age [13]. In predisposed patients of fair complexion the lesions can appear as early as in their thirties. Studies from Australia [13–17], were AKs are more common than anywhere else in the world, estimated the prevalence of AKs to be 40–50% in people over 40 years of age, but even in studies in northern hemisphere populations prevalence rates range from 11 to 26% [18–20].

12.3 Clinical appearance

AK lesions usually appear on chronically sun exposed skin of elderly people: forehead, nose, cheeks, and especially in male patients ear helices and scalp. Apart from the head the dorsum of the hands and forearms are often affected. Usually other signs of chronic sun damage, such as freckled pigmentation and elastosis of the surrounding skin can be seen. The lesions themselves can first present as pink papules or patches (Fig. 1) from a few mm to sometimes several cm in diameter. In the beginning hyperkeratosis can be mild and the lesions can often better be felt than seen (Fig. 2). Later stages can present as rather thick hyperkeratotic tumors (Fig. 3) clinically indistinguishable from invasive disease. Often histological examination is the only way to determine whether a lesion is still superficial or already invasive. The color spectrum ranges from red to skin-colored to grayish-brown depending on the amount of hyperkeratosis. Special variants include pigmented AKs, cutaneous horns (Fig. 4), or actinic cheilitis of the (lower) lip. AKs may occur as single or multiple, sometimes confluent lesions [21]. AKs thus are the beginning of a spectrum that leads to invasive squamous cell carcinoma (Figs. 5 & 6) and even metastatic [22] and life-threatening disease.

Actinic Keratoses can appear in a variety of clinical constellations which have implications in the choice of the treatment modality. Size, localization and number of lesions, thickness and probability of invasive disease, are the lesions very hyperkeratotic or erosive, age and general status of the patient, underlying disease or pharmacological therapy (anticoagulation, immunosuppression). Because of this spectrum there cannot be a standard therapy for actinic keratoses but the choice has to be made on an individual basis. Furthermore cosmetic outcome, time off from work, time and costs, as well as the pain of the procedure have influence on the decision which form of treatment is the most suitable in a certain situation.

12.4 Treatment options

Photodynamic therapy competes with many methods of treatment of AKs (Table 1) [23,24]. Some of them are being widely used, some have a more investigational character, and others are performed only by a smaller number of physicians due to the specialization or special equipment needed. For just a few well demarcated lesions

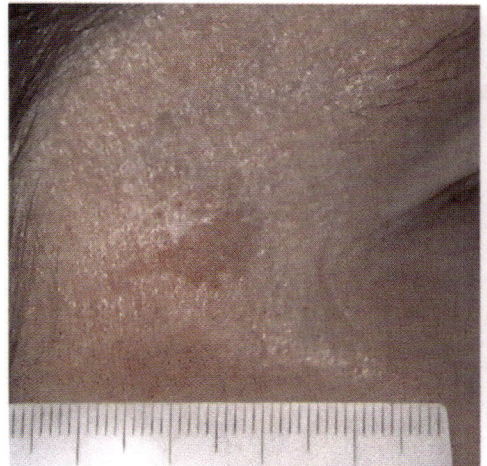

Figure 1. very mild actinic keratosis on the temple of a 37 year old female patient.

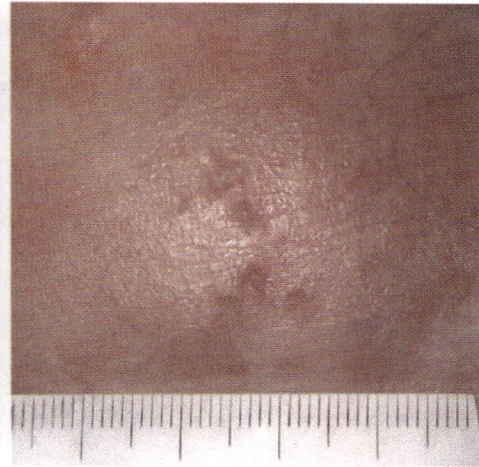

Figure 2. mildly hyperkeratotic actinic keratosis on the scalp of a 66 year old male patient.

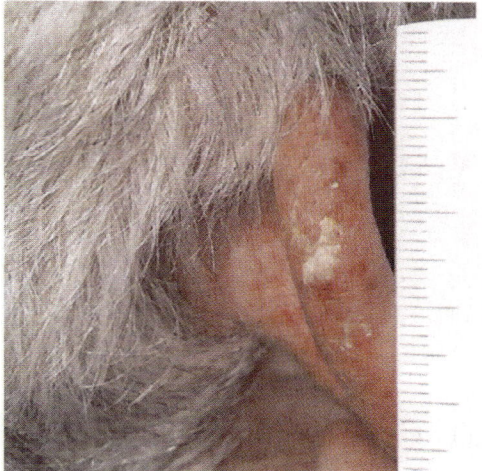

Figure 3. more pronounced and hyperkeratotic actinic keratosis on the helix of a 70 year old male patient.

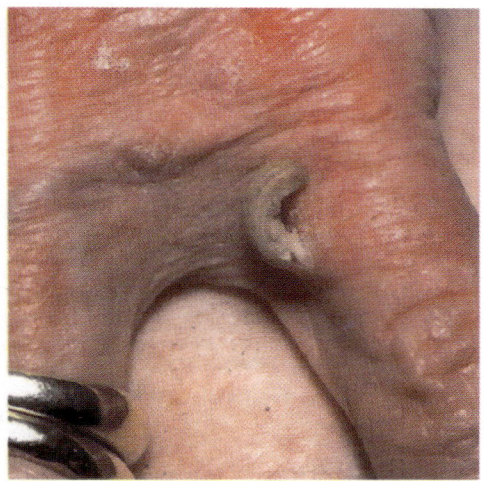

Figure 4. digital cutaneous horn in a 90 year old female patient.

cryotherapy with liquid nitrogen – either with cotton tip swabs or with a spraying device – is most commonly used [25,26]. The procedure is quick, easy, and cost-effective. Some authors also suggest cryotherapy for the treatment of more extensive areas (cryopeeling) [27,28]. Other mechanic possibilities of treatment include curettage, either with or without electrosurgery, dermabrasion [29], laser-ablation (CO_2-laser [30,31], Er:YAG-laser [32,33]) or surgical excision.

For more widespread lesions topical application of 5-fluorouracil (5-FU) has become a standard treatment modality. The methylation reaction of deoxyuridylic acid to

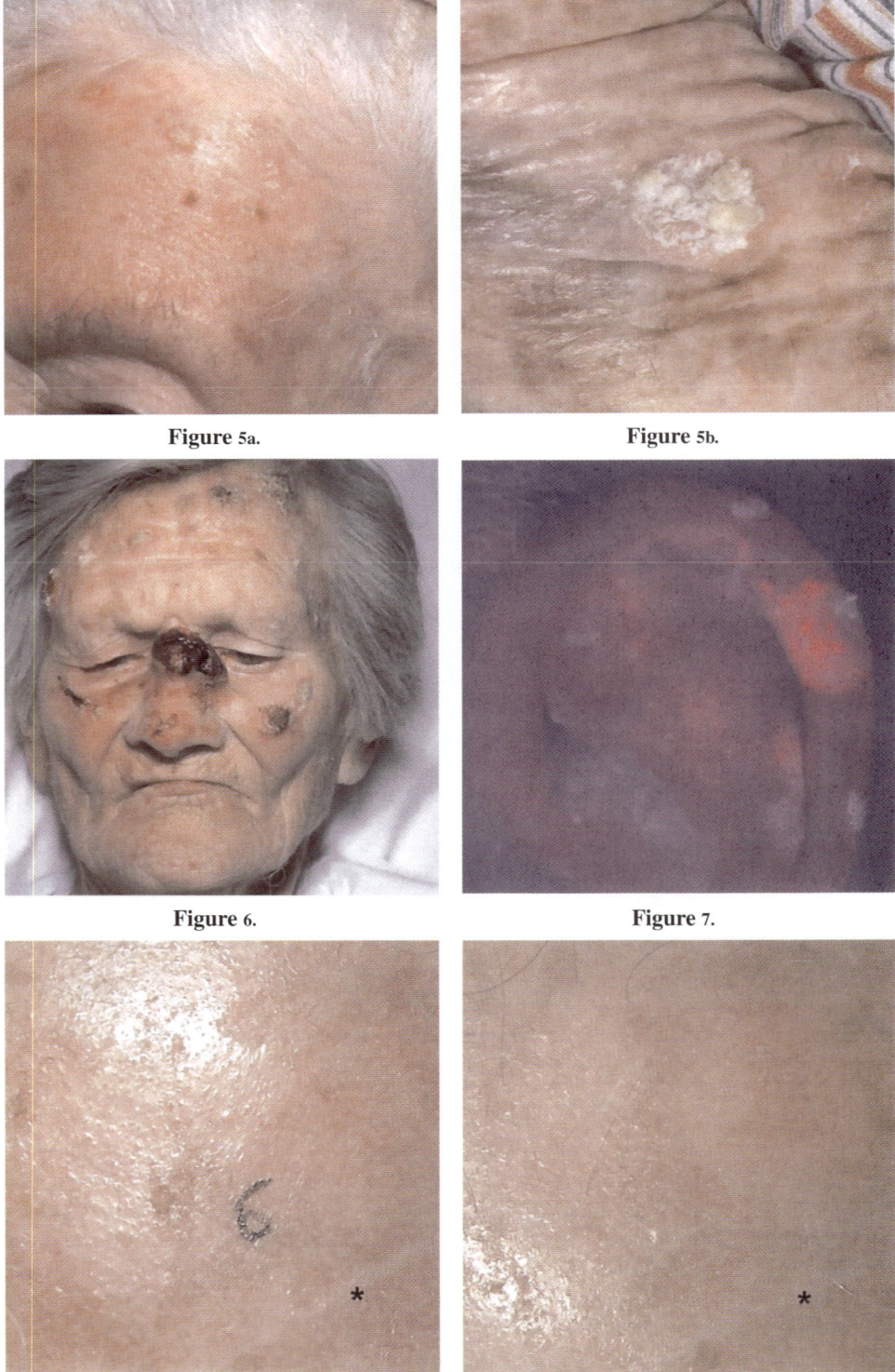

Figure 5a.

Figure 5b.

Figure 6.

Figure 7.

Figure 8a.

Figure 8b.

* hypopigmented line as reference for localization

Figure 5a–b. the spectrum from actinic keratosis (a) to invasive squamous cell carcinoma (b) in a 90 years old female patient.

Figure 6. large destructive squamous cell carcinoma in the face of a 88 year old female patient.

Figure 7. 75-year old male patient: pink fluorescence of actinic keratoses on the ear helix after 5 hours of incubation with a 20% o/w ALA formulation and illumination with a Wood's light.

Figure 8a–b. therapeutic effect of PDT with ALA methyl-ester: (a) before treatment; (b) 3 months after treatment.

thymidylic acid is blocked by this cytotoxic agent [34,35]. 5-FU in an o/w basis is applied twice daily on the affected area until it becomes erosive which may take two to four weeks. This regimen is sometimes modified either by intensifying it (4 applications daily) [36] or changing it to a pulse therapy where 5-FU is only applied once or twice a week for several months [37,38]. The response rate of topical 5-FU treatment is very much dependent on the compliance of the patients and the accuracy with which the treatment is being carried out. The use of masoprocol, another topical antineoplastic

Table 1. Options for the topical treatment of actinic keratoses

Prevention
Cryotherapy/cryopeeling
Curettage
Electrosurgery
Laserablation
Dermabrasion
Excisional surgery

Photodynamic therapy

Topical 5-fluorouracil
Chemical peels
Salicylic acid
Topical retinoids (e.g. Tretinoin)
Alpha hydroxy acids
Topical masoprocol
Interferon
Topical tubercidin
Diclofenac in hyaluronic acid gel
Solasodine glycosides
Calcitriol and isotretinoin

agent [39], has been discontinued because it has a potential of allergic sensitization [40].

Chemical peels with various substances like trichloracetic acid (TCA) [41], α-hydroxy acids, phenol [42] – alone or in combination [43,44] – have also been used and/or are still under investigation. Other treatment trategies include the intralesional [45,46] or topical [47] application of interferon alpha 2b, the topical use of retinoids [48,49], a mixture of topical 3% diclofenac in a 2.5% hyaluronic acid gel [50], topical tubercidin [51], or solasodine glycosides [52].

These options can even be expanded by the possibility of combining topical treatment modalities [53,54] or adding systemic treatment with retinoids [55–57].

Last but not least it has to be stated that prevention of UVR avoidance is a of utmost importance in the effort towards fighting non-melanoma skin cancer.

12.5 Photodynamic therapy of actinic keratosis using topical sensitizers

Photodynamic therapy is a modality in which by (either systemic or topical) administration of non-psoralen photosensitizing drugs and consecutive exposure to light tumors and inflammatory diseases can be treated by oxygen-dependent mechanisms. The major drawback of systemic PDT is that it results in a prolonged cutaneous light sensitivity. For many of the easily accessible cutaneous diseases topical application of the sensitizer is the more desirable alternative. Besides superficial basal cell carcinoma (BCC) and Bowen's disease AKs are one of the most common dermatological indications for photodynamic therapy using topical sensitizers and the treatment procedures are comparable in these diseases. As a consequence the same basic considerations concerning the sensitizers and light sources used apply and lead to a variety of regimens with the use of different sensitizers, sensitizer concentrations and preparations, incubation times, light sources, illumination times, number of treatments and treatment intervals (Table 2).

12.5.1 Sensitization

For sensitization topical preparations of δ-aminolevulinc acid (5-ALA) are most commonly used [58, 59]. 5-ALA is not a sensitizer itself but a metabolite in the heme-pathway which induces the synthesis of protoporphyrin IX (PpIX). For reasons described earlier in this book there is quite a selective accumulation of PpIX in tumor-tissue (about 10fold the concentration found in the surrounding tissue). Until recently no topical sensitizer was commercially available so that in many centers individual preparations were and are still made by the local pharmacists. 5-ALA-HCl concentra-tions used vary from 10 to 30% in indifferent o/w emulsions. For a better penetration DMSO or EDTA is sometimes added [60] or used to pre-treat the target area and also the addition of 3% of the iron chelator desferrioxamine (Desferal®, Ciba-Geigy, Basle, Switzerland) or ferrochelatase inhibitors [61,62] to stimulate the photosensitization has been reported [63,64]. Also to enhance penetration hyperkeratotic lesions can be

Table 2. Variables in topical PDT of actinic keratoses.

type of lesion treated
localization
O$_2$ in tissue (e.g. perfusion, use of antioxidants)
geometry of target area
sensitizer
sensitizer preparation
additives
application procedure
incubation time
local analgesia
light source
light intensity (mW/cm^2)
fluence (J/cm^2)
illumination regimen (e.g. fractionating)
number of treatments
treatment intervals

cautiously curetted before the application of the sensitizer. The lesion is then covered with an approximately 1mm thick layer of the ALA cream with an overlap to clinically normal appearing skin. An occlusive dressing (e.g. Tegaderm™; 3M Healthcare Ltd. or Opsite™; Smith and Nephew Medical Ltd.) is applied and topped by a swab, a gauze bandage, and/or aluminum foil for light protection. Application times described vary depending on the formulations used but commonly range from 3 to 6 hours to grant a sufficient PpIX concentration in the target tissue and to prevent too high levels in the unaffected surrounding tissue by unspecific enrichment. The only ALA preparation approved in the US for topical use is a solution (Levulan®). A special application device called Kerastick™ allows precise targeting of the lesions to be treated and the sensitizer is supposed to be applied approximately 14 to 18 hours before exposure to (blue) light.

Porphyrin molecules themselves are too large to penetrate when applied topically, but a few other sensitizers have been used.

Phase II clinical studies are currently being performed with ATMPn (9-acetoxy-2,7,12,17-tetrakis-(β-methoxyethyl)-porphycene) a member of the porphycene family. Porphycenes are synthetically generated porphin isomers whose properties seem to be very promising for systemic and topical PDT [65–70]. Another porphin isomer that has been investigated for topical PDT is Tetrasodium-meso-tetraphenylporphinesulfonate (TPPS4) [71]. Unfortunately the compound penetrates unselectively into the subcutaneous fat [72] and is potentially neurotoxic [73], which is why investigations have been discontinued.

12.5.2 Fluorescence diagnosis

After removing the dressing and gently wiping off the remaining ALA cream from the surface of the lesion with a soft tissue, fluorescence diagnosis can be performed. The term photodynamic diagnosis (PDD) should not be used, as the fluorescence is not caused by photodynamic effects. As a light source an ultraviolet – blue light (Wood's light, $\lambda = 340–400$ nm) is used and a bright pink fluorescence shows the localization of PpIX and thus the borders of the lesion (Fig. 7) often revealing clinically inconspicuous areas to be affected as well.

12.5.3 Light exposure

Basically two types of light sources can be used to cause the desired photodynamic effects: lasers producing monochromatic, coherent light and incoherent light sources providing light of a broader wavelength spectrum. The principal differences and the question whether there is a clinically relevant difference in the photodynamic effect between the type of light used are discussed in chapters 2 and 6. Basic considerations for the choice of a light source are, that the emitted wavelength should match an absorption peak of the photosensitizer used and that longer wavelengths penetrate deeper into the tissue. The accessibility of cutaneous lesions makes coupling of light into a fiberoptic system as it is used to treat other organs (e.g. urinary bladder, gastrointestinal tract, trachea and bronchi) unnecessary. Although spreading of the laser beam is possible with diffuser-lenses and fiber-splitters allow the simultaneous treatment of a couple of lesions with lasers, incoherent light sources can usually produce a larger field of illumination. The broader spectrum of these lamps also makes it possible to use them for different sensitizers with different absorption spectra. Thus for ALA PDT in dermatology preference is often given to the cheaper and easier to maintain incoherent light sources.

The spectrum of incoherent light sources used ranges from high output slide projectors with special filters [74,75] over prototypes to commercially available lamps like the Waldmann PDT 1200 ($\lambda = 600$ to 800 nm, 30 to 200 mW/cm^2) [76], or the Curelight lamp from PhotoCure (150W halogen lamp, $\lambda = 570–680$ nm, 70–200 mW/cm^2). The Paterson PDT by Photo Therapeutics Limited (UK) offers a variable wavelength spectrum and ESC Medical Systems is marketing the VersaLight™, a multiple band continuous wave light source with a treatment band from 580–720 nm providing an intensity of 100–150 mW/cm^2, a fluorescence excitation light ($\lambda = 400–450$ nm) and a built-in CCD camera. Recently LED light sources have been developed like the Fireplace Video-3 by JSC BioSpec (LED, 50mW/cm^2, $\lambda = 632$ nm) with an integrated camera device or the PRP 100 by PRP Optoelectronics (LED array, $\lambda = 635$ nm, 150W/cm^2), just to name a few in the red spectrum. Controversy exists on whether the infrared portion of the lightsource should be filtered out (as it increases the risk of hyperpigmantation) or not. The only treatment regimen that got approval from the FDA in the US and in which the sensitizer is the Levulan® Kerastick™ uses a blue light source (BLU-U™, $\lambda = 417$ nm).

When using monochromatic light sources particular attention has to be taken that the emitted wavelength matches an absorption peak of the sensitizer. For 5-ALA induced PpIX Szeimies and co-workers could demonstrate that a wavelength of 635 nm had the best cell killing capabilities [77]. Lasers often used in PDT such as the argon ion pumped dye laser or the copper vapor laser, thus do not provide the optimal wavelength for ALA-PDT. The Laserscope PDT System that uses a KTP/YAG laser as pump system and a dye module can operate from 624 to 670 nm (732 nm). Dedicated Diode Lasers for PDT, which are currently being developed and marketed by various manufacturers, have the advantage of being relatively inexpensive, more compact in size and need less maintenance.

12.5.4 Treatment effects and side effects

During illumination patients have a burning, stinging sensation, which without anesthesia can be quite severe and even lead to the interruption of the treatment. This pain is not caused solely by the heat of the light source but also by the photodynamic process itself. In particular pain is reported in cases of extensive and/or erosive lesions. If infiltrative local anesthesia is used, no vasoconstrictive agents should be added as the presence of oxygen is necessary for the photodynamic processes. The possibility of adding topical local anesthetics [78] (e.g. EMLA®-Cream; AstraZeneca) to the topical sensitizer is controversial as, due to the pH of 9, condensation of 5-ALA is possible. After illumination erythema of the treated area is usually seen and edema can occur. A steroid cream can be used to reduce this acute inflammatory reaction. After a few hours erosions and crusts develop. Healing usually takes about 1 to 3 weeks and can be supported by application of indifferent ointments or topical antibiotics to prevent bacterial infection. Transient hyperpigmentation can be observed, especially when DMSO is added to the sensitizer.

Phototherapy with ultraviolet radiation (PUVA, UVB) has been demonstrated to be carcinogenic [79–81]. There has been one case report of a 82 year old patient who developed a malignant melanoma on the site of repeated PDT for AKs [82], but this of course does not imply a causative connection and UVA was also used in this patient. Apart from this report there has been no clinical evidence that ALA PDT has caused secondary malignancies. In a review of this topic Fuchs et al. [83] point out that further studies are needed to evaluate the carcinogenic risk of PDT but available data and the fact that porphyria patients do not seem to develop more malignancies than others suggest that this risk seems to be very low.

12.5.5 Outcome

The first reports on the effectiveness of topical PDT of epithelial tumors were published in the early 20th century by von Tappeiner and Jesionek [84]: a 5% eosin solution was applied on a SCC in the face of 70-year-old woman and exposed to light which led to healing of the tumor. But it wasn't until the 1990s that topical PDT with ALA was systematically investigated and published. All 6 lesions labeled as in situ or early invasive SCC and 9 of 10 AKs incubated for 3 to 6 hours with a 20% o/w ALA formulation and irradiated with a slide projector as a light source showed remission in a study published by Kennedy and co-workers in 1990 [75]. Although most of the consecutive studies concerning PDT with topically applied sensitizers deal mainly with the treatment of BCCs or Bowen's disease some data on the response of AKs or SCCs to this treatment exist: In a study performed by Wolf and co-workers all 9 AKs and 5 of 6 superficial SCCs responded to a therapy with a 20% ALA emulsion and irradiation with a slide projector [85]. In another study presented by Hürlimann in 1994 4 SCCs were treated with a 20% ALA formulation and in all a complete remission could be achieved [86]. Also in Harth's study [60], where 2% DMSO or 2% EDTA were added to 20% ALA cream to enhance penetration 4 of 5 superficial SCC showed a complete response. A 20% ALA emulsion and light from an argon-ion-pumped dye laser were used in a study by Calzavara Pinton. 1–3 treatments were performed in 12 superficial BCCs (response rate 84%) and 50 AKs (long term response rate 84%) [87]. An incoherent light source (Waldmann PDT 1200, 160 mW/cm^2, 150 J/cm^2) was used to irradiate 36 AKs in a phase II study by Szeimies et al. The lesions were incubated for 5–6 h with a 10% ALA emulsion and the response rate after 3 months was 71% for lesions on the head [88]. A similar PDT modality (20% 5-ALA cream, 150 J/cm^2 with the Waldmann PDT 1200 incoherent light source) was compared to topical 5-FU by Kurawa and co-workers [89]. In the relatively small group of 17 patients one episode of PDT treatment seemed as effective as 5-FU applied twice daily over a period of 3 weeks in a hand to hand comparison. Different concentrations of ALA (0%, 10%, 20% or 30%) were investigated by Jeffes and co-workers on 40 patients [90]. With an incubation time of 3 hrs and illumination with an argon pumped dye laser (630 nm, 10–150 J/cm^2) the response rate was 91% for AK lesions on the scalp and 45% for lesions on the trunk or extremities and did not differ significantly with the different ALA-concentrations. At present – November 2000 – ALA is the only topical sensitizer to have approval for the treatment of actinic keratoses (nonhyperkeratotic lesions on the face or scalp) in the U.S. The results of the clinical trials initiated by DUSA Pharmaceuticals (Toronto, Canada) were summarized by Ormrod and Jarvis [91]. A 20% ALA solution (Levulan®) was applied topically with an applicator device (Kerastick™) 14–18 hrs before exposure to blue light (BLU-U™) at 417 nm for 1000 seconds (10 mW/cm^2, 10 J/cm^2). Response rates in the verum groups were 86% and 81% compared to 32% and 20% in the placebo groups. Across all patients 82.8% of the 1403 lesions treated responded with a recurrence rate of 5% between 8 and 12 weeks after treatment. A phase III clinical trial comparing an ALA methyl-ester cream (160 mg/g) to cryotherapy has recently been completed showing similar response rates in both treatment arms (submitted for publication) (Fig. 8).

A treatment modality that uses a long pulsed flash lamp pumped dye laser (LPDL, pulse duration 1.5 ms; $\lambda = 585$ nm; 16 J/cm^2) as light source with the intention of reducing pain during illumination showed a response rate of 79% 4 weeks after PDT (20% ALA emulsion, incubation time 6hrs) [92].

Etiological variants of non-melanoma skin cancer (NMSC) in which topical PDT is also applicable are arsenic keratoses [93] and NMSC arising in organ transplant patients who have to undergo long term immunosuppressive medication.

In summary topical PDT has been documented to be an efficient treatment of AKs but rather poor long term results have also been reported. In a study by Fink-Puches et al. [94] only 8% of the treated SCCs did not recur. Histological examinations showed post treatment fibrosis in the dermis so that insufficient penetration of the treatment did not seem to be the reason for the unsatisfactory efficacy. As an alternative explanation insufficient marking of the target cells was proposed [94].

In our routine procedure for the treatment of AKs with topical PDT we use 20% 5-ALA in a o/w cream (Doritin®; AstaMedica, Vienna, Austria) with an incubation time of 5–6 hrs under an occlusive dressing and illumination with an incoherent light source ($\lambda = 600$–800 nm) at 100–150 mW/cm^2 with 100–150 J/cm^2. This method has established itself as a reliable alternative for example in instances, where routine cryotherapy is not feasible because of the extent of the lesions and topical 5-FU treatment could only be performed on an inpatient basis. Cosmetic results are very satisfactory. As with any treatment of AKs a regular follow-up is of utmost importance, not only because treatment failures are possible but also because of the nature of the disease itself which makes the emergence of new lesions probable.

12.6 Perspectives

PDT of AKs with topical sensitizers is effective. But, as already mentioned, the efficacy of the treatment is dependent on many variables (Table 2). Small modifications of the treatment procedure may have quite a significant impact on the outcome and are often not sufficiently described in publications. This makes it difficult to compare the published reports and to determine which of topical PDT procedures performed in the different centers is currently the best. Furthermore phase III clinical trials initiated by the pharmaceutical industry either use modified sensitizers, like the ALA-methyl-ester Metvix® (Photocure ASA) or use other ALA preparations and light sources like the procedure based on the Levulan® Kerastick™ (Schering AG) and the BLU-U™ blue light. Therefore modalities that already are or will be approved in the near future differ quite significantly from those that have been investigated and used for years in specialized research centers. Direct comparative studies or data that sufficiently allow to compare the different modalities are missing. One of the next steps that should to be taken is the standardization of the "classical" topical PDT with 5-ALA and red light and its approval by the regulatory authorities. However, the long term interest – not only for the treatment of AKs – will be the development of new sensitizers and light sources, both, to raise the efficacy of the treatment and to make PDT for the patients as convenient and painless as possible. Where sensitizers are concerned, properties optimizing penetration, tumor targeting, and enhancement of the photodynamic mechanisms like high quantum yield of singlet oxygen are desired. Light sources should be easy to handle and maintain, affordable, and suitable for different illumination situations. Larger fields of illumination would for example be helpful in the treatment of widespread NMSC in transplant patients.

In our view topical PDT should already be classified as an equivalent alternative in the treatment of AKs in certain clinical situations. On the other hand there is no doubt that it has the potential to be further improved and thus to be established as a standard procedure in the therapy of the growing number of non-melanoma skin cancer.

References

1. R. Marks, G, Rennie, T.S. Selwood (1988). Malignant transformation of solar keratoses to squamous cell carcinoma. *Lancet*, **1**(8589), 795–797.

2. J.H. Graham (1976). Precancerous lesions of the skin. *Prim. Care*, 2, 699–716.

3. R.G. Glogau (2000). The risk of progression to invasive disease. *J. Am. Acad. Dermatol.*, **42**(1 Pt 2), 23–24.

4. P. Ng, A.B. Ackermann (1999). The major types of squamous-cell carcinoma. *Dermatopathol: Pract. and Conc.*, **5**(3), 250–252.

5. A.B. Ackermann (1997). Respect at last for solar keratosis (Editiorial). *Dermatopathol: Pract. and Conc.*, **3**(2), 101–103.

6. A. Ziegler, A.S. Jonason, D.J. Leffell, J.A. Simon, H.W. Sharma, J. Kimmelman et al. (1994). Sunburn and p53 in the onset of skin cancer [see comments]. *Nature*, **372**(6508), 773–776.

7. C. Frost, G. Williams, A. Green (2000). High incidence and regression rates of solar keratoses in a queensland community. *J. Invest. Dermatol.*, **115**(2), 273–277.

8. B. Vogelstein, K.W. Kinzler (1993). The multistep nature of cancer. *Trends Genet.*, **9**(4), 138–141.

9. A. Ziegler, D.J. Leffell, S. Kunala, H.W. Sharma, M. Gailani, J.A. Simon et al. (1993). Mutation hotspots due to sunlight in the p53 gene of nonmelanoma skin cancers. *Proc. Natl. Acad. Sci. USA*, **90**(9), 4216–4220.

10. D.E. Brash, A. Ziegler, A.S. Jonason, J.A. Simon, S. Kunala, D.J. Leffell (1996). Sunlight and sunburn in human skin cancer: p53, apoptosis, and tumor promotion. *J. Investig. Dermatol. Symp. Proc.*, **1**(2), 136–142.

11. B.C. Vitasa, H.R. Taylor, P.T. Strickland, F.S. Rosenthal, S. West, H. Abbey et al. (1990). Association of nonmelanoma skin cancer and actinic keratosis with cumulative solar ultraviolet exposure in Maryland watermen. *Cancer*, **65**(12), 2811–2817.

12. R. Marks. (1990). Solar keratoses. *Br. J. Dermatol.* (Suppl), **122**(35), 49–54.

13. A. Green, G. Beardmore, V. Hart, D. Leslie, R. Marks, D. Staines (1988). Skin cancer in a Queensland population. *J. Am. Acad. Dermatol.*, **19**(6), 1045–1052.

14. R. Marks, D. Jolley, A.P. Dorevitch, T.S. Selwood (1989). The incidence of non-melanocytic skin cancers in an Australian population: results of a five-year prospective study. *Med. J. Aust.*, **150**(9), 475–478.

15. R. Marks, M. Staples, G.G. Giles (1993). Trends in non-melanocytic skin cancer treated in Australia: the second national survey. *Int. J. Cancer*, **53**(4), 585–590.

16. R. Marks (1995). An overview of skin cancers. Incidence and causation. *Cancer* (Suppl), **75**(2), 607–612.

17. C.A. Frost, A.C. Green (1994). Epidemiology of solar keratoses. *Br. J. Dermatol.*, **131**(4), 455–464.

18. I. Harvey, S. Frankel, R. Marks, D. Shalom (1996). Nolan-Farrell M. Non-melanoma skin cancer and solar keratoses. I. Methods and descriptive results of the South Wales Skin Cancer Study. *Br. J. Cancer*, **74**(8), 1302–1307.

19. I. Harvey, S. Frankel, R. Marks, D. Shalom, M. Nolan-Farrell (1996). Non-melanoma skin cancer and solar keratoses II analytical results of the South Wales Skin Cancer Study. *Br. J. Cancer*, **74**(8), 1308–1312.

20. M.T. Johnson, J. Roberts (1978). Skin conditions and related need for medical care among persons 1–74 years. United States, 1971–1974. *Vital Health Stat.*, **11**(212), i-v, 1–72.

21. R.L. Moy (2000). Clinical presentation of actinic keratoses and squamous cell carcinoma. *J. Am. Acad. Dermatol.*, **42**(1 Pt 2), 8–10.

22. S.M. Dinehart, P. Nelson-Adesokan, C. Cockerell, S. Russell, R. Brown (1997). Metastatic cutaneous squamous cell carcinoma derived from actinic keratosis. *Cancer*, **79**(5), 920–923.

23. L.A. Drake, R.I. Ceilley, R.L. Cornelison, W.L. Dobes, W. Dorner, R.W. Goltz et al. (1995). Guidelines of care for actinic keratoses. Committee on Guidelines of Care. *J. Am. Acad.*

Dermatol., **32**(1), 95–98.

24. S.R. Feldman, A.B. Fleischer, Jr., P.M. Williford, J.L. Jorizzo (1999). Destructive procedures are the standard of care for treatment of actinic keratoses. *J. Am. Acad. Dermatol.*, **40**(1), 43–47.

25. P.J. Holt (1988). Cryotherapy for skin cancer: results over a 5-year period using liquid nitrogen spray cryosurgery. *Br. J. Dermatol.*, **119**(2), 231–240.

26. R.R. Lubritz, S.A. Smolewski (1982). Cryosurgery cure rate of actinic keratoses. *J. Am. Acad. Dermatol.*, **7**(5), 631–632.

27. S.E. Chiarello (1992). Full-face cryo- (liquid nitrogen) peel. *J. Dermatol. Surg. Oncol.*, **18**(4), 329–332.

28. S.E. Chiarello (2000). Cryopeeling (extensive cryosurgery) for treatment of actinic keratoses: an update and comparison. *Dermatol. Surg.*, **26**(8), 728–732.

29. W.P. Coleman, 3rd, J.M. Yarborough, S.H. Mandy (1996). Dermabrasion for prophylaxis and treatment of actinic keratoses. *Dermatol. Surg.*, **22**(1), 17–21.

30. S.J. Trimas, D.A. Ellis, R.D. Metz (1997). The carbon dioxide laser. An alternative for the treatment of actinically damaged skin [see comments]. *Dermatol. Surg.*, **23**(10), 885–889.

31. J.E. Fulton, A.D. Rahimi, P. Helton, K. Dahlberg, A.G. Kelly (1999). Disappointing results following resurfacing of facial skin with CO_2 lasers for prophylaxis of keratoses and cancers [see comments]. *Dermatol. Surg.*, **25**(9), 729–732.

32. B. Dmovsek-Olup, B. Vedlin (1997). Use of Er:YAG laser for benign skin disorders. *Lasers Surg. Med.*, **21**(1), 13–19.

33. S.B. Jiang, V.J. Levine, K.S. Nehal, M. Baldassano, H. Kamino, R.A. Ashinoff (2000). Er:YAG laser for the treatment of actinic keratoses. *Dermatol. Surg.*, **26**(5), 437–440.

34. W.H. Eaglstein, G.D. Weinstein, P. Frost (1970). Fluorouracil: mechanism of action in human skin and actinic keratoses. I. Effect on DNA synthesis in vivo. *Arch. Dermatol.*, **101**(2), 132–139.

35. C.J. Dillaha, G.T. Jansen, W.M. Honeycutt, A.C. Bradford (1983). Selective cytotoxic effect of topical 5-fluorouracil. *Arch. Dermatol.*, **119**(9), 774–783.

36. M.E. Unis (1995). Short-term intensive 5-fluorouracil treatment of actinic keratoses [see comments]. *Dermatol. Surg.*, **21**(2), 162–163.

37. D.L. Pearlman (1991). Weekly pulse dosing: effective and comfortable topical 5-fluorouracil treatment of multiple facial actinic keratoses. *J. Am. Acad. Dermatol.*, **25**(4), 665–667.

38. E. Epstein (1998). Does intermittent "pulse" topical 5-fluorouracil therapy allow destruction of actinic keratoses without significant inflammation? *J. Am. Acad. Dermatol.*, **38**(1), 77–80.

39. E.A. Olsen, M.L. Abernethy, C. Kulp-Shorten, J.P. Callen, S.D. Glazer, A. Huntley et al. (1991). A double-blind, vehicle-controlled study evaluating masoprocol cream in the treatment of actinic keratoses on the head and neck [see comments]. *J. Am. Acad. Dermatol.*, **24**(5 Pt 1), 738–743.

40. E. Epstein (1994). Warning! Masoprocol is a potent sensitizer [letter; comment]. *J. Am. Acad. Dermatol.*, **31**(2 Pt 1), 295–297.

41. D.G. Brodland, R.K. Roenigk (1988). Trichloroacetic acid chemexfoliation (chemical peel) for extensive premalignant actinic damage of the face and scalp [published erratum appears in *Mayo Clin. Proc.*, **63**(11, Nov. 1988), 1122]. *Mayo Clin. Proc.*, **63**(9), 887–896.

42. P.A. Stone (1998). The use of modified phenol for chemical face peeling. *Clin. Plast. Surg.*, **25**(1), 21–44.

43. Y. Tse, A. Ostad, H.S. Lee, V.J. Levine, K. Koenig, H. Kamino et al. (1996). A clinical and histologic evaluation of two medium-depth peels. Glycolic acid versus Jessner's trichloroacetic acid. *Dermatol. Surg.*, **22**(9), 781–786.

44. N. Lawrence, S.E. Cox, C.J. Cockerell, R.G. Freeman, P.D. Cruz, Jr. (1995). A comparison

fluorouracil in the treatment of widespread facial actinic keratoses. *Arch. Dermatol.*, **131**(2), 176–181.

45. L. Edwards, N. Levine, M. Weidner, M. Piepkorn, K. Smiles (1986). Effect of intralesional alpha 2-interferon on actinic keratoses. *Arch. Dermatol.*, **122**(7), 779–782.

46. D. Shuttleworth, R. Marks (1989). A comparison of the effects of intralesional interferon a-2b and topical 5% fluorouracil cream in the treatment of solar keratoses and Bowen's disease. *J. Dermatol. Treat.*, (1), 91–93.

47. L. Edwards, N. Levine, K.A. Smiles (1990). The effect of topical interferon alpha 2b on actinic keratoses. *J. Dermatol. Surg. Oncol.*, **16**(5), 446–449.

48. M. Alirezai, P. Dupuy, P. Amblard, B. Kalis, P. Souteyrand, A. Frappaz et al. (1994). Clinical evaluation of topical isotretinoin in the treatment of actinic keratoses. *J. Am. Acad. Dermatol.*, **30**(3), 447–451.

49. W. Bollag, F. Ott (1970). Retinoic acid: topical treatment of senile or actinic keratoses and basal cell carcinomas. *Agents Actions*, **1**(4), 172–175.

50. J.K. Rivers, D.I. McLean (1997). An open study to assess the efficacy and safety of topical 3% diclofenac in a 2.5% hyaluronic acid gel for the treatment of actinic keratoses. *Arch. Dermatol.*, **133**(10), 1239–1242.

51. G.H. Burgess, A. Bloch, H. Stoll, H. Milgrom, F. Helm, E. Klein (1974). Effect of topical tubercidin on basal cell carcinomas and actinic keratoses. *Cancer*, **34**(2), 250–253.

52. B.E. Cham, B. Daunter, R.A. Evans (1991). Topical treatment of malignant and premalignant skin lesions by very low concentrations of a standard mixture (BEC) of solasodine glycosides. *Cancer Lett.*, **59**(3), 183–192.

53. T.D. Griffin, E.J. Van Scott (1991). Use of pyruvic acid in the treatment of actinic keratoses: a clinical and histopathologic study. *Cutis*, **47**(5), 325–329.

54. G.M. Marrero, B.E. Katz (1998). The new fluor-hydroxy pulse peel. A combination of 5-fluorouracil and glycolic acid. *Dermatol. Surg.*, **24**(9), 973–978.

55. L. Bercovitch (1987). Topical chemotherapy of actinic keratoses of the upper extremity with tretinoin and 5-fluorouracil: a double-blind controlled study. *Br. J. Dermatol.*, **116**(4), 549–552.

56. T.R. Humphreys, V. Werth, L. Dzubow, A. Kligman (1996). Treatment of photodamaged skin with trichloroacetic acid and topical tretinoin. *J. Am. Acad. Dermatol.*, **34**(4), 638–644.

57. S. Majewski, M. Skopinska, W. Bollag, S. Jablonska (1994). Combination of isotretinoin and calcitriol for precancerous and cancerous skin lesions [letter]. *Lancet*, **344**(8935), 1510–1511.

58. R.M. Szeimies, P. Calzavara-Pinton, S. Karrer, B. Ortel, M. Landthaler (1996). Topical photodynamic therapy in dermatology. *J. Photochem. Photobiol. B*, **36**(2), 213–219.

59. C. Fritsch, G. Goerz, T. Ruzicka (1998). Photodynamic therapy in dermatology. *Arch. Dermatol.*, **134**(2), 207–214.

60. Y. Harth, B. Hirshowitz, B. Kaplan (1998). Modified topical photodynamic therapy of superficial skin tumors, utilizing aminolevulinic acid, penetration enhancers, red light, and hyperthermia. *Dermatol. Surg.*, **24**(7), 723–726.

61. N. Rebeiz, C.C. Rebeiz, S. Arkins, K.W. Kelley, C.A. Rebeiz (1992). Photodestruction of tumor cells by induction of endogenous accumulation of protoporphyrin IX: enhancement by 1,10-phenanthroline. *Photochem. Photobiol.*, **55**(3), 431–435.

62. N. Rebeiz, S. Arkins, C.A. Rebeiz, J. Simon, J.F. Zachary, K.W. Kelley (1996). Induction of tumor necrosis by delta-aminolevulinic acid and 1,10- phenanthroline photodynamic therapy. *Cancer Res.*, **56**(2), 339–344.

63. B. Ortel, A. Tanew, H. Honigsmann (1993). Lethal photosensitization by endogenous porphyrins of PAM cells – modification by desferrioxamine. *J. Photochem. Photobiol. B*, **17**(3), 273–278.

64. S. Fijan, H. Honigsmann, B. Ortel (1995). Photodynamic therapy of epithelial skin tumours using delta- aminolaevulinic acid and desferrioxamine. *Br. J. Dermatol.*, **133**(2), 282–288.

65. C. Abels, R.M. Szeimies, P. Steinbach, C. Richert, A.E. Goetz (1997). Targeting of the tumor microcirculation by photodynamic therapy with a synthetic porphycene. *J. Photochem. Photobiol. B*, **40**(3), 305–312.

66. P.F. Aramendia, R.W. Redmond, S. Nonell, W. Schuster, S.E. Braslavsky, K. Schaffner et al. (1986). The photophysical properties of porphycenes: potential photodynamic therapy agents. *Photochem. Photobiol.*, **44**(5), 555–559.

67. M. Guardiano, R. Biolo, G. Jori, K. Schaffner (1989). Tetra-n-propylporphycene as a tumour localizer: pharmacokinetic and phototherapeutic studies in mice. *Cancer Lett.*, **44**(1), 1–6.

68. S. Karrer, C. Abels, R.M. Szeimies, W. Baumler, M. Dellian, U. Hohenleutner et al. (1997). Topical application of a first porphycene dye for photodynamic therapy – penetration studies in human perilesional skin and basal cell carcinoma. *Arch. Dermatol. Res.*, **289**(3), 132–137.

69. C. Richert, J.M. Wessels, M. Muller, M. Kisters, T. Benninghaus, A.E. Goetz (1994). Photodynamic antitumor agents: beta-methoxyethyl groups give access to functionalized porphycenes and enhance cellular uptake and activity. *J. Med. Chem.*, **37**(17), 2797–2807.

70. R.M. Szeimies, S. Karrer, C. Abels, P. Steinbach, S. Fickweiler, H. Messmann et al. (1996). 9-Acetoxy-2,7,12,17-tetrakis-(beta-methoxyethyl)-porphycene (ATMPn), a novel photo-sensitizer for photodynamic therapy: uptake kinetics and intracellular localization. *J. Photochem. Photobiol. B*, **34**(1), 67–72.

71. V. Sacchini, E. Melloni, R. Marchesini, T. Fabrizio, N. Cascinelli, O. Santoro et al. (1987). Topical administration of tetrasodium-meso-tetraphenylporphinesulfonate (TPPS) and red light irradiation for the treatment of superficial neoplastic lesions. *Tumori*, **73**(1), 19–23.

72. U. Hohenleutner, R.M. Szeimies, M. Landthaler (1993). Photodynamische Therapie zur Behandlung oberflächlicher Basaliome. In: O. Braun-Falco, G. Plewig, M. Meurer (Eds), *Fortschritte der praktischen Dermatologie und Venerologie* (pp. 472–474). Berlin, Heidelberg, New York: Springer.

73. J.W. Winkelman, G.H. Collins (1987). Neurotoxicity of tetraphenylporphinesulfonate TPPS4 and its relation to photodynamic therapy. *Photochem. Photobiol.*, **46**(5), 801–807.

74. I.J. Forbes, P.A. Cowled, A.S. Leong, A.D. Ward, R.B. Black, A.J. Blake et al. (1980). Phototherapy of human tumours using haematoporphyrin derivative. *Med. J. Aust.*, **2**(9), 489–493.

75. J.C. Kennedy, R.H. Pottier, D.C. Pross (1990). Photodynamic therapy with endogenous protoporphyrin IX: basic principles and present clinical experience. *J. Photochem. Photobiol. B*, **6**(1–2), 143–148.

76. R.M. Szeimies, R. Hein, W. Baumler, A. Heine, M. Landthaler (1994). A possible new incoherent lamp for photodynamic treatment of superficial skin lesions. *Acta Derm. Venereol.*, **74**(2), 117–119.

77. R.M. Szeimies, C. Abels, C. Fritsch, S. Karrer, P. Steinbach, W. Baumler et al. (1995). Wavelength dependency of photodynamic effects after sensitization with 5-aminolevulinic acid in vitro and in vivo. *J. Invest. Dermatol.*, **105**(5), 672–677.

78. F. Cairnduff, M.R. Stringer, E.J. Hudson, D.V. Ash, S.B. Brown (1994). Superficial photodynamic therapy with topical 5-aminolaevulinic acid for superficial primary and secondary skin cancer. *Br. J. Cancer*, **69**(3), 605–608.

79. R.S. Stern, N. Laird (1994). The carcinogenic risk of treatments for severe psoriasis. Photochemotherapy Follow-up Study. *Cancer*, **73**(11), 2759–2764.

80. R. Stern, S. Zierler, J.A. Parrish (1982). Psoriasis and the risk of cancer. *J. Invest. Dermatol.*, **78**(2), 147–149.

81. T.M. Runger (1995). [DNA damage to the skin caused by ultraviolet irradiation.

Development, significance, modification]. *Med. Klin.* (Suppl), **90**(1), 22–26.

82. P. Wolf, R. Fink-Puches, A. Reimann-Weber, H. Kerl (1997). Development of malignant melanoma after repeated topical photodynamic therapy with 5-aminolevulinic acid at the exposed site. *Dermatology*, **194**(1), 53–54.

83. J. Fuchs, S. Weber, R. Kaufmann (2000). Genotoxic potential of porphyrin type photosensitizers with particular emphasis on 5-aminolevulinic acid: implications for clinical photodynamic therapy. *Free Radic. Biol. Med.*, **28**(4), 537–548.

84. H. von Tappeiner, A. Jesionek (1903). Therapeutische Versuche mit fluorescierenden Stoffen. *Münch. Med. Wochenschr.*, **47**, 2042–2044.

85. P. Wolf, E. Rieger, H. Kerl (1993). Topical photodynamic therapy with endogenous porphyrins after application of 5-aminolevulinic acid. An alternative treatment modality for solar keratoses, superficial squamous cell carcinomas, and basal cell carcinomas? [published erratum appears in *J. Am. Acad. Dermatol.*, Jul. 29(1), 41]. *J. Am. Acad. Dermatol.*, **28**(1), 17–21.

86. A.F. Hurlimann, R.G. Panizzon, G. Burg (1994). Topical photodynamic treatment of tumors and dermatoses. *Dermatology*, (3), 327.

87. P.G. Calzavara-Pinton (1995). Repetitive photodynamic therapy with topical delta-aminolaevulinic acid as an appropriate approach to the routine treatment of superficial non-melanoma skin tumours. *J. Photochem. Photobiol. B*, **29**(1), 53–57.

88. R.M. Szeimies, S. Karrer, A. Sauerwald, M. Landthaler (1996). Photodynamic therapy with topical application of 5-aminolevulinic acid in the treatment of actinic keratoses: an initial clinical study. *Dermatology*, **192**(3), 246–251.

89. H.A. Kurwa, S.A. Yong-Gee, P.T. Seed, A.C. Markey, R.J. Barlow (1999). A randomized paired comparison of photodynamic therapy and topical 5-fluorouracil in the treatment of actinic keratoses. *J. Am. Acad. Dermatol.*, **41**(3 Pt 1), 414–418.

90. E.W. Jeffes, J.L. McCullough, G.D. Weinstein, P.E. Fergin, J.S. Nelson, T.F. Shull et al. (1997). Photodynamic therapy of actinic keratosis with topical 5-aminolevulinic acid. A pilot dose-ranging study. Arch. Dermatol., **133**(6), 727–732.

91. D. Ormrod, B. Jarvis (2000). Topical Aminolevulinic Acid HCL Photodynamic Therapy. *Am. J. Clin. Dermatol.*, **1**(2), 133–139.

92. S. Karrer, W. Baumler, C. Abels, U. Hohenleutner, M. Landthaler, R.M. Szeimies (1999). Long-pulse dye laser for photodynamic therapy: investigations in vitro and in vivo. *Lasers Surg. Med.*, **25**(1), 51–59.

93. R.M. Szeimies, S. Karrer, A. Heine, U. Hohenleutner, M. Landthaler (1995). Topical photodynamic therapy with 5-aminolevulinic acid in the treatment of arsenic-induced skin tumors. *Eur. J. Dermatol.*, (5), 208–211.

94. R. Fink-Puches, H.P. Soyer, A. Hofer, H. Kerl, P. Wolf (1998). Long-term follow-up and histological changes of superficial nonmelanoma skin cancers treated with topical delta-aminolevulinic acid photodynamic therapy. *Arch. Dermatol.*, **134**(7), 821–826.

Photodynamic Therapy and Fluorescence Diagnosis in Dermatology
P.-G. Calzavara-Pinton, R.-M. Szeimies and B. Ortel, editors.

Chapter 13

Topical sensitization – oncologic indications – Bowen's disease

Colin A. Morton

Table of contents

Abstract

Overview. Bowen's disease, squamous cell carcinoma in situ, is a relatively common pre-malignant lesion usually treated by surgery, curettage or cryotherapy, although no modality is ideal. Topical ALA-PDT has also been shown to be effective with clearance rates of 90–100%. This chapter assesses the evidence and compares ALA-PDT with conventional therapies.

Summary. Despite variation in the treatment parameters of reported studies, improved outcome is associated with 3–6 hour application times for 20% 5-ALA and the use of narrowband red light. Randomised comparison studies indicate ALA-PDT to be at least as effective as cryotherapy and topical 5-fluorouracil, but with fewer adverse reactions. In addition to achieving high clearance rates, tissue preservation, absence of ulceration and failure to heal, along with a uniformly high standard of cosmesis, support the use of ALA-PDT in clinical practice. Recurrence rates of 0–10% over 12–36 months compare favourably with historical data for conventional therapies.

Implications. Although all patches of Bowen's can respond, ALA-PDT is probably best suited to large and multiple lesions, and single patches in sites where standard therapies risk complications.

Future directions. Further studies are required to confirm the optimal treatment parameters, but ALA-PDT is now proposed as a good choice in therapy guidelines for Bowen's disease. The development of 5-ALA esters and refinement of light sources aim to improve the efficacy of single treatment ALA-PDT in Bowen's disease.

13.1 Introduction

Bowen's disease, intra-epidermal (in situ) squamous cell carcinoma of the skin, is usually persistent and progressive, although there has been the occasional report of spontaneous regression. The risk for invasive malignancy is around 3%, with up to one-third of these lesions demonstrating metastases. Bowen's disease typically presents as a slowly enlarging, well demarcated, erythematous plaque with an irregular border, and surface crust and scale. Lesions are usually solitary, but are multiple in 10–20% of patients. Around three-quarters of lesions develop on the lower leg, although any body site can be affected.

The peak age of presentation is the seventh decade, indicative of a relationship to chronic solar damage. Bowen's disease is also linked to previous arsenic exposure, immunosuppression and ionizing radiation. In comparison with the actinic keratoses which demonstrate focal histological atypia usually sparing adnexae, the epidermis in Bowen's disease is replaced by abnormal keratinocytes with loss of polarity and disordered maturation, with dysplastic cells extending down ducts up to 3 mm from the surface.

Bowen's disease can be treated with cryotherapy, curettage and electrodessication, topical 5-fluorouracil or surgical excision although no therapy is completely curative and each has potential complications. Radiotherapy has also been used, although full tumour doses are required. These therapies may be impractical for patients with numerous or large lesions, or lesions in anatomically difficult areas. As photodynamic therapy is tissue-sparing, it is potentially very useful in these circumstances.

13.2 The clinical efficacy of PDT in Bowen's disease

Initial studies of photodynamic therapy in Bowen's disease utilised a systemic photosensitiser. Warlow et al. [1], Jones et al. [2], and Petrelli et al. [3], each used Photofrin (dihematoporphyrin) to treat 3, 8 and 2 lesions respectively with complete clearance following a single treatment. Robinson et al. [4] also used Photofrin to treat two patients with unusually high numbers of lesions of Bowen's disease to clear 81/87 patches following a single treatment. The opportunity to avoid the generalized skin photosensitivity induced by systemic agents, combined with the usually low patient number of lesions, suggested that Kennedy's description in 1990 of topical 5-ALA-derived PDT could be particularly well-suited to the treatment of Bowen's disease.

Table 1 summarizes the treatment parameters and outcome of 11 studies reporting an efficacy of 50–100% for ALA-PDT in Bowen's disease. Kennedy et al. [5] used a slide projector (containing a 500W tungsten lamp) modified only by the inclusion of a long-wave pass colour filter to eliminate wavelengths less than 600 nm. He cleared the 3 patches of Bowen's disease in 1 visit although with only a 3 month follow-up reported. High irradiances between 150–300 mW/cm^2 suggest that heating of superficial tissue may have occurred with an unknown contribution to the PDT treatment. Hyperthermia has been reported as an effective monotherapy for Bowen's disease [6] and thus may have contributed to the therapeutic effect as 150 mW/cm^2 is considered to be the threshold intensity above which hyperthermic injury may occur [7]. Infra-red emissions from these projector sources, despite the incorporation of heat rejection filters, may also contribute to tissue heating independent of the irradiance intensity.

Cairnduff et al. [8]. and Svanberg et al. [10] cleared 35/36 and 9/10 lesions respectively following a single treatment using 20% 5-ALA applied for 3–6 hours and 630 nm laser light. Svanberg evaluated the pre-treatment surface (PpIX) fluorescence at 630 nm and demonstrated a demarcation ratio of 15:1 between tumour and normal skin, with almost undetectable values on completion of the laser treatment. Calzavara-Pinton [9] also used a 630 nm laser source and cleared all 6 lesions in 1–3 visits. He excised 3 of the treatment sites with histological examination showing no residual disease. Irradiance was less than 150 mW/cm^2 in each study and only 3 lesions recurred between these studies during review periods of 6–36 months. These 3 independent studies represent a persuasive argument for the high efficacy of ALA-PDT in Bowen's disease.

Fijan et al. [11] applied 5-ALA for 20 hours in combination with the iron chelator desferrioxamine, prior to illumination with an incandescent source (filtered output 540–720 nm). Histological confirmation of clearance was pursued for all lesions of Bowen's disease. A single treatment cleared only 5/10 sites of Bowen's disease with 2 'recurrences' after 1 and 4 months.

Like Kennedy [5], Lui et al. [12] used a tungsten slide projector source with light <570 nm filtered out. The 2/3 response was confirmed by histology, but with only 4/11 tumours (BCC and SCC) clearing in the same study, a low efficacy of ALA-PDT using the projector source was implied.

Table 1. Summary of treatment conditions and results of studies using 20% ALA for PDT in Bowen's disease

Reference	No. of lesions	Lesion size (mm)	Interval (hrs.)	Light source	Light dose (J/cm²)	Irradiance (mW/cm²)	No. of Visits	CCR% on 1 visit	CCR (%)	Recurrence – no. (%)	Follow-up (months)
Kennedy et al. [5]	6	NK	3–6	Tungsten	54–540	150–300	1	100	100	0	3
Cairnduff et al. [8]	36	5–75 (med. 20)	3–6	CuVDL	125–150	<150	1	97	97	3 (9%)	18
Calzavara-Pinton [9]	6	10–50 (med. 30)	6–8	ArPDL	60–80	100	1–3	NK	100	0	24–36
Svanberg et al. [10]	10	20–50 (med. 25)	4–6	Nd:YAG	60	<110	1	90	90	0	6–14
Fijan et al. [11]	10	10–45	20	Tungsten	300	50–300	1	50	50	2 (20%)	6
Lui et al. [12]	3	NK	3	Tungsten	100	19–44	1	67	67	0	3
Morton et al. [13]	20	3–20	4	Xenon	94–156	55–158	1–2	60	100	2 (10%)	12
Wennberg et al. [14]	18	7–15	3	Xenon	75–100	125–166	1	78	78	0	6
Morton et al. [15]	20	5–20	4	Xenon	125	70	1–2	75	100	0	12
Fritsch et al. [16]	8	18–158	6	Tungsten	180	150	1–3	50	75	0	12–24
Varma et al. [17]	50	NK	4	Tungsten	105	NK	1–2	NK	90	14 (32%)	12

NK = Not Known, med. = median

Fritsch et al. [16] and Varma et al. [17] utilised a halogen source (Waldmann 1200, 580–740 nm) to clinically clear 75% and 90% of lesions, but with 14 recurrences bringing the response rate in Varma's study down to 62% at 1 year.

The three studies that utilised a narrowband filtered xenon source (Table 1), 620–670 nm, achieved a single treatment response of 60–78% increasing to 100% in the two studies which permitted a second treatment [13,15].

We (Morton et al. [15]) have described the importance of lesion size in predicting response in Bowen's disease, with larger lesions likely to require repeat treatments. This has been observed not only in ALA-PDT, but also following cryotherapy. Fritsch et al. [16] treated lesions up to 16 cm in diameter, which may account for lower response rates in comparison with other studies. The successful treatment of 3 further large patches of Bowen's disease was reported by Stables et al. [18] with lesions 4×3, 7×6 and 8×6 cm clearing after only 2 treatments with ALA-PDT (20% 5-ALA for 4 hours, projector tungsten-halogen lamp 400–700 nm, 125 J/cm^2 at < 150 mW/cm^2).

The response of larger patches of Bowen's disease to ALA-PDT has recently been confirmed [19] in a study where 40 lesions, 2–5 cm diameter, were treated using a xenon source filtered to 630 ± 15 nm, 100 J/cm^2 at 18–55 mW/cm^2. Eighty-eight percent (35/40) lesions cleared after 1–3 treatments (Fig. 1) at 6 weekly intervals (17 on 1, 13 on 2, 5 on 3). Larger lesions were again statistically more likely to require more treatments to clear suggesting that a superior response may have been achieved if additional treatments had been permitted in the protocol.

5-ALA has been the predominant topical agent utilised in PDT for Bowen's disease. Although a topical formulation of meso-tetraphenylporphinesulphonate tetrasodium has been used with a 94% initial CCR in superficial BCC [20], only topical meso-tetra(hydroxyphenyl) chlorin (mTHPC) has been studied as an alternative agent in Bowen's disease [21]. A 64% clinical clearance rate for the topical 2% mTHPC preparation was reduced to a 32% pathological clearance rate following biopsy of lesions presumed clear. This was a dose-ranging pilot study suggesting further study is required to clarify the potential of this agent.

Four patients with Erythroplasia of Queyrat, histologically also intra-epithelial carcinoma in situ, have been treated with 1–2 treatments of ALA-PDT (125 J/cm^2, 630 nm copper vapour pumped dye laser, 20% 5-ALA for 3 hours, then local anaesthetic gel for 1 hour) [22]. Clearance was achieved in 2 (but recurrence at 18 months in 1) and significant improvement observed in the remaining 2 patients with more extensive disease, allowing easier subsequent laser vaporisation.

13.3 Treatment parameters for ALA-PDT in Bowen's disease

13.3.1 Surface preparation

Surface crust removal and gentle abrasion prior to 5-ALA application is described in 5/11 studies [8–10,13,15] in Table 1. The edge of a scalpel blade or forceps may be used, but debulking of lesions in Bowen's disease only extends to removing a thick piece of crust.

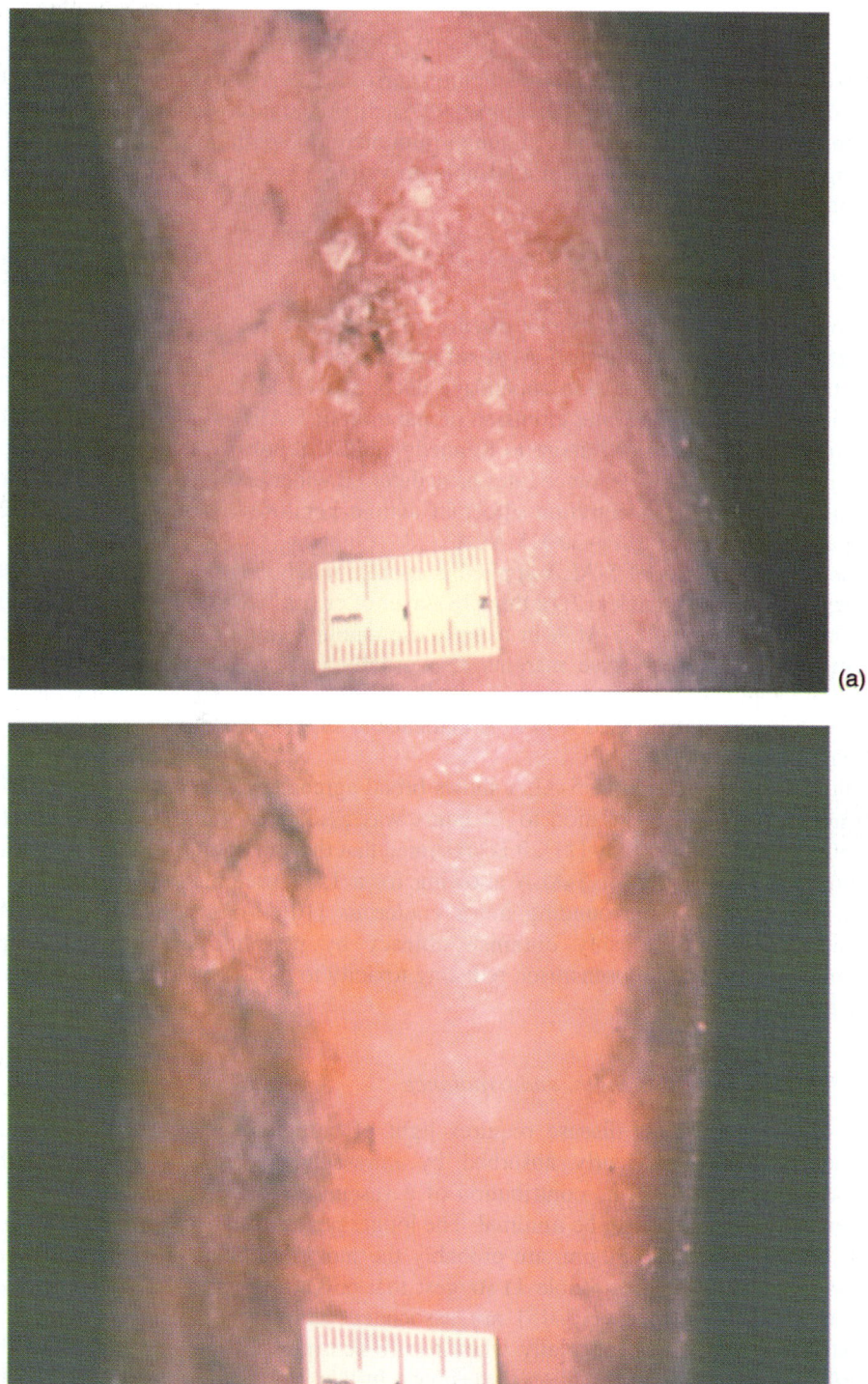

Figure 1. Large patch of Bowen's disease; 40×40mm, (a) before and (b) the same site 3 months following a single treatment with ALA-PDT.

No direct comparison of lesion preparation versus no de-crusting/abrasion exists. Due to additional study variables the impact of such preparation on efficacy cannot be determined from current studies, although minimizing potential obstruction of 5-ALA and light penetration would seem advisable providing a simple technique that requires no anaesthesia is pursued.

13.3.2 Photosensitiser

The optimal response of Bowen's disease to ALA-PDT requires ensuring adequate delivery of the topical pro-drug 5-ALA sufficient to be absorbed by the dysplastic cells, with the need to penetrate skin appendage ducts as well as the entire epidermis. The adoption of a 20% oil in water emulsion preparation of 5-ALA as the standard by most investigators appears to owe more to Kennedy's initial report [5] than to detailed dose ranging study. The high clearance rates have deterred wider exploration, although, there was an absence of concentration effect in actinic keratoses when Jeffes et al. [23] compared outcome using 10–30% 5-ALA.

The typical quantity of 5-ALA applied is approximately 50 mg/cm^2 although certain studies refer only to a thick [12] or thin [18] layer of cream. We presume that the tumour is saturated by available 5-ALA in using these amounts, although formal evidence is lacking. Clinical disease free margins also usually receive topical 5-ALA, with the deliberate application of cream described to a rim of 5–20 mm. Oil in water emulsions of 5-ALA are uniformly used with little evidence to suggest one preparation is superior to another.

Duration of application of 5-ALA shows broad agreement with a typical interval of 3–6 hours. Application duration to 20 hours was associated with a poor 50% initial response rate in Fijan's study [11]. It is presumed that the abnormal dysplastic epidermis facilitates the absorption of 5-ALA allowing for a 15:1 preferential accumulation of PpIX compared with surrounding intact epidermis [10]. Long application intervals beyond 3–6 hours probably diminish efficacy by permitting wider and deeper dissemination of the photosensitiser, reducing toxicity to the target epidermis as well as reducing lesion specificity.

13.3.3 Light dose, wavelength, and irradiance

ALA-PDT for Bowen's disease requires light of sufficient energy and appropriate wavelength to be efficiently absorbed by intra-cellular PpIX to effect a lethal photodynamic reaction. The contribution of a vascular effect of topical ALA-PDT in Bowen's disease is likely to be minimal. The longer peaks of the absorption spectrum of PpIX, at 540, 580 and 630 nm, are probably the most important in treating Bowen's Disease. Published studies (Table 1) suggest that 630 nm laser [8–10] or narrowband (620–670) xenon sources [13–15] are associated with higher response rates than tungsten-halogen lamps, especially when recurrence rates are included. The possible contribution of infrared irradiation to response further clouds the interpretation of the efficiency of tungsten sources for promoting photodynamic therapy.

Total light dosage can only meaningfully be compared for similar light sources with limited data on comparison of dose with response for individual lamps. A dose as low as 60 J/cm^2 for laser appears as effective as 125–150 J/cm^2 (Table 1), with 100 J/cm^2 as effective as higher doses in the quoted studies using a xenon source. A dose response comparison we performed using the xenon source [13,15] (unpublished data) compared the response of similar small (up to 3 cm diameter) patches of Bowen's disease within a narrow irradiance range of 55–86 mW/cm^2 and otherwise identical parameters. Light doses of 50,75,100 or 125 J/cm^2 were randomly allocated with a total of 116 lesions subsequently available for assessment. Whilst 125 and 100 J/cm^2 achieved initial complete response rates of 100 and 93% respectively, during the 12 months of follow-up, three recurrences in the 125 J/cm^2 group and one recurrence in the 100 J/cm^2 group reduced the clearance rates to 90 and 89% respectively. Seven recurrences in the 75 J/cm^2 group and five recurrences in the 50 J/cm^2 group reduced overall clearance rates in these groups from 90% in both groups to 66 and 70% respectively. Statistical analysis of these results confirmed a significant reduction in the probability of successful clearance at 1 year for lesions receiving less than 10 J/cm^2 ($p < 0.001$). In view of the positive response of the majority of lesions to 50 J/cm^{2}m a further six lesions received 25 J/cm^2 but no complete response was observed although a partial response was noted in three of the lesions. This study, whilst requiring confirmation by others, underlines the importance of follow-up as well as correction for lesion size, in deriving even a dose-response profile. The high recurrence rate at the lower doses is presumed to suggest insufficient light dose in the deepest parts of the lesions.

In a randomised comparison study of red with green light [24] in Bowen's disease, we compared 630 ± 15 nm with 540 ± 15 nm. At a fluence rate of 86 mW/cm^2, lesions received 125 J/cm^2 of red light or 62.5 J/cm^2 of green light. This dose of green light was chosen to normalize the quantity of protoporphyrin IX produced as quantum yield at 540 nm is approximately twice that at 630 nm [25]. The initial clearance rate for lesions treated by red light was 94% (30/32) in comparison with 72% (21/29) for those lesions receiving green light ($p = 0.002$). Over the following 12 months, there were 2 recurrences in the red light group and 7 in the green light group reducing the clearance rates to 88% and 48% respectively. The frequency and severity of pain experienced were similar between the two treatment groups. Green light is thus less effective than red light, at a theoretically equivalent dose, in the treatment of Bowen's disease by topical ALA-PDT. A model of optical distribution in Caucasian skin indicates that 635 nm light peaks in the upper dermis (due to the addition of scattered light to the incident fluence) in comparison with 514 nm light, which peaks in the stratum corneum [26]. Despite the superficial nature of Bowen's disease, this difference in penetration appears to be of therapeutic importance when red and green light is compared. The high recurrence rate with green light suggests that the deepest dysplastic cells may have been inadequately treated by green light.

Fritsch et al. [27] reported a half-side red/green comparison study of ALA-PDT for 6 patients with extensive facial actinic keratoses, with green light as effective as red. The difference between the studies is likely to reflect the pathological differences between actinic keratoses, characteristically a focal disease sparing adnexal structures, and Bowen's disease, which commonly involves the appendage ducts.

The influence of irradiance on the response of Bowen's disease is also difficult to assess from published studies in view of the small number of lesions treated and additional variables. Lui et al. [12] used a lower irradiance (19–44 mW/cm^2) than other groups with clearance of 2/3 patches with a tungsten source on 1 treatment. We have compared the effect of fluence on outcome when using the xenon source in Bowen's disease in 90 lesions (unpublished data). Irradiance varied between 18–158 mW/cm^2, with lesions ranging from 5–50 mm in diameter having received 125 J/cm^2 of 630 ± 15 nm light. Lesion size and fluence rate are linked variables when using this light source as irradiance reduces as the fibre bundle is pulled away from the lesion to accommodate a larger treatment field. After correcting for the influence of size on clearance, a significant improvement in response was apparent at fluence rates at or below 48 mW/cm^2 (p = 0.0097, by multivariate analysis). 35/90 lesions received this irradiance with a clearance at 12 months of 71% despite this group containing the largest lesions, compared with 85% of the smaller lesions clearing having received higher iradiances. Lower irradiance values may promote a more effective photodynamic process with oxygen depletion observed to affect outcome at levels above 50 mW/cm^2 in a study assessing the effects of systemically administered 5-ALA [28].

We have also demonstrated (unpublished data), using an infrared probe, that for irradiances of 18–158 mW/cm^2 no hyperthermia could be demonstrated, with red light inducing a peak increase of 3.1°C (1.5–5.6°C) with a maximum value of 34.3°C (29.7–37.1°C).

There is no published data on the use of fractionated light in Bowen's disease although Calzavara-Pinton et al. [9] repeated treatments after 48 hours until treatment sites appeared uniformly eroded in order to maximise effect. Other investigators wait 4 [16] or 8 weeks [13,15] to assess clearance before repeating treatment.

13.3.4 Adjuvant therapeutic strategies

Only Fijan et al. [11] reports using the iron chelator desferrioxamine in Bowen's disease to promote greater accumulation of PpIX. The poor efficacy of PDT, clearing only 50% of lesions after a 20 hour 5-ALA application prevents specific assessment of the influence of this chelator.

Harth et al. [29] used the penetration enhancers ethyleneamine tetra acetic acid (EDTA) and dimethylsulphoxide (DMSO) to enhance ALA-PDT in clearing 4/5 very superficial squamous cell carcinomas and a solitary patch of Bowen's disease on the penis. The light they used has 2 bandwidths at 585–720 nm and 1.25–1.6 mm, producing 25% of its output as infrared to deliberately promote hyperthermia. No comparison data is identified to show difference with and without these potential adjuvant strategies.

13.4 Comparison of PDT with conventional therapy for Bowen's disease

13.4.1 ALA-PDT vs. cryotherapy

Cryotherapy is suggested to be the optimal therapy for small, single lesions in good healing sites and a good or fair choice even for multiple, facial or penile lesions and those at sites of poor healing in recent guidelines on the management of Bowen's disease [30]. Although Holt [31] describes clearing all 128 lesions of Bowen's disease following a single 30s freeze-thaw cycle with only one recurrence, local anaesthesia was used and lesions on the calf and shin were associated with slow healing times of up to 6 months. Whilst effective, with recurrence rates < 10% at 12 months, cryotherapy is associated with many potential complications including delayed healing, scar formation and even nerve damage [32].

In a randomised comparison study, 40 histologically-proven lesions (≤ 2 cm in diameter) of Bowen's disease received either PDT (20% 5-ALA for 4 hours, xenon lamp filtered to 630 ± 15 nm, 125 J/cm^2 at 70 mW/cm^2) or cryotherapy (one freeze-thaw cycle of 20 seconds) [15]. Cryotherapy produced clearance in 10/20 lesions after one treatment, the remaining 10 lesions requiring two or three treatment applications. PDT resulted in clearance of 15/20 lesions after one and of the remaining 5 lesions after a second treatment. The probability that a lesion cleared after one treatment was greater with PDT than cryotherapy (p < 0.01). Cryotherapy was associated with ulceration (n = 5), infection (n = 2) and recurrent disease (n = 2); no such complications occurred following PDT.

Pain during cryotherapy was present in 19 lesions and described as mild in 12 and moderate in 7 lesions. Pain during PDT was present in only 11 lesions and was mild in six and moderate in the five lesions. This difference was statistically significant (p = 0.01). 12 months following clearance, a visible scar in the treatment field, outwith diagnostic biopsy sites, was observed in 4 lesions treated by cryotherapy, whilst absent in all lesions treated by PDT.

ALA-PDT was thus at least as effective as cryotherapy for Bowen's disease and associated in this series with fewer adverse effects and a lower recurrence rate. A longer 30s duration of cryotherapy may have improved response, however as approximately 50% of lesions were situated on the legs in each group, this probably would have resulted in more adverse effects. The low incidence of ulceration and absence of infection following PDT is of particular importance in the elderly population developing Bowen's disease. The absence of clinically obvious scar formation is consistent with the good cosmetic results reported following ALA-PDT for Bowen's disease [5,8–10,13,14].

13.4.2 ALA-PDT vs. topical 5-fluorouracil

Topical 5-fluorouracil has the attraction of a cream formulation that patients can apply at home. Several regimes are described with application commonly of 5% cream usually once or twice daily for 3–4 weeks. Although irritation is a recognised adverse effect, it is thought important for efficacy [33].

Kurwa et al. [34] reported that a single treatment of ALA-PDT was as effective as 3 weeks of twice daily topical 5% 5-fluorouracil in a left/right comparison study of patients with extensive actinic keratoses on the backs of their hands. A single treatment probably did not optimise efficacy with hyperkeratotic actinic keratoses probably less responsive to ALA-PDT than Bowen's disease.

Salim [35] has reported the interim results on a randomised comparison study of PDT with topical 5-fluorouracil, in 40 patients with biopsy-proven Bowen's disease. Patients received either 5% 5-fluorouracil once daily for 7 days, then twice daily for 21 days, or topical ALA-PDT (xenon lamp, filtered to 630 ± 15 nm, 100 J/cm^2 at 48mW/cm^2, 20% 5-ALA applied for 4 hours). Thirty-three lesions received PDT with clearance of 88% (22 on 1st and 7 on 2nd treatment), and 33 lesions received 5-FU, although 3 patients (with 5 lesions) withdrew due to severe eczematous reactions during treatment. 5-fluorouracil initially cleared 67% of lesions (12 on 1st and 10 on 2nd treatment cycle). After adjustment for the influence of lesion size on response, this difference in initial clearance rates was not significant. In the 5-FU group, 1 further patient developed widespread eczema but completed therapy, whilst 3 lesions ulcerated following treatment. No such reactions occurred following PDT. Two lesions recurred in each group after a mean follow-up of 8 months. PDT is at least as effective as topical 5-FU in the treatment of Bowen's disease, but with fewer adverse reactions observed.

13.4.3 ALA-PDT vs. surgery

No comparison of these modalities exists and the literature is limited on the efficacy of surgery despite its high frequency of use in practice. Thestrup-Pederson et al. [36] reviewed their management of 617 patients with proven Bowen's disease in Denmark with 35% followed up in excess of 5 years. They report recurrence rates for surgery in this non-randomised study, with physicians choice of modality, as the lowest at 4.6% (3/65) in comparison with radiotherapy (6%, 6/97), topical 5-fluorouracil (14%, 3/21), curettage followed by curettage (19%, 65/345) and cryotherapy (34%, 19/56). No comments on adverse treatment reactions nor cosmetic outcome were made in this study.

Graham and Helwig [37] observed a 19% recurrence rate following simple excision during 5 years of review of 96 patients. Moritz and Lynch [38] proposed Mohs micrographic surgery to minimise tissue loss in penile shaft lesions that were virtually circumferential. Topical ALA-PDT would appear now a good and doubtless much less expensive option in such a situation.

PDT offers the advantages of being non-invasive, well tolerated in slow healing sites, and tissue sparing, leaving the skin surrounding the tumour intact and functional. Surgery has the advantage of histological confirmation of clearance, but at the risk of slow healing, grafting in large lesions, and a permanent usually visible scar, frequently on exposed body sites.

13.4.4 ALA-PDT vs. curettage and cautery

Although no direct comparison of ALA-PDT with curettage exists, Ahmed et al. [39] recently compared cryotherapy (2 freeze-thaw cycles, 5–10 sec beyond iceball formation) with curettage and electrocautery. Seventy-four percent of lesions were situated on the lower leg and significantly longer healing times were observed for lesions treated by cryotherapy compared with curettage (means: 90 vs. 39 days). Following initial clinical clearance of all lesions (after 1 treatment in all cases except 2 of the cryotherapy group) there was infection requiring antibiotics in 6 patients, and recurrence rates after 24 months of 36% following cryotherapy and 9% after curettage.

Curettage, but not cryotherapy, was performed under local anaesthetic. Whilst not severe, pain was 10 times more likely to be reported during the 24 hours following cryotherapy rather than after curettage. Half of all patients reported pain in the week following each therapy and 12% still described pain after 6 weeks. Although pain during and following ALA-PDT is an important adverse reaction to consider, this study reminds us of the considerable morbidity due to pain following conventional therapies.

Recurrence rates following curettage vary from 0% at 2 years (n = 33) [40], 20% over 5 years (65/345), [36]to 73% [37] with differences in treatment regimes as well as follow-up duration likely to be relevant.

13.4.5 ALA-PDT vs. radiotherapy

With no comparison data, a review of the success of radiotherapy in Bowen's disease reveals widely divergent results, from as low as 12% in a small study of 12 patients [37], to a 100% initial response in 77 patients [41]. Cox and Dyson [42] treated only lower leg patches of Bowen's by either cryotherapy (n = 82) or external beam radiotherapy (n = 59). All lesions cleared, although in contrast to a single radiotherapy visit, up to 3 staged cryotherapy treatments (two 20 sec freeze-thaw cycles each) were used so no lesion had > 2cm diameter field at a single visit. 20% of radiotherapy wounds failed to heal after a minimum of 3 months in comparison with 2% after cryotherapy, although the only recurrences were in the latter group (6%). A residual ulcer or salvage surgery was performed on the non-healing wounds suggesting that radiotherapy, whilst effective, has significant morbidity that is not observed following ALA-PDT. The cosmetic outcome is not commented upon, but radiotherapy commonly leaves obvious skin change and concerns remain over the mutagenicity of this therapy. ALA-PDT, on the basis of current knowledge, offers several advantages, especially for larger lesions where the dermatologist is most likely to consider radiotherapy – high clearance rates in addition to good healing and cosmesis [19]. If PDT fails, all alternative therapies remain, in comparison with radiotherapy, where surgical excision of non-healing wounds larger than the original lesion may be required [42].

13.5 Indications for the use of ALA-PDT in Bowen's disease

A recent review of the management of all 68 patients with Bowen's disease who presented in one year to a UK dermatology department reported that up to 8 clinic visits (median 4) were required [43]. The median age of patients was 71 years, 32% had multiple lesions and in 73%, the Bowen's disease was located on the lower leg. This suggests that Bowen's disease represents a significant clinical management burden.

Recently prepared guidelines on the management of Bowen's disease [30] suggests PDT as a good or fair choice for most presentations of Bowen's disease. No treatment is currently recommended as first choice for large lesions of Bowen's disease, although other therapies are considered to be inferior to PDT in this indication. Cryotherapy is suggested as a preferable therapy to PDT for multiple lesions in good healing sites although, in practice, such multiple lesions usually occur on poor healing sites.

The optimal therapy in Bowen's disease should combine high clearance rates with low morbidity, acknowledging the poor healing of the lower leg, and a good cosmetic outcome. ALA-PDT satisfies these criteria, although it is more time-consuming and more likely to require repeat visits, especially for larger lesions >2 cm in diameter. Current evidence suggests that 20% 5-ALA in an oil in water emulsion applied approximately 4 hours pre-illumination with a narrowband red light source, delivering 60–125 J/cm^2 at <150mW/cm^2 can clear 90–100% of patches of Bowen's disease.

ALA-PDT is an effective, well tolerated therapy at sites of poor healing, that compares favourably to existing modalities. Pain during ALA-PDT may necessitate local anaesthesia in the minority of larger lesions, but this is no different to excision surgery and curettage. For patches <2 cm in diameter, ALA-PDT is significantly less painful than cryotherapy, although the latter modality remains a quick and effective treatment for small lesions in good healing sites. Topical ALA-PDT is therefore probably as effective as current routine therapies in small, single patches of Bowen's and provides an increased choice to the dermatologist. The treatment of digital, genital and perianal lesions are all feasible by ALA-PDT with the advantages of tissue preservation and avoidance of loss of function of adjacent tissue, e.g. no urethral meatal obstruction.

The cosmetic outcome is widely reported as generally good, although temporary pigmentary change with a residual faint erythematous hue and localized hair loss is occasionally noted. Visible scar formation has not been seen following PDT to Bowen's disease, although histological evidence of mild scarring is occasionally observed following PDT for basal cell carcinoma and superficial invasive squamous cell carcinoma, probably reflecting their dermal involvement. ALA-PDT would thus seem indicated for lesions in sites of particular cosmetic importance such as the face (Fig. 2).

ALA-PDT has certain advantages over existing therapies for large (>2 cm) and multiple lesions, where an effective, tissue sparing therapy that achieves good healing and cosmesis, makes PDT a first line therapy.

Long term recurrence rates are awaited, but the high rates occasionally observed following ALA-PDT in basal cell carcinoma would seem unlikely for the intra-epithelial disease of Bowen's. Current evidence suggests rates of 0–10% after follow-up periods of 12–36 months, with late results likely to reflect new disease.

13.6 Future directions

Bowen's disease is already a good indication for ALA-PDT, but multi-centre studies to confirm the optimal parameters are required. The few studies that report a poor response of Bowen's disease indicate that careful dosimetry is essential even for such an intra-epidermal lesion. Ensuring adequate penetration of photosensitiser and light appears important for optimal response. The development of ALA methylester and the refinement of non-laser light sources to maximise the photodynamic reaction throughout

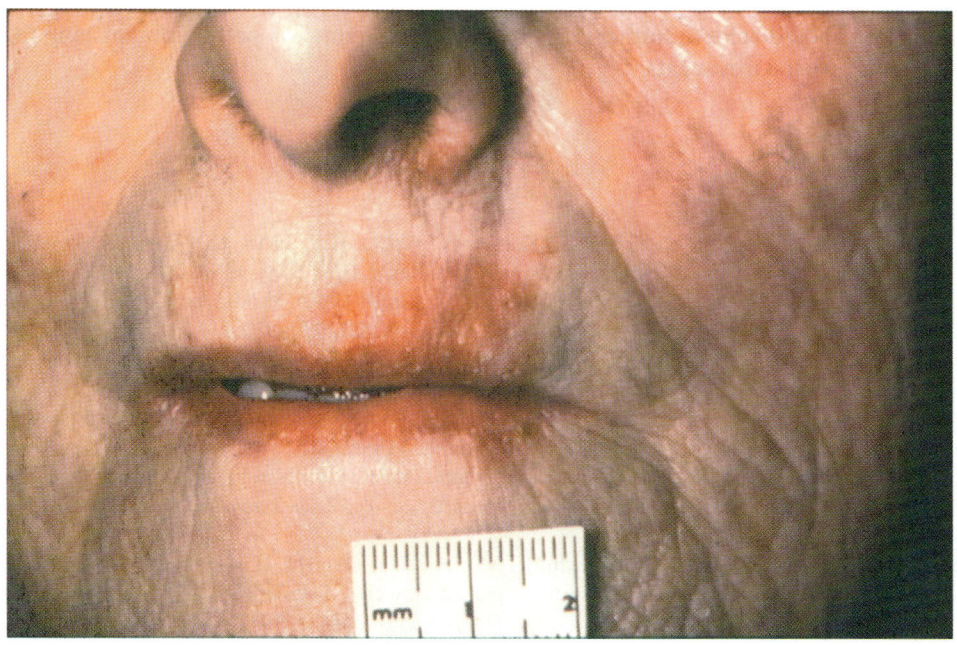

(a)

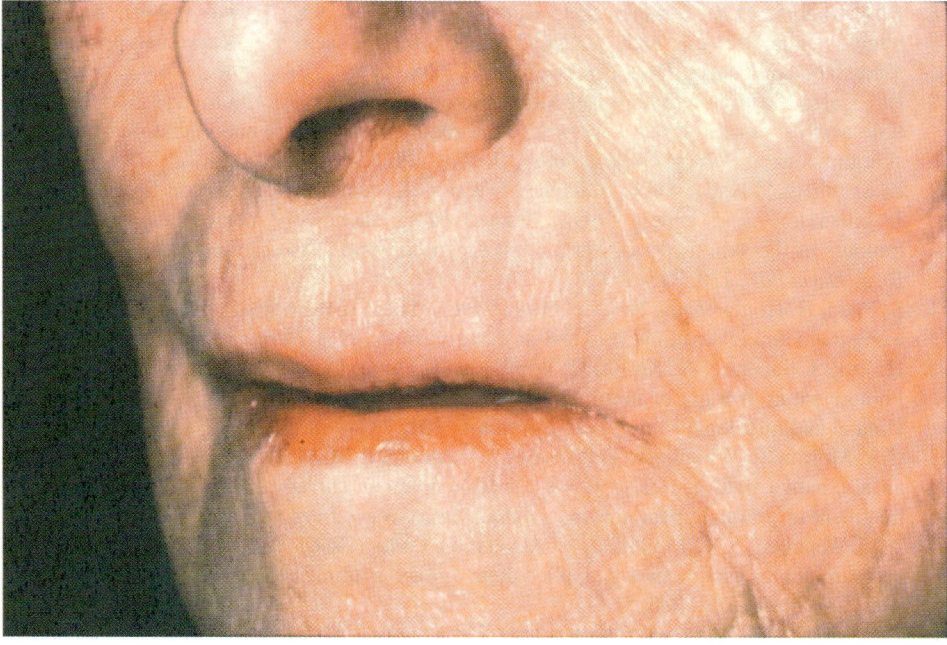

(b)

Figure 2. Bowen's disease on the upper and lower lip (a) before and (b) the same site 12 months following a single treatment with ALA-PDT.

the lesion should achieve further improvements in response, hopefully diminishing the need for repeated treatments.

References

1. S.M. Waldow, R.V. Lobraico, I.K. Kohler, S. Wallk, H.T. Fritts (1987). Photodynamic therapy for treatment of malignant cutaneous lesions. *Lasers Surg. Med.*, **7**, 451–456.
2. C.M. Jones, T. Mang, M. Cooper, D. Wilson, H.L. Stoll (1992). Photodynamic therapy in the treatment of Bowen's disease. *J. Am. Acad. Dermatol.*, **27**, 979–982.
3. N.J. Petrelli, J.A. Cebollero, M. Rodriguez-Bigas, T. Mang (1992). Photodynamic therapy in the management of neoplasms of the perianal skin. *Arch. Surg.*, **127**, 1436–1438.
4. P.J. Robinson, J.A.S. Carruth, G.M. Fairris (1988). Photodynamic therapy: a better treatment for widespread Bowen's disease. *Br. J. Dermatol.*, **119**, 59–61.
5. J.C. Kennedy, R.H. Pottier (1990). Photodynamic therapy with endogenous protoporphyrin IX: basic principles and present clinical experience. *J. Photochem. Photobiol. B: Biol.*, **6**, 143–148.
6. M. Hiruma, A. Kawada, H. Noguchi, K. Morimoto, Y. Ohnishi, H. Takahashi, A. Ishibashi, M. Yoshida (1994). Hyperthermic treatment of Bowen's disease with disposable chemical pocket warmers: report of three cases. *J. Dermatol. Treatment*, **5**, 37–41.
7. L.O. Svaasand (1984). Thermal and optical dosimetry for photoradiation therapy of malignant tumours. In: A. Andreoni & R. Cubeddu (Eds), *Porphyrins in Tumour Phototherapy* (pp. 261–279). Plenum, New York.
8. F. Cairnduff, M.R. Stringer, E.J. Hudson, D.V. Ash, S.B. Brown (1994). Superficial photodynamic therapy with topical 5-aminolaevulinic acid for superficial primary and secondary skin cancer. *Br. J. Cancer*, **69**, 605–608.
9. P.G. Calzavara-Pinton (1995). Repetitive photodynamic therapy with topical 5-Aminolaevulinic acid as an appropriate approach to the routine treatment of superficial non-melanoma skin tumours. *J. Photochem. Photobiol. B: Biol.*, **29**, 53–57.
10. K. Svanberg, T. Anderson, D. Killander, I. Wang, U. Stenram, S. Andersson-Engels, R. Berg, J. Johansson, S. Svanberg (1994). Photodynamic therapy of non-melanoma malignant tumours of the skin using topical 5-aminolaevulinic acid sensitisation and laser irradiation. *Br. J. Dermatol.*, **130**, 743–751.
11. S. Fijan, H. Honigsmann, B. Ortel (1995). Photodynamic therapy of epithelial skin tumours using delta aminolaevulinic acid and desferrioxamine. *Br. J. Dermatol.*, **133**, 282–288.
12. H. Lui, S. Salasche, N. Kollias, J. Wimberly, T. Flotte, D. Mclean, R.R. Anderson (1995). Photodynamic therapy of non-melanoma skin cancer with topical aminolaevulinic acid: A clinical and histologic study. *Arch. Dermatol.*, **131**, 737–738.
13. C.A. Morton, C. Whitehurst, H. Moseley, J.V. Moore, R.M. MacKie (1995). Development of an alternative light source to lasers for photodynamic therapy: Clinical evaluation in the treatment of pre-malignant non-melanoma skin cancer. *Lasers Med. Sci.*, **10**, 165–171.
14. A.M. Wennberg, L-E. Lindholm, M. Alpsten, O. Larko (1996). Treatment of superficial basal cell carcinomas using topically applied delta-aminolaevulinic acid and a filtered xenon lamp. *Arch. Dermatol. Res.*, **288**, 561–564.
15. C.A. Morton, C. Whitehurst, H. Moseley, J.V. Moore, R.M. MacKie (1996). Comparison of photodynamic therapy with cryotherapy in the treatment of Bowen's disease. *Br. J. Dermatol.*, **135**, 766–771.
16. C. Fritsch, G. Goerz, T. Ruzicka (1998). Photodynamic therapy in Dermatology. *Arch. Dermatol.*, **134**, 207–214.

17. S. Varma, H. Wilson, H.A. Kurwa, C. Charman, B. Gambles, A. Anstey (1999). One year relapse rates for Bowen's disease, basal cell carcinomas and solar keratoses treated by photodynamic therapy: analysis of 189 lesions. *Br. J. Dermatol.*, **141**(suppl. 55), 114 (abstract).

18. G.I. Stables, M.R. Stringer, D.J. Robnson, D.V. Ash (1997). Large patches of Bowen's disease treated by topical aminolaevulinic acid Photodynamic therapy. *Br. J. Dermatol.*, **136**, 957–960.

19. C.A. Morton, C. Whitehurst, H. Moseley, J.V. Moore, R.M. MacKie (2001). Photodynamic therapy for large or multiple plaques of Bowen's disease and basal cell carcinoma. *Arch. Dermatol.*, **137**, 319–324.

20. O. Santoro, G. Bandieramonte, E. Melloni, R. Marchesini, F. Zunino, P. Lepera, G. De Palo (1990). Photodynamic therapy by topical meso-Tetraphenylporphinesulfonate tetrasodium salt administration in superficial basal cell carcinomas. *Cancer Research*, **50**, 4501–4503.

21. G. Gupta, C.A. Morton, C. Whitehurst, J.V. Moore, R.M. MacKie (1999). Photodynamic therapy with meso-tetra(hydroxyphenyl) chlorin in the tpical treatment of Bowen's disease and basal cell carcinoma. *Br. J. Dermatol.*, **141**, 385–386.

22. G.I. Stables, M.R. Stringer, D.J. Robinson, D.V. Ash (1999). Erythroplasia of Queyrat treated by topical aminolaevulinic acid photodynamic therapy. *Br. J. Dermatol.*, **140**, 514–517.

23. E.W. Jeffes, J.L. McCullough, G.D. Weinstein, P.E. Fergin, J.S. Nelson, T.F. Shull, K.R. Simpson, L.M. Bukaty, W.L. Hoffman, N.L. Fong (1997). Photodynamic therapy of actinic keratoses with topical 5-aminolaevulinic acid. *Arch. Dermatol.*, **133**, 727–732.

24. C.A. Morton, C. Whitehurst, J.V. Moore, R.M. MacKie (2000). Comparison of red and green light in the treatment of Bowen's disease by photodynamic therapy. *Br. J. Dermatol.*, **143**, 767–772.

25. K. Konig, S. Auchter (1991). Testing der photodynamischen wirksamkeit von farbstoffen. *Biomed. Technik.*, **36**, 201–205.

26. L.O. Svaasand, B.J. Tromberg, P. Wyss, M.-T. Wyss-Desserich, Y. Tadir, M.W. Berns (1996). Light and drug administration with topically applied photosensitizers. *Lasers Med. Sci.*, **11**, 261–265.

27. C. Fritsch, H. Stege, G. Saalmann, G. Goerz, T. Ruzicka, J. Krutmann (1997). Green light is effective and less painful than red light in photodynamic therapy of facial solar keratoses. *Photoderm. Photoimmunol. Photomed.*, **13**, 181–185.

28. Q. Peng, T. Warloe, K. Berg, J. Moan, M. Kongshaug, K.E. Giercksky, J.M. Nesland (1997). 5-ALA based photodynamic therapy. *Cancer*, **79**(12), 2282–2308.

29. Y. Harth, B. Hirshowitz, B. Kaplan (1998). Modified topical photodynamic therapy of superficial skin tumours, utilising aminolaevulinic acid, penetration enhancers, red light, and hyperthermia. *Dermatol. Surg.*, **24**(7), 723–726.

30. N.H. Cox, D.J. Eady, C.A. Morton (1999). Guidelines for the management of Bowen's Disease. *Br. J. Dermatol.*, **141**, 633–641.

31. P.J.A. Holt (1988). Cryotherapy for skin cancer: results over a 5-year period using liquid nitrogen spray cryosurgery. *Br. J. Dermatol.*, **119**, 231–240.

32. R.F. Elton (1983). Complications of cutaneous cryosurgery. *J. Am. Acad. Dermatol.*, **8**, 513–519.

33. H.M. Sturm (1979). Bowen's disease and 5-fluorouracil. *J. Am. Acad. Dermatol.*, **1**, 513–522.

34. H.A. Kurwa, S.A. Yong-Gee, P.T. Seed, A.C. Markey, R.J. Barlow (1999). A randomised paired comparison of photodynamic therapy with topical 5-fluorouracil in the treatment of actinic keratoses. *J. Am. Acad. Dermatol.*, **41**, 414–418.

35. A. Salim, C.A. Morton (2000). Comparison of photodynamic therapy with topical 5-fluorouracil in Bowen's disease. *Br. J. Dermatol.*, **143**(suppl 57), 114.

36. K. Thestrup-Pedersen, L. Ravnborg, F. Reymann (1998). Morbus Bowen. *Acta Derm. Venereol. (Stockh.).*, **68**, 236–239.

37. J.H. Graham, E.B. Helwig (1961). Bowen's disease and its relationship to systemic cancer. *Arch. Dermatol.*, **83**, 76–96.

38. D.L. Moritz, W.S. Lynch (1991). Extensive Bowen's disease of the penile shaft treated with fresh tissue Mohs micrographic surgery in two separate operations. *J. Dermatol. Surg. Oncol.*, **17**, 374–378.

39. I. Ahmed, J. Berth-Jones, S. Charles-Holmes, C.J.O. Callaghan, A. Ilchyshyn (2000). Comparison of cryotherapy with curettage in the treatment of Bowen's disease: a prospective study. *Br. J. Dermatol.*, **143**, 759–766.

40. S.K. Veien, N.K. Veien, T. Hattel, G. Laurberg (1996). Results of treatment of non-melanoma skin cancer in a dermatologic practice. A prospective study. *Ugeskrirt Laeger*, **158**, 7213–7215.

41. A.A. Blank, U.W. Schnyder (1985). Soft X-ray therapy in Bowen's disease and erthroplasia of Queyrat. *Dermatologica*, **171**, 89–94.

42. N.H. Cox, P. Dyson (1995). Wound healing on the lower leg after radiotherapy or cryotherapy of Bowen's disease and other malignant skin lesions. *Br. J. Dermatol.*, **133**, 60–65.

43. H.K. Bell, L.E. Rhodes (1999). Bowen's Disease – a retrospective review of clinical management. *Clin. Exp. Dermatol.*, **24**, 336–337.

Chapter 14

Photodynamic therapy of basal cell carcinoma

Ann-Marie Wennberg and Olle Larkö

Table of contents

14.1 Introduction

Basal cell carcinoma (BCC) is a malignant tumour of the skin that arises from basal cells in the epidermis. It requires a stroma for proper growth. It rarely metastasises.

14.2 Clinical manifestations

BCCs are most common in the face and most BCCs appear in the head and neck region. Approximately 25% occur on the nose alone, the most common site [1,2]. They are commonly divided into three main forms:

Nodular basal cell carcinomas (NBCC) have well defined borders and are often located in the face. Pigmented forms of BCCs are seen and pose differential diagnostic problems versus malignant melanoma. NBCCs are often excised. In special regions, such as the nose, ear and eyelid, cryotherapy may be the therapy of choice [3,4]. Nodular tumours can be very large and destroy deeper structures.

Superficial basal cell carcinomas (SBCC) usually occur on the trunk. In many instances, they are multiple. They can mimic guttate psoriasis. Treatment consists of curettage and electrodesiccation, cryosurgery and sometimes excision. Recently, photodynamic therapy (PDT) has evolved as a therapeutic alternative.

Aggressive BCCs are of either intermediate or morfeiform type and grow with diffuse borders. These tumours are often much larger than expected from the clinical expression. An average of 7.2 mm of subclinical tumour extension was found in 51 morfeiform BCCs, as compared with 2.1 mm in 138 well circumscribed nodular lesions [5]. Morfeiform BCCs are often located in the face and especially on the nose, forehead and temple. This is generally called the "H-zone" with a high risk of recurrence [2]. The embryonic fusion planes, such as the nasolabial fold, medial canthus and the area behind the ears are especially troublesome. In these regions, BCCs can grow as "icebergs" with only the top of the tumour visible. Treatment of aggressive BCCs is usually surgical. Mohs micrographic surgery has been advocated since this technique offers the best possible evaluation of the excision margins [6]. This is especially important on the eyelids [7,8].

14.3 Photodynamic therapy (PDT) for basal cell carcinomas

Photodynamic therapy involves the destruction of tumour cells achieved by a photochemical reaction producing toxic substances, mainly singlet oxygen. An exogenous photosensitiser, which can be administered topically or systemically, usually intravenously, is concentrated in tumour tissue. A photochemical reaction occurs when the tissue is irradiated at a wavelength matching the absorption spectrum of the photosensitiser. Molecular oxygen, present in the tissue is converted into cytotoxic singlet oxygen which causes the destruction of the malignant cells while surrounding normal tissue remains unharmed [9]. Delta-aminolevulinic acid (ALA) is being used clinically for the treatment of skin cancer [10–15]. ALA can be delivered topically or systemically and can also be used for diagnostic purposes [16,17].

ALA is the precursor in the biosynthesis of haeme. Most cells in the human body are able to synthesise haeme. ALA is formed from glycine and succinyl CoA by the enzyme

ALA-synthetase (ALAS). This step is regulated by the amount of haeme in the cell. At the end of the synthesis, iron is incorporated into the photosensitive protoporphyrin IX molecule by the enzyme ferrochelatase and haeme is formed. The first and the last step of the haeme biosynthesis take place in the mitochondria of the cell. The formation of ALA and haeme by the enzymes ALAS and ferrochelatase are the rate-limiting steps. Exogenous ALA bypasses the first rate-limiting reaction. The activity of ferrochelatase in tumour tissue is lower than in normal skin [18–20]. Consequently, protoporphyrin IX (Pp IX) is accumulated in tumour tissue. Another reason might be the increased ALA permeability through tumour skin. It is known that the hydrophilic ALA penetrates intact healthy skin poorly [21–23].

PDT is suitable mainly for superficial basal cell carcinomas, actinic keratoses and early stages of squamous cell carcinomas [12,13,21,24–26]. The use of immunosuppressants have made organ transplantations more common. However, the risk of developing basal and squamous cell skin cancers is increased dramatically [27,28]. Often the patients have large areas of precancerous lesions. PDT seems promising for this type of lesions.

14.4 Light sources for photodynamic therapy

A variety of different light sources can be used, even slide projectors! [21]. The light source for ALA-PDT has to deliver the right amount of light at the proper wavelength. This can be achieved either by a laser or a non-coherent light source [11,29]. Although light in the Soret band (410 nm) would give the highest absorption, it is seldom used because of the poor penetration of blue light [30]. For skin tumours there is basically no need for lasers as their unique optical properties – coherence and non-divergence – are not required [21,24]. It might even be of advantage to use a broadband light source since photosensitising photoproducts with other absorption peaks are formed during PDT and a broadband lamp could cover their spectrum [31].

14.5 Recurrence rates

Mohs micrographic surgery has the highest cure rate [32]. However, topical PDT has shown promising results when treating superficial BCCs with a clearance rate of 80–100% in several studies [9–11,33]. Most studies show some decrease in the cure rate after long-term follow-up. Warloe et al have published a study with a large number of lesions and achieved a short-term cure rate of 94% [33]. Fink-Punches et al had an initial cure rate of 86% that decreased to 56% after a median follow-up of 19 months [34]. The cure rate for nodular BCCs after a single PDT session is rather low with an average cure rate below 50% [10,24]. Repeated treatments seem to improve the cure rate [11].

The use of EDTA, DFO and DMSO can increase efficacy [25,33].

14.6 PDT procedure

Before treatment, the tumour outlines is marked with ink. Anaesthesia etc can distort the tumour borders. A slight curettage without major bleeding is then sometimes performed and the tumour area is covered with a plastic sheet under occlusion for three to six hours. The application time varies between different centres. Our experience is that EMLA anaesthesia is usually necessary to minimise the discomfort for the patient. EMLA causes vasoconstriction and it is not known yet whether this contributes to a good effect or decreases the efficacy of the treatment.

The area is then illuminated with a device emitting red light (Fig. 1). The amount of light delivered varies between different centres and using meters with different spectral sensitivity, the readout can be different. In our setting, 70–110 J/cm^2 is used. The lower legs and the face are most sensitive. The exact dose-response relationship is not known. Also, the role of the dose rate should be further investigated.

During illumination some patients experience pain. It is usually impossible to tell which patient that will feel pain. Pain can be minimised by a fan or by pouring water to the treated area. It may be that a photoproduct of Pp IX absorbing at 670 nm is responsible for part of the pain. However, omitting this by using a laser has not resulted in total pain relief [35]. Also, the importance of the dose rate for pain development is not known.

In our hands, we agree that only superficial lesions are suitable for PDT due to the relatively poor penetration of ALA [24]. Recently, it has been shown that a prior curettage may be helpful when treating thicker lesions [36,37]. Possibly repeated or fractionated PDT sessions give a more favourable result. When PDT is applied to superficial lesions the long-term results seem good.

Immediately after treatment an erythema develops, partly due to heat. A couple of days post treatment a scaling reaction takes place which subsides within a week. A postinflammatory reaction, most often a transient hyperpigmentation, is sometimes seen. It is important to protect the treated area with a sunscreen protecting both against UVA and UVB. UV protection should be advocated whenever the patient is outdoors even in cloudy weather as a relatively large amount of UV passes through the atmosphere even under these conditions.

We usually see the patients 6–8 weeks after treatment. The later follow-up is individualised as many patients are severely sun damaged and require treatment for precancerous or cancer lesions frequently. The cosmetic results are usually excellent. In many patients it is impossible to locate the exact origin of the previous tumour. This is a crucial aspect in regions of the body which have a high tendency for keloid formation. A typical case is seen in Fig. 2a, b and c.

In our material the cure rate is approximately 80% during a follow-up period of 4 years concerning superficial basal cell carcinomas [38].

14.7 Future improvements of PDT for basal cell carcinomas

Measuring the fluorescence of Pp IX immediately before therapy may be a way of optimising therapy in order to avoid under- or overdose.Optimising the time of topical application could increase the penetration of ALA into tumour tissue [39].

Warloe has tried curettage before ALA application in order to reduce the tumour volume and remove the surface structure in 152 nodular tumours before PDT and achieved an 85% cure rate [15].

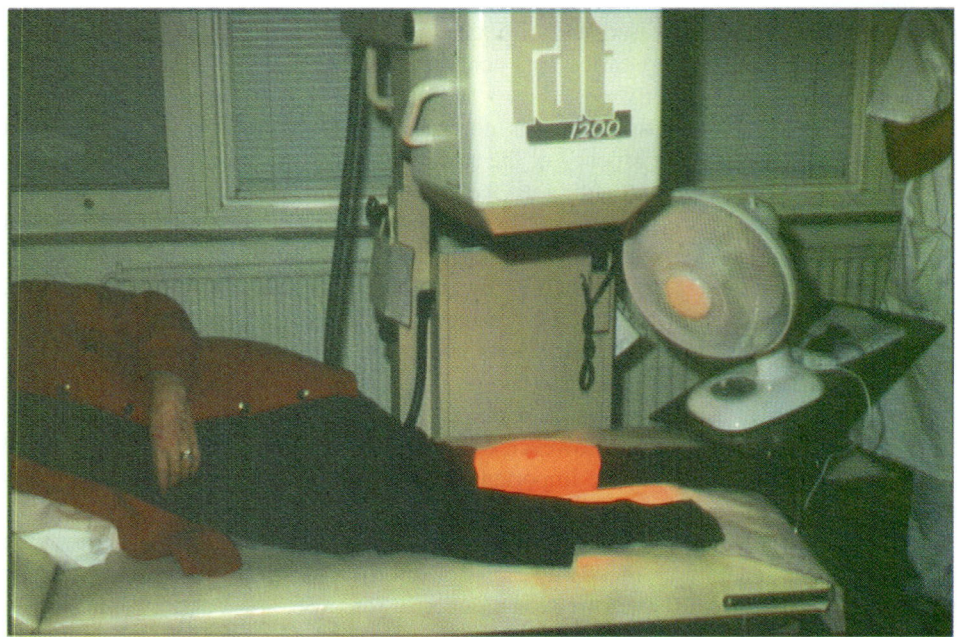

Figure 1. Photodynamic treatment of a basal cell carcinoma on the leg.

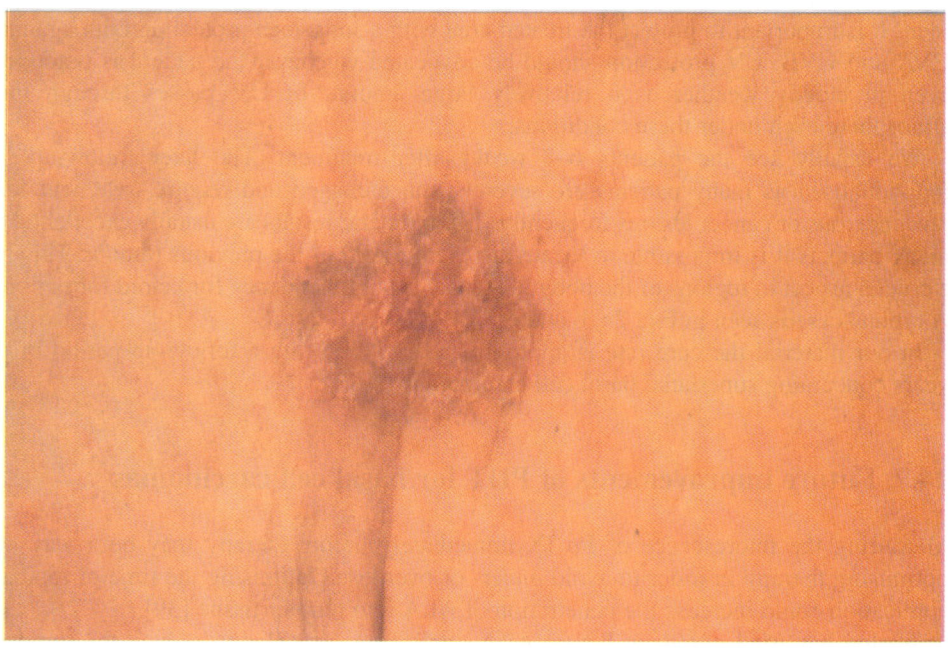

Figure 2a. A superficial basal cell carcinoma (1.5 cm largest diameter) localised on the chest. Before PDT.

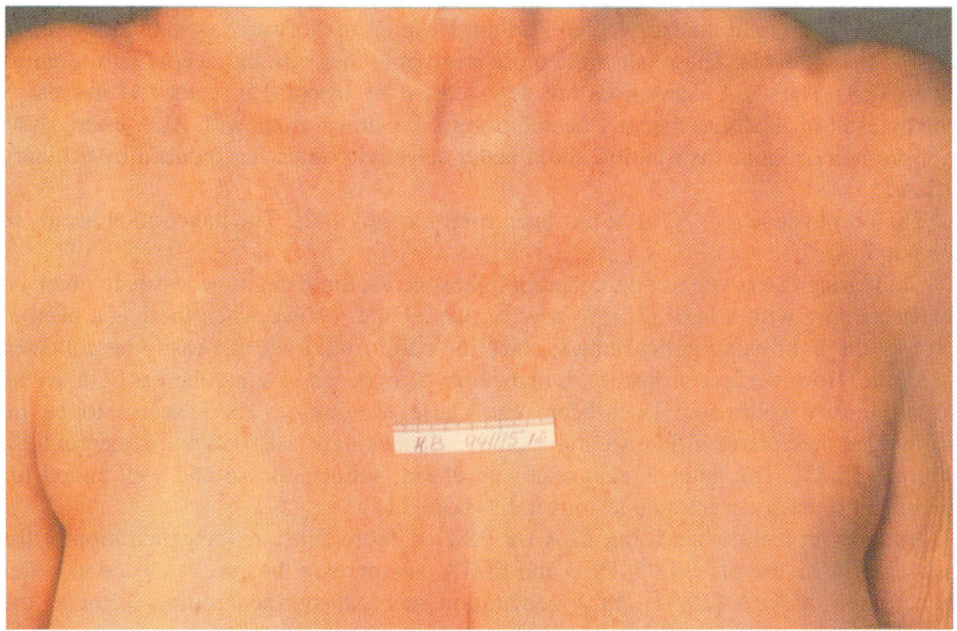

Figure 2b. An over-view of the chest area described in fig 2a, one year after PDT.

Figure 2c. A close-up of the same area as in fig. 2a, one year after PDT.

The PDT-group at the Norwegian Radium Hospital in Norway has studied lipophilic ALA-esters [40,41]. These penetrate more easily into the cells and deeper into tumours than ALA. The ALA derivatives are de-esterified by esterase in cells and tissues. A methylester of ALA has recently been introduced. Clinical trials with ALA-esters show a more homogenous distribution and a better selectivity than that induced by ordinary ALA [15].

The porphycene dye ATMPn has been tried recently [42]. The penetration seems to be good.

Comparing lasers versus non-coherent light sources, there might be less pain involved when treating with a laser [35]. A laser is also a better choice when there is a need to convey the light using optical fibres, since the collimated light can easily be collected in a fibre. However, eye protection is mandatory as opposed to non-coherent light, as the high-intensity collimated laser beam can cause eye injury. There seems to be no significant therapeutic difference in the effect when using a laser or a non-coherent light source [15,43]. The latter is also relatively cheap, simpler to use and well suited for treatment of easily accessible lesions in the skin [24].

A major problem when using ALA for PDT or fluorescence is the penetration of the substance. The use of DMSO, DFO and EDTA can increase the cure rates due to better penetration of ALA [25,33]. ALA penetration seems to be the limiting factor as the penetration of red light is fairly good. Thus, measurements of the ALA concentration in tissue are of great interest. In recent years microdialysis has evolved as a new elegant way of doing this. It turns out that the penetration in tumour areas is rapid and after 15 minutes the concentration is high and stable [23]. On the other hand, virtually no ALA penetrates healthy skin. However, it should be remembered that light curettage causing no bleeding was performed in tumour areas. This probably affects the result but, on the other hand, we mimic the clinical situation. The ALA uptake in skin after curettage would be enhanced as crusts are removed and the stratum corneum is diminished. The light curettage would also ensure a more even uptake of ALA. The exact consequence of curettage needs to be investigated further. The penetration kinetics at various depths also need to be clarified in further studies.

Laser Doppler-imaging technique reveal that the perfusion in BCCs is 2.5-fold higher than in normal skin [23]. This could lead to a situation where topically applied ALA is washed away, reducing the tissue concentration of ALA and Pp IX. On the other hand, this can affect the kinetics of ALA penetration in tumour and healthy skin with an earlier uptake in tumour tissue. Wang and co-workers have investigated the perfusion in superficial basal cell carcinomas and there was an increased blood perfusion after topical PDT [44].

Due to the pain during therapy, EMLA is often applied. As EMLA causes vasoconstriction, this can probably increase the tissue concentration of ALA and Pp IX and lead to a better therapeutic response. This has to be investigated further.

14.8 Photodynamic diagnosis (PDD) of basal cell carcinoma

Chromophores such as porphyrins have the ability to absorb energy when excited by light of a certain wavelength. Positioned at their excitation levels, the molecules strive to return to their non-excited level, giving away part of the energy as fluorescence when doing so. Imaging spectroscopy using digital cameras (CCD) can be used to study such physical processes. The fluorescence imaging technique extends our ability to see patterns that can not be preserved with the naked eye [11,17,45–47].

Today's imaging spectroscopy is a field gaining in importance due to the use of digital cameras and the possibility of handling large quantities of information by computer technology. The equipment needed for fluorescence imaging is a light source for excitation and, a digital camera for detection and the software for data processing. Many different types of digital cameras are available on the market. The special range of the camera determines which parts of the spectra that can be recorded. The CCD chip has a spectral response, which is in itself limited by the band gap in silicon and is thus limited to the visible part of the spectral range, 400–800 nm. Outside this spectral range one has to use other techniques. By coating the CCD a possibility to extend the sensitivity to UV is given. In experiments with fluorescence imaging it is often desirable to record a smaller part of the incident spectrum. A way of doing this is to use glass filters.

When using fluorescence imaging to determine spatial variation of the fluorescent species it is important to control the light distribution. One way of doing this is to ensure a homogenous illumination of the investigation area. Pp IX has the ability to fluoresce when illuminated with photoactivating light. This property can be used for detecting tumour tissue. It is of importance to choose the proper light source and detection device so that a clear signal from the Pp IX fluorescence is given as opposed to other fluorescing substances in the skin. Several groups work with this [48–50].

In our set-up a bandpass glass filtered mercury lamp is used as a light source with three peaks as emission spectrum. These three peaks at 365, 366 and 405 nm. The set up is shown in Fig. 3. The CCD camera recording the fluorescence images has a longpass filter letting only wavelengths longer than 590 nm through. The light source and the CCD device are fitted with the filters to match the emission and absorption of Pp IX. The fluorescence images are analysed with a computer program using the threshold criterion for tumour fluorescence as 40% or more of the mean fluorescence in the ALA area treated.

There are many pitfalls when using fluorescence spectroscopy.

The filters on the CCD-camera should be chosen to match the excitation and emission wavelengths of Pp IX. The cut-off point for collecting fluorescence light was at 590 nm. The sensitivity of a CDD-camera is low above 750 nm. The exposure time used for the images is 2 seconds and with the patients lying on a bed there is no problems with unsharp pictures. A typical fluorescence case is shown in Fig. 4.

The area has to be evenly illuminated and especially in the face, curvatures etc can affect the result. Often, fluorescence is at its maximum on the vertex of an investigated area. This is important on the cheek, the eyelids and the nose. Of course, this plays a minor role on the trunk. The uneven illumination is important in the normalisation process.

In our work we used the mean fluorescence intensity in the entire ALA-treated area as the normalisation value. This is a simple way to normalise but it supposes that the tumour area is small compared to the whole ALA-treated area. If the tumour is large relative to the total area applied with ALA, the sensitivity of the method declines. Another approach can be to compare the fluorescence of the tumour area with that from

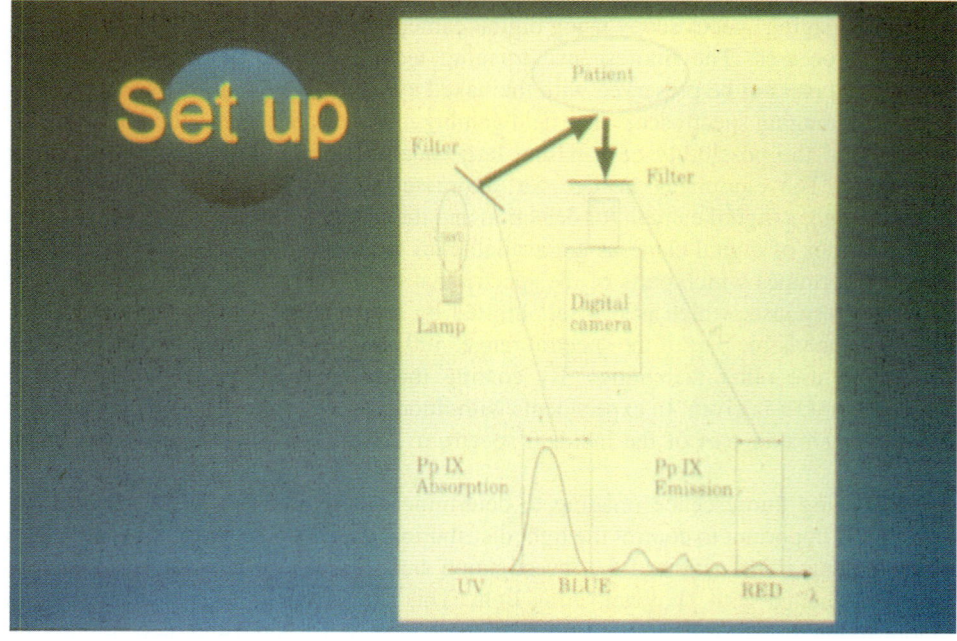

Figure 3. The set-up of the mercury lamp, the patient and the CCD device.

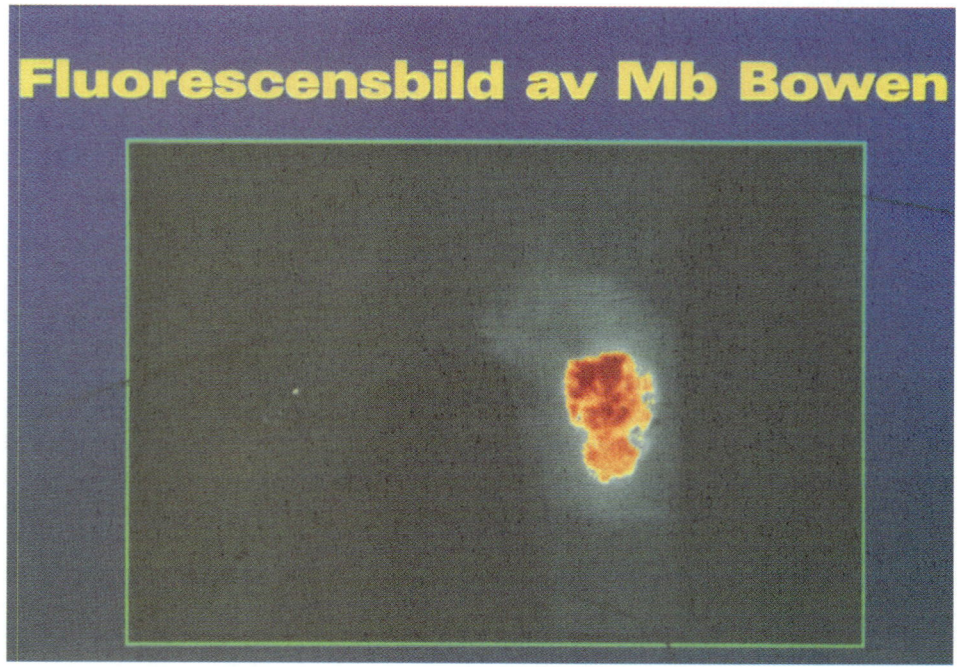

Figure 4. A fluorescence image of a Mb. Bowen on a finger.

a healthy area using the same angle of incident light. This method may have a higher degree of precision. Autofluorescence can also be used.

Sebaceous glands fluoresce and this is a disturbing factor especially on the nose [51]. Actinic keratoses and Mb. Bowen also fluoresce, as do areas with sun damage. However, an experienced observer can often tell the difference between an area of solar damage and a BCC. On the other hand, necrotic material in wounds of ulcerative BCCs does not produce any Pp IX and no fluorescence is observed here. In approximately 75% of morfeiform BCCs, the fluorescence technique is of clinical value for demarcating this type of lesion. The fluorescence technique is convenient and promising but the observer needs to be trained and experienced.

A major advantage of the fluorescence technique is of course that it can be used in sensitive areas of the face, i.e. the H-zone, as the tissue sparing is of utmost importance in these delicate areas. Fluorescence measurements can also be made before treatment with cryosurgery and radiotherapy. The method cannot replace Mohs micrographic surgery but it can be a complement for complete tumour eradication. The detection of tumour tissue by imaging Pp IX fluorescence is based on the fact that topical ALA-application results in tissue specific photosensitization where BCC is one type of lesion that gives rise of increased Pp IX fluorescence [9,21,45,52]. The question as to whether the Pp IX concentration is higher in cancer cells than in normal cells has been addressed by several authors. Martin et al. [53] reported that there was no selectivity for Pp IX fluorescence in tumour tissue vs. normal skin as seen with fluorescence microscopy.

Also, the penetration of light into different tissues varies. It is reasonable to believe that light penetration in dense tumours such as morpheiform BCCs is less than in superficial BCCs.

The technique needs further improvement and other derivatives than ALA can possibly be more useful as the penetration of ALA is not optimal. Only the superficial extension of a BCC can be visualised. The ALA ester p-1202 seems to be more specific and penetrates more deeply.

14.9 Conclusions and outlook for the future

Photodynamic therapy (PDT) is a good alternative to other treatments for superficial basal cell carcinomas, actinic keratoses and Mb Bowen. The cosmetic outcome is generally very good and the side effects are few. Nodular basal cell carcinomas are not suited for PDT as the penetration of delta-aminolevulinic acid is not deep enough. New esterified ALA derivatives look promising.

The selectivity is of ALA-induced tumour fluorescence can be used for tumour demarcation. This implies a safer delineation of morpheiform and other BCCs with ill-defined borders before therapy. Mohs micrographic surgery can be facilitated. However, the method has to be developed further as there is reasonable agreement between fluorescence and histology in only 75% of the cases.

References

1. S.J. Miller (1991). Biology of basal cell carcinoma (Part I). *J. Am. Acad. Dermatol.*, **24**(1), 1–13.

2. R.K. Roenigk, J.L. Ratz, P.L. Bailin, R.G. Wheeland (1986). Trends in the presentation and treatment of basal cell carcinomas. *J. Dermatol. Surg. Oncol.*, **12**(8), 860–865.

3. P. Nordin, O. Larko, B. Stenquist (1997). Five-year results of curettage-cryosurgery of selected large primary basal cell carcinomas on the nose: an alternative treatment in a geographical area underserved by Mohs' surgery. *Br. J. Dermatol.*, **136**(2), 180–183.

4. P. Nordin (1999). Curettage-cryosurgery for non-melanoma skin cancer of the external ear: excellent 5-year results. *Br. J. Dermatol.*, **140**(2), 291–293.

5. S.J. Salasche, R.A. Amonette (1981). Morpheaform basal-cell epitheliomas. A study of subclinical extensions in a series of 51 cases. *J. Dermatol. Surg. Oncol.*, **7**(5), 387–394.

6. A.M. Wennberg, O. Larko, B. Stenquist (1999). Five-year results of Mohs' micrographic surgery for aggressive facial basal cell carcinoma in Sweden. *Acta Derm. Venereol.*, **79**(5), 370–372.

7. G. Lindgren, B. Lindblom, O. Larko (2000). Mohs' micrographic surgery for basal cell carcinomas on the eyelids and medial canthal area. II. Reconstruction and follow-up [In Process Citation]. *Acta Ophthalmol. Scand.*, **78**(4), 430–436.

8. G. Lindgren, B. Lindblom, A.T. Bratel, L. Molne, O. Larko (2000). Mohs' micrographic surgery for basal cell carcinomas on the eyelids and medial canthal area. I. Characteristics of the tumours and details of the procedure [In Process Citation]. *Acta Ophthalmol. Scand.*, **78**(4), 425–429.

9. J.C. Kennedy, R.H. Pottier (1992). Endogenous protoporphyrin IX, a clinically useful photosensitizer for photodynamic therapy. *J. Photochem. Photobiol. B*, **14**(4), 275–292.

10. P. Wolf, E. Rieger, H. Kerl (1993). Topical photodynamic therapy with endogenous porphyrins after application of 5-aminolevulinic acid. An alternative treatment modality for solar keratoses, superficial squamous cell carcinomas, and basal cell carcinomas? [published erratum appears in *J. Am. Acad. Dermatol.*, 1993 Jul. 29(1), 41]. *J. Am. Acad. Dermatol.*, **28**(1), 17–21.

11. K. Svanberg, T. Andersson, D. Killander, I. Wang, U. Stenram, S. Andersson-Engels et al. (1994). Photodynamic therapy of non-melanoma malignant tumours of the skin using topical delta-amino levulinic acid sensitization and laser irradiation. *Br. J. Dermatol.*, **130**(6), 743–751.

12. F. Cairnduff, M.R. Stringer, E.J. Hudson, D.V. Ash, S.B. Brown (1994). Superficial photodynamic therapy with topical 5-aminolaevulinic acid for superficial primary and secondary skin cancer. *Br. J. Cancer*, **69**(3), 605–608.

13. P.G. Calzavara-Pinton (1995). Repetitive photodynamic therapy with topical delta-aminolaevulinic acid as an appropriate approach to the routine treatment of superficial non-melanoma skin tumours. *J. Photochem. Photobiol. B*, **29**(1), 53–57.

14. R.M. Szeimies, P. Calzavara-Pinton, S. Karrer, B. Ortel, M. Landthaler (1996). Topical photodynamic therapy in dermatology. *J. Photochem. Photobiol. B*, **36**(2), 213–219.

15. Q. Peng, T. Warloe, K. Berg, J. Moan, M Kongshaug, K.E. Giercksky et al. (1997). 5-Aminolevulinic acid-based photodynamic therapy. Clinical research and future challenges. *Cancer*, **79**(12), 2282–2308.

16. T.J. Dougherty (1987). Photosensitizers: therapy and detection of malignant tumors. *Photochem. Photobiol.*, **45**(6), 879–889.

17. K. Svanberg, I. Wang, S. Colleen, I. Idvall, C. Ingvar, R. Rydell et al. (1998). Clinical multicolour fluorescence imaging of malignant tumours – initial experience. *Acta Radiol.*, **39**(1), 2–9.

18. N. Schoenfeld, O. Epstein, M. Lahav, R. Mamet, M. Shaklai, A. Atsmon (1988). The heme biosynthetic pathway in lymphocytes of patients with malignant lymphoproliferative disorders. *Cancer Lett.*, **43**(1–2), 43–48.

19. M.M. el-Sharabasy, A.M. el-Waseef, M.M. Hafez, S.A. Salim (1992). Porphyrin metabolism in some malignant diseases. *Br. J. Cancer*, **65**(3), 409–412.
20. M. Kondo, N. Hirota, T. Takaoka, M. Kajiwara (1993). Heme-biosynthetic enzyme activities and porphyrin accumulation in normal liver and hepatoma cell lines of rat. *Cell Biol. Toxicol.*, **9**(1), 95–105.
21. J.C. Kennedy, R.H. Pottier, D.C. Pross (1990). Photodynamic therapy with endogenous protoporphyrin IX: basic principles and present clinical experience. *J. Photochem. Photobiol. B*, **6**(1–2), 143–148.
22. B.A. Goff, R. Bachor, N. Kollias, T. Hasan (1992). Effects of photodynamic therapy with topical application of 5- aminolevulinic acid on normal skin of hairless guinea pigs. *J. Photochem. Photobiol. B*, **15**(3), 239–251.
23. A. Wennberg, O. Larko, P. Lonnroth, G. Larson, A. Krogstad (2000). Delta-aminolevulinic acid in superficial basal cell carcinomas and normal skin-a microdialysis and perfusion study [In Process Citation]. *Clin. Exp. Dermatol.*, **25**(4), 317–322.
24. A.M. Wennberg, L.E. Lindholm, M. Alpsten, O. Larko (1996). Treatment of superficial basal cell carcinomas using topically applied delta- aminolaevulinic acid and a filtered xenon lamp. *Arch. Dermatol. Res.*, **288**(10), 561–564.
25. S. Fijan, H. Honigsmann, B. Ortel (1995). Photodynamic therapy of epithelial skin tumours using delta- aminolaevulinic acid and desferrioxamine. *Br. J. Dermatol.*, **133**(2), 282–288.
26. R.M. Szeimies, S. Karrer, A. Sauerwald, M. Landthaler (1996). Photodynamic therapy with topical application of 5-aminolevulinic acid in the treatment of actinic keratoses: an initial clinical study. *Dermatology*, **192**(3), 246–251.
27. L. Bergfelt, O. Larko, I. Blohme (1993). Skin disease in immunosuppressed patients in relation to epidermal Langerhans' cells. *Acta Derm. Venereol.*, **73**(5), 330–334.
28. P. Jensen, S. Hansen, B. Moller, T. Leivestad, P. Pfeffer, O. Geiran et al. (1999). Skin cancer in kidney and heart transplant recipients and different long-term immunosuppressive therapy regimens [see comments]. *J. Am. Acad. Dermatol.*, **40**(2 Pt 1), 177–186.
29. R.M. Szeimies, R. Hein, W. Baumler, A. Heine, M. Landthaler (1994). A possible new incoherent lamp for photodynamic treatment of superficial skin lesions. *Acta Derm. Venereol.*, **74**(2), 117–119.
30. J.I.V. Moan, L.W. Ma (1996). Choice of proper wavelength for photochemotherapy. In: *Proc. SPIE* (pp. 544–549).
31. E.F. Gudgin Dickson, R.H. Pottier (1995). On the role of protoporphyrin IX photoproducts in photodynamic therapy [news]. *J. Photochem. Photobiol. B*, **29**(1), 91–93.
32. D.L. Shriner, D.K. McCoy, D.J. Goldberg, R.F. Wagner, Jr. (1998). Mohs micrographic surgery. *J. Am. Acad. Dermatol.*, **39**(1), 79–97.
33. T. Warloe (1995). *Photodynamic therapy of human malignant tumors*. Oslo: University of Oslo, Norway.
34. R. Fink-Puches, H.P. Soyer, A. Hofer, H. Kerl, P. Wolf (1998). Long-term follow-up and histological changes of superficial nonmelanoma skin cancers treated with topical delta-aminolevulinic acid photodynamic therapy. *Arch. Dermatol.*, **134**(7), 821–826.
35. S. Karrer, W. Baumler, C. Abels, U. Hohenleutner, M. Landthaler, R.M. Szeimies (1999). Long-pulse dye laser for photodynamic therapy: investigations in vitro and in vivo. *Lasers Surg. Med.*, **25**(1), 51–59.
36. A.M. Soler, T. Warloe, J. Tausjo, A. Berner (1999). Photodynamic therapy by topical aminolevulinic acid, dimethylsulphoxide and curettage in nodular basal cell carcinoma: a one-year follow-up study. *Acta Derm. Venereol.*, **79**(3), 204–206.
37. M.R. Thissen, C.A. Schroeter, H.A. Neumann (2000). Photodynamic therapy with delta-aminolaevulinic acid for nodular basal cell carcinomas using a prior debulking technique. *Br. J. Dermatol.*, **142**(2), 338–339.

38. A.M. Wennberg (2000). Basal cell carcinoma – new aspects of diagnosis and treatment [In Process Citation]. *Acta Derm. Venereol. Suppl.*, **209**, 5–25.
39. R.M. Szeimies, T. Sassy, M. Landthaler (1994). Penetration potency of topical applied delta-aminolevulinic acid for photodynamic therapy of basal cell carcinoma. *Photochem. Photobiol.*, **59**(1), 73–76.
40. Q. Peng, J. Moan, T. Warloe, V. Iani, H.B. Steen, A. Bjorseth et al. (1996). Build-up of esterified aminolevulinic-acid-derivative-induced porphyrin fluorescence in normal mouse skin. *J. Photochem. Photobiol. B*, **34**(1), 95–96.
41. J.M. Gaullier, K. Berg, Q. Peng, H. Anholt, P.K. Selbo, L.W. Ma et al. (1997). Use of 5-aminolevulinic acid esters to improve photodynamic therapy on cells in culture. *Cancer Res.*, **57**(8), 1481–1486.
42. S. Karrer, C. Abels, R.M. Szeimies, W. Baumler, M. Dellian, U. Hohenleutner et al. (1997). Topical application of a first porphycene dye for photodynamic therapy – penetration studies in human perilesional skin and basal cell carcinoma. *Arch. Dermatol. Res.*, **289**(3), 132–137.
43. A.M. Soler, E. Angell-Petersen, T. Warloe, J. Tausjo, H.B. Steen, J. Moan et al. (2000). Photodynamic therapy of superficial basal cell carcinoma with 5- aminolevulinic acid with dimethylsulfoxide and ethylendiaminetetraacetic acid: a comparison of two light sources. *Photochem. Photobiol.*, **71**(6), 724–729.
44. I. Wang, S. Andersson-Engels, G.E. Nilsson, K. Wardell, K. Svanberg (1997). Superficial blood flow following photodynamic therapy of malignant non-melanoma skin tumours measured by laser Doppler perfusion imaging. *Br. J. Dermatol.*, **136**(2), 184–189.
45. Q. Peng, T. Warloe, J. Moan, H. Heyerdahl, H.B. Steen, J.M. Nesland et al. (1995). Distribution of 5-aminolevulinic acid-induced porphyrins in noduloulcerative basal cell carcinoma. *Photochem. Photobiol.*, **62**(5), 906–913.
46. H. Heyerdahl, I. Wang, D.L. Liu, R. Berg, S. Andersson-Engels, Q. Peng et al. (1997). Pharmacokinetic studies on 5-aminolevulinic acid-induced protoporphyrin IX accumulation in tumours and normal tissues. *Cancer Lett.*, **112**(2), 225–231.
47. J. Moan, O. Bech, Q. Peng, K. Berg (1998). [Use of 5-aminolevulinic acid in photochemotherapy and fluorescence diagnostics]. *Tidsskr Nor Laegeforen*, **118**(8), 1206–1211.
48. G. Ackermann, C. Abels, W. Baumler, S. Langer, M. Landthaler, E.W. Lang et al. (1998). Simulations on the selectivity of 5-aminolaevulinic acid-induced fluorescence in vivo. *J. Photochem. Photobiol. B*, **47**(2–3), 121–128.
49. A.M. Wennberg, F. Gudmundson, B. Stenquist, A. Ternesten, L. Molne, A. Rosen et al. (1999). In vivo detection of basal cell carcinoma using imaging spectroscopy. *Acta Derm. Venereol.*, **79**(1), 54–61.
50. S. Andersson-Engels, G. Canti, R. Cubeddu, C. Eker, C. af Klinteberg, A. Pifferi et al. (2000). Preliminary evaluation of two fluorescence imaging methods for the detection and the delineation of basal cell carcinomas of the skin. *Lasers Surg. Med.*, **26**(1), 76–82.
51. D.X. Divaris, J.C. Kennedy, R.H. Pottier (1990). Phototoxic damage to sebaceous glands and hair follicles of mice after systemic administration of 5-aminolevulinic acid correlates with localized protoporphyrin IX fluorescence. *Am. J. Pathol.*, **136**(4), 891–897.
52. A. Orenstein, G. Kostenich, Z. Malik (1997). The kinetics of protoporphyrin fluorescence during ALA-PDT in human malignant skin tumors. *Cancer Lett.*, **120**(2), 229–234.
53. A. Martin, W.D. Tope, J.M. Grevelink, J.C. Starr, J.L. Fewkes, T.J. Flotte et al. (1995). Lack of selectivity of protoporphyrin IX fluorescence for basal cell carcinoma after topical application of 5-aminolevulinic acid: implications for photodynamic treatment. *Arch. Dermatol. Res.*, **287**(7), 665–674.

Photodynamic Therapy and Fluorescence Diagnosis in Dermatology
P.-G. Calzavara-Pinton, R.-M. Szeimies and B. Ortel, editors.

Chapter 15

Topical sensitization – oncologic indications – others (lymphoma)

Regina Fink-Puches and Peter Wolf

Table of contents

Abstract

In recent years the therapeutic search for cutaneous T-cell lymphoma (CTCL) and especially for mycosis fungoides (MF) has continued and several innovative therapies are under study. Photodynamic therapy (PDT) after topical photosensitization with 5-aminolevulinic acid (ALA) might be an alternative treatment modality. The following results of research in the last years led to the clinical application of ALA-PDT in patients with CTCL: ALA and light inactivated erythrocytic, lymphocytic and myelocytic leukemic cells in vitro; ALA-PDT was effective in T-cell-derived cell lines; selective PpIX accumulation could be demonstrated in CD71 positive lymphocytes; and selective clinical CTCL photosensitization after topical ALA application was observed. Indeed, clinical studies in a limited number of patients revealed complete, partial or no response of CTCL lesions treated by different ALA-PDT treatment protocols. However, further studies are required in larger numbers of CTCL patients to determine the optimal ALA and light dose combination, frequency of treatment, overall response rate, and duration of disease remission.

15.1 Mycosis fungoides – Clinical and histologic features – Standard management

Mycosis fungoides (MF) is an epidermotropic cutaneous T-cell lymphoma (CTCL) characterized by proliferation of small or medium-sized neoplastic T-lymphocytes with cerebriform nuclei [1]. It classically presents with patch or plaque stage lesions which may be present for many years or even decades, before more aggressive disease develops. The progression of cutaneous disease is mirrored pathologically in that tumor stage lesions may lose their mature T-cell phenotype, show blastic transformation, a higher mitotic rate and little or no epidermotropism. Lymph node and internal organs may become involved. Tumor stage disease is associated with bad prognosis. For instance studies from the U.S. indicated that tumor stage, lymph node involvement or erythroderma gives a median survival of 24 to 36 months [2]. However, the estimated 5-year survival of 278 patients with MF of all stages from the Dutch Cutaneous Lymphoma Working Group registry was 87% [1]. In a study from our Department we found an estimated 5-year-survival of 89% for 335 consecutive patients with MF (Fink-Puches et al. unpublished data).

The treatment of CTCL has been the subject of many reviews [3]. Treatment strategies can be divided into two categories: skin-directed and systemic therapy. Skin-directed therapy includes psoralens plus ultraviolet light (PUVA), topical chemotherapy and radiation therapy. Systemic therapy includes chemotherapy and other cytotoxic agents, photopheresis, interferons (IFN), retinoids, as well as other investigational biologic response modifiers. However, due to the long-term, natural course of the disease it remains unclear at present whether definite cure of CTCL can be obtained by any type of therapy. For instance, Kaye et al. evaluated the use of early, aggressive therapy with total-skin electron beam therapy and chemotherapy versus sequential topical therapies [4]. Although the aggressively treated patients had a higher complete remission rate, they also experienced significantly more toxicity. On the other hand, topical nitrogen mustard application can result in atypical histologic changes in the skin and radiotherapy and PUVA have been shown to be carcinogenic [5]. More importantly, there were no significant durable differences between disease-free and overall survival between the patients treated with aggressive therapy and those who received only local therapy [4]. No studies have demonstrated superiority of any topical therapy, and the decision of which therapy to choose usually depends on physician's and patient's preference. Several novel therapies for CTCL are under study and topical PDT might be an alternative treatment modality.

15.2 Rationality for ALA-PDT in CTCL – Experimental studies

Reasons for applying ALA-PDT in CTCL are certain experimental studies, which have been published in the recent years [6–9]. Malik et al. have demonstrated in-vitro studies that fast-dividing lymphoma cells, can intrinsically synthesize endogenous porphyrins after exogenous administration of ALA [7]. The same studies also have shown, that inactivation of ALA-sensitized T cell lymphoma Eb-Esb cells occurred after photoexcitation by visible light [7]. Grebenova et al. recently investigated the cytotoxic effects of ALA-PDT on promyeloid, erythroid, and T and B leukemia/lymphoma cell lines and compared the effects with those on normal human lymphocytes (quiescent and activated) and bone marrow progenitor cells. They demonstrated that ALA-PDT reduced the number and/or suppressed proliferation of T-cell-derived cell lines by an order of 2 log magnitude, while the effect on quiescent human lymphocytes was small [6].

15.3 Clinical studies

15.3.1

Despite increasing experience with ALA-PDT in the treatment of non-melanoma skin cancers there are only a few reports at present about treatment of patients with CTCL. Svanberg et al. reported on complete remission in two of four CTCL lesions after one or two treatment sessions using 20% ALA cream and laser photoirradiation with a light dose of 60 J/cm^2 [9]. Importantly, the same group of investigators demonstrated by laser-induced fluorescence that the Pp IX distribution after topical ALA-treatment showed a demarcation between CTCL lesional skin and normal skin by an approximate 5 : 1 ratio [9].

15.3.2

Conti et al. investigated clinical and histological responses to topical ALA-PDT on 24 CTCL lesions in 6 patients with patch and plaque stage CTCL and on more than 15 tumors of CTCL up to 1 cm thickness in 3 patients. ALA was topically applied for 4–5 hours or 18–24 hours, respectively, before irradiation. Short application times and single treatments induced clinical and histological remissions lasting more than 91 weeks in patch and plaque stage disease; overnight ALA applications yielded significant, light-dose dependent patch and tumor reductions with minimal or no epidermal damage [10].

15.3.3

Babich et al. reported on the treatment of CTCL with fractionated PDT using topical δ–aminolevulinic acid [11]. They treated more than 50 lesions in eight MF patients of different stage. Two to 20% ALA was applied under occlusion before exposure to 50–150 J/cm^2 of 630 nm dye laser light at 30–150 mW/cm^2 or light from a non-coherent 590–700 nm lamp. PDT treatments were repeated every 2–4 weeks. The number of treatments necessary to reach complete or partial responses depended on the lesion's thickness and the stage of the disease. CTCL patches and thin plaques resolved after one or two treatments, whereas CTCL tumors required five to seven treatments for complete remission. Taken together the complete clinical remission rate for all types of lesions was approximately 60%.

15.3.4

We have up to now treated more than 60 CTCL lesions with a diameter from 2.0 to 18.0 cm in eight patients with MF (patch/plaque stage IA-IIB). The treatment involved the topical administration of 20% ALA dissolved in Doritin®, a proprietary oil-in-water emulsion. The ALA-containing emulsion was applied to the CTCL lesions and to adjacent skin under occlusive and light-shielding dressing and left for 4 to 24 hours to allow penetration of ALA into the tissue and synthesis of porphyrin. Subsequently, visual evaluation of the ALA-treated areas under Wood's light showed in many cases the typical red fluorescence of porphyrin limited to the MF lesions (Fig. 1). These ALA-treated areas were then exposed repeatedly in 1- to 3-week intervals to different light doses ranging from 0.5 to 60 J/cm^2 of UVA (320–400 nm) or visible light. In two of those patients complete clinical and histological remission was achieved after 5 PDT sessions within 18 weeks and 4 PDT sessions within 7 weeks, respectively [12]. However, in the remaining six patients we noted partial or no response of CTCL lesions to topical ALA-PDT although we applied in single cases up to 10 PDT exposures [13].

15.3.5

Orenstein et al. reported on the successful treatment of two MF patients in different disease stages by topical ALA-PDT [14]. Twenty percent ALA-Decoderm® cream was applied to the CTCL lesions of those patients for 16 hours. One 20-minute session of photoirradiation [total light dose 170 J/cm^2) was performed to treat a 5 × 9-cm MF patch in patient 1. For nodular CTCL lesions in patient 2, the treatment protocol consisted of one 30-minute session of photoirradiation, a 1-hour-dark period, followed by a second

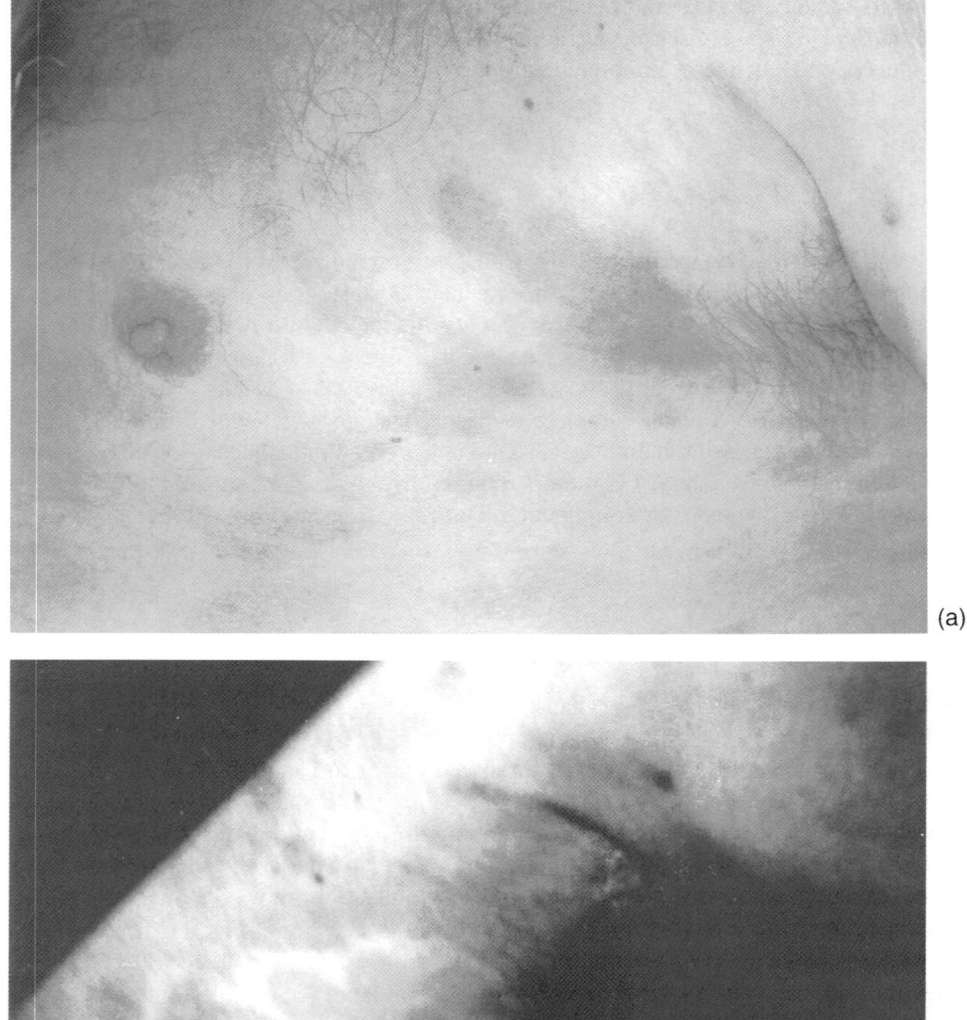

(a)

(b)

Figure 1. Photodynamic therapy of MF with topical 5-aminolevulinic acid photosensitization. (a) MF plaques in a male patient on the right chest. (b) Photograph taken under Wood's light 4 hours after topical application of 5-aminolevulinic acid. A Schott OG 530 long-wave-pass color glass filter was employed to enhance the contrast of fluorescence between normal skin and MF patches.

session of 10–15 minutes of photoirradiation (total light doses 340 and 380 J/cm^2, respectively) until fluorescence (detected by fluorescence measurements) had permanently diminished. In both patients complete remission was observed and the follow-up revealed that they were free of recurrence at 27 and 24 months after therapy, respectively. The cosmetic results of the treatment in both patients were excellent [14].

15.3.6

Eich et al. reported on the successful treatment of MF in initial tumor stage by topical ALA-PDT to the eyebrow and side of the right foot [15]. The treatment involved the topical administration of 12% ALA. The ALA-containing emulsion was left for 6 hours to allow penetration of ALA before exposure for 20 minutes to light with the wavelength of 550 nm at an intensity of 60 mW/cm^2. They observed after 12 PDT treatment sessions complete clinical and histological response of MF and no signs of recurrence at 24 months of follow-up.

15.3.7

Eich et al. reported also on histologically confirmed partial remission to ALA-PDT of a rare type of CD8 + CTCL with primary involvement of the ear [15]. However, besides this report of treatment of this CD8 + CTCL, to the best of our knowledge no further reports of topical PDT on other types of CTCL are available.

15.4 Possible mechanisms for efficiency of ALA-PDT in CTCL

Several factors such as stage of the disease, photosensitizer concentration, light dose, and oxygen supply could influence the result of PDT in the treatment of CTCL. The mechanism of selective ALA photosensitization in CTCL is not completely understood at present. However, Oseroff and Rittenhouse-Diakun suggested a possible mechanism for the selectivity of ALA-PDT for activated CD71 + cells [16, 8]: PpIX is the last but one step in the biosynthetic pathway of heme [17]. If the rate of conversion from ALA to PpIX is greater than that of PpIX to heme, PpIX will accumulate. Intracellular iron is essential for the conversion of PpIX to heme on the one hand [18], and for cell replication on the other. The transferrin receptor CD71 (OKT 9) – necessary for uptake of iron into the cell [19] – is upregulated in proliferating and activated cells and low intracellular iron upregulates membrane levels of transferrin receptor (CD 71). Thus, activated lymphocytes, expressing high levels of CD71, will accumulate PpIX, because of low intracellular iron, which is needed for ALA-dependent heme production and cell proliferation, respectively [16]. Thus CD71 positive cells might serve as targets for ALA-PDT. Rittenhouse-Diakun et al. showed that activated lymphocytes expressing high levels of surface CD71 transferrin receptors after activation with mitogens (phytohemagglutinin A, concanavalin A and pokeweed prior to incubation with ALA) generated more PpIX than those with low CD71 expression. They also showed that malignant CD71 + lymphocytes from a patient with Sezary's syndrome had higher levels of PpIX when compared to normal lymphocytes. After ALA-PDT activated lymphocytes and Sezary cells showed lower survival fractions compared to normal, unstimulated CD71 + cells [8]. Taken together, these findings suggest a possible mechanism for selective ALA photosensitization and the efficiency of ALA-PDT in CTCL.

15.5 Conclusions

In conclusion, the potential efficacy of topical ALA-PDT in CTCL has been shown in a limited number of patients. Further studies are required in larger numbers of CTCL patients to determine the optimal ALA and light dose combination, frequency of treatment, overall response rate, and duration of disease remission. For instance, potential advantages of topical ALA-PDT compared to PUVA photochemotherapy in CTCL might be: (1) the possibility to treat effectively thick lesions of up to 1 cm in thickness with the former modality, and (2) that in contrast to PUVA ALA-PDT does not seem to be mutagenic. Moreover, compared to other therapeutic modalities ALA-PDT is not invasive, can be used to treat multiple lesions in short treatment sessions, produces excellent cosmetic results and it is well accepted by patients [20].

References

1. R. Willemze, H. Kerl, W. Sterry, L. Berti, L. Cerroni, S. Chimenti, J.L. Diaz-Perez, M.L. Geerts, M. Goos, R. Knobler, E. Ralfkiaer, M. Santucci, J. Smith, W.A. van Vloten, C.J.L.M. Meijer (1997). EORTC classification for primary cutaneous lymphomas: a proposal from the cutaneous lymphoma study group of the European Organization for Research and Treatment of Cancer. *Blood*, **90**, 354–371.
2. M.A. Weinstock, J.W. Horm (1997). Mycosis fungoides in the United States: increasing incidence and descriptive epidemiology. *JAMA.*, **260**, 42–46.
3. P.A. Bunn, S.J. Hoffmann, D. Norris (1994). Systemic therapy of cutaneous T-cell lymphoma (mycosis fungoides and the Sezary syndrome). *Annals of Internal Medicine*, **121**, 592–602.
4. F.J. Kaye, P.A. Bunn, S.M. Steinberg (1989). A randomized trial comparing combination electron-beam radiation and chemotherapy with topical therapy in the initial treatment of mycosis fungoides. *N. Engl. J. Med.*, **321**, 1784–1790.
5. V.B. Reddy, D. Ramsay, J.A. Garola, H. Kamino (1996). Atypical cutaneous changes after topical treatment with nitrogen mustard in patients with mycosis fungoides. *Am. J. Dermatopathol.*, **18**, 19–23.
6. D. Grebenova, H. Cathamlova, J. Bartosova, J. Marinov, H. Klamova, O. Fuchs, Z. Hrkal (1998). Selective destruction of leukaemic cells by photo-activation of 5-aminolevulinic acid-induced protoporphyrin-IX. *J. Photochem. Photobiol. B*, **47**, 74–81.
7. Z. Malik, B. Ehrenberg, A. Faraggi (1989). Inactivation of erythrocytic, lymphocytic, and myelocytic leukemic cells by photoexcitation of endogenous porphyrins, *J. Photochem. Photobiol. B*, **4**, 195–205.
8. K. Rittenhouse-Diakun, H. van Leengoed, J. Morgan, E. Hryhorenko, G. Paskiewicz, J.E. Whitaker, A.R. Oseroff (1995). The role of transferrin receptor (CD71) in photodynamic

therapy of activated and malignant lymphocytes using the heme precursor δ-aminolevulinic acid (ALA). *Photochem. Photobiol.*, **61**, 523–528.

9. K. Svanberg, T. Andersson, D. Killander, I. Wang, U. Stenram, S. Andersson-Engels, R. Berg, J. Johansson, S. Svanberg (1994). Photodynamic therapy of non-melanoma malignant tumours of the skin using topical δ-amino levulinic acid sensitization and laser irradiation. *Br. J. Dermatol.*, **130**, 743–751.

10. C.M. Conti, S. Shanler, C. Sommer, H.L.L.M. van Leengoed, S. Dozier, J.E Whitaker, M. Barcos, H.L. Stoll, A.R. Oseroff (1995). Fractionated dose PDT with topical δ-aminolevulinic acid is effective for patch, plaque and tumor stage cutaneous T-cell lymphomas. *J. Invest. Dermatol.*, **104**, 575 (abstract).

11. D. Babich, J. Whitaker, C. Conti, D. Blaird-Wagner, H.L. Stoll, S. Dozier, A.R. Oseroff (1996). Treatment of all stages of cutaneous T-cell lymphomas with fractionated photodynamic therapy using topical δ-aminolevulinic acid (ALA-PDT). *J. Invest. Dermatol.*, **106**, 840 (abstract).

12. P. Wolf, R. Fink-Puches, L. Cerroni, H. Kerl (1994). Photodynamic therapy for mycosis fungoides after topical photosensitization with 5-aminolevulinic acid. *J. Am. Acad. Dermatol.*, **31**, 678–680.

13. P. Wolf, R. Fink-Puches, L. Cerroni, H. Kerl (1994). Aminolevulinic acid in photodynamic therapy of cutaneous T-cell lymphoma. *J. Invest. Dermatol.*, **102**, 602 (abstract).

14. A. Orenstein, J. Haik, J. Tamir, E. Winkler, H. Trau, Z. Malik, G. Kostenich (2000). Photodynamic therapy of cutaneous lymphoma using 5-aminolevulinic acid topical application. *Dermatol. Surg.*, **26**, 765–769.

15. D. Eich, H.T. Eich, H.-G. Otte, V. Ghilescu, C.E. Orfanos (1999). Photodynamische Therapie kutaner T-Zell-Lymphome in besonderer Lokalisation. 2 Falldarstellungen. *Hautarzt*, **50**, 109–114.

16. A.R. Oseroff, E. Hryhorenko, H. van Leengoed, J. Morgan, K. Rittenhose-Diakun (1995). Intracellular iron levels regulate ALA-induced protoporphyrin IX accumulation in activated and malignant lymphocytes: implications for selective therapies. *J. Invest. Dermatol.*, **104**, 575 (abstract).

17. Z. Malik, H. Lugavi (1987). Selective destruction of erythroleukemic cells by photo-activation of endogenous porphyrins. *Br. J. Cancer*, **56**, 389–395.

18. S. Iinuma, S.S. Farshi, B. Ortel, T. Hasan (1994). A mechanistic study of cellular photodestruction with 5-aminolevulinic acid-induced porphyrin. *Br. J. Cancer*, **70**, 21–28.

19. U. Testa, E. Pelosi, C. Peschle (1993). The transferrin receptor. *Critical Reviews in Oncogenesis*, **4**, 241–276.

20. P. Wolf, H. Kerl (1995). Photodynamic therapy with 5-aminolevulinic acid: a promising concept for the treatment of cutaneous tumors. *Dermatology*, **190**, 183–185.

Photodynamic Therapy and Fluorescence Diagnosis in Dermatology
P.-G. Calzavara-Pinton, R.-M. Szeimies and B. Ortel, editors.

Chapter 16

Topical photodynamic therapy for psoriasis

Wolf-Henning Boehncke

Table of contents

Abstract

Photochemotherapy using oral psoralen and ultraviolet A radiation (PUVA) is a highly effective treatment for severe psoriasis but has carcinogenic potential. Photodynamic therapy (PDT) combining application of a photosensitizing porphyrin derivative and subsequent irradiation with red light may represent an alternative photochemotherapeutic modality with potentially a lower carcinogenic risk.

In an attempt to limit potential side effects the concept of topical PDT has been persued for treating this inflammatory disease. In this context, the parameters determining susceptibility and selectivity of the targeted cell types as well as the biological effects induced in these cells could be defined. Following this pioneering work several photosensitizers were shown to exhibit good clinical efficacy in the topical PDT of psoriasis.

Since the general feasibility of this approach is now well documented future studies will have to investigate the effectiveness of this modality in comparison to established regimen. Furthermore, the concept of PDT as antimicrobial regimen will have to be evaluated in the light of psoriasis being a disease triggered by bacterial superantigens.

16.1 The rationale for topical PDT in psoriasis

Photochemotherapies combining application of photosensitizing agents and subsequent irradiation with light of activating wavelengths are considered valuable treatment modalities in several clinical fields. In dermatology, oral administration of 8-methoxypsoralen and irradiation with ultraviolet A radiation (PUVA) is a well established regimen for the treatment of a variety of dermatoses, most importantly moderate to severe psoriasis [1]. An increased risk of nonmelanoma skin cancer [2,3] as well as malignant melanoma [4] has been reported in patients receiving this form of treatment. Moreover, epithelial malignancies in other organs also appear more frequently than expected in this population [5]. Since psoriasis is essentially a benign inflammatory dermatosis evidence for an increased risk of malignant melanoma prompted a dialogue as to whether PUVA should be abandoned for this indication [6]. A strategy that reduces the carcinogenic risk of PUVA is to abolish systemic application of the photosensitizer and administer it topically instead, e.g. in the form of bath PUVA [7] therapy.

From a theoretical point of view, PDT may well be a suitable alternative photochemotherapy for the following reasons:

- Topical application of the photosensitizer: PDT has already been widely used as treatment modality for skin tumours. In these clinical settings the respective photosensitizers have preferentially been administered topically (see Chaps 13–16). This mode of application allows prevention of long lasting photosensitization as well as systemic immunosuppressive effects (see Section 16.2 of this chap.).
- Biological effects: PDT induces a wide variety of effects as described in detail in Chaps. 3–5. Many of them are suitable to alter the functional status of the skin immune system and thus are potentially relevant for treating psoriasis.
- Tissue penetration: Due to the longer wave length of red light most frequently used for PDT in comparison to ultraviolet light applied in conventional phototherapies and

photochemotherapies the former exhibits a better tissue penetration. Currently, the papillary dermis is looked at as being the hot spot of the psoriatic inflammatory reaction, and exocytic lymphocytes might be the most relevant effector cell population. But given the profound epidermal thickening and the pronounced dermal inflammatory infiltrate it might still be advantageous to reach the complete dermal compartment with the photochemotherapy of choice.

• Cancerogenicity: At the molecular level, DNA modifications by psoralens activated with UV-A light are well documented [8,9]. Similarly, DNA strand breaks have been demonstrated after porphyrin-based PDT [10,11]. However, DNA is not the main site of damage, and significant transformation and mutagenesis have not been demonstrated in mammalian culture systems exposed to porphyrins and light [12]. Still, occurence of a malignant melanoma at a site previously treated with PDT using 5-aminolevulinic acid has been reported [13]. But large epidemiologic studies to cover this point are still lacking.

16.2 The effects of PDT relevant to the treatment of psoriasis

Psoriasis currently is considered to be a T-cell mediated immune disease possibly triggered by superantigens. Consequently, PDT-induced effects on the function of the skin immune system are likely to be of relevance with regard to achieving therapeutic success. The T-cells themselves along with professional and non-professional antigen-presenting cells (e.g. keratinocytes) therefore may be the most relevant targets. Since recent data suggest a role for angiogenesis in the pathogenesis of psoriasis, endothelial cells should also be included as candidate target cells. Contrary to treating malignancies, PDT does not necessarily have to achieve necrosis or apoptosis but might only need to induce functional alterations of the targeted cell types in order to yield good clinical responses in psoriasis. Previous chapters give a full description of the PDT-mediated effects at the molecular and cellular level (see Chaps. 3–5). This section briefly summarizes those effects presumably most meanigful in the treatment of psoriasis.

From the experiences with treating tumours it is clear that PDT utilizing porphyrin photosensitizers can exhibit direct cytotoxic effects. These are at least in part related to the in situ generation of reactive oxygen species, which could induce local changes of the redox state of the cell [14]. Consequently, cells can either undergo necrosis or apoptosis [15]. The latter effect is believed to be directly related to the localization of porphyrin photosensitizers to mitochondria representing a primary site for the regulation of apoptotic stimuli [16].

Sub-lethal PDT is known to alter the expression pattern of surface molecules on immune cells. Human blood mononuclear cells exhibit a markedly reduced level of HLA class II molecules. This effect might be due to either induced internalization or altered recycling kinetics and results in loss of stimulatory capacity to T-cells [17]. Moreover, downregulation of the expression of adhesion molecules as well as co-stimulatory molecules also resulting in reduced T-cell stimulatory capacity has been described [18]. Besides surface molecule expression, cytokine secretion can be influenced by PDT. Treating peripheral mononuclear cells from psoriatic patients secretion of pro-inflammatory cytokines could be inhibited by PDT in a fashion comparable to PUVA in vitro [19]. An in vivo immunosuppressive effect of PDT was demonstrated in animal models with regard to delayed-type hypersensitivity [20] as well as autoimmunity [21].

The effects triggered by PDT depend largely on the uptake of the respective photosensitizer. Cell type and state of activation and differentiation have been defined as the key parameters determining the susceptibility of a target to PDT. Resident cells of the skin such as keratinocytes and fibroblasts are much less susceptible to PDT when compared to lymphocytes [22]. Susceptibility of cells increases when they are stimulated [23–25]. With regard to light sources that can potentially be used for PDT it was important to establish that polychromatic light was at least as effective at inducing photodynamic reaction as monochromatic laser light [22].

Another interesting feature of PDT is its ability to kill bacteria [26], an effect that can be applied to the treatment of acne [27]. Since bacterial superantigens are among the potential trigger factors for psoriasis [28], this strategy might be yet another promising track to follow.Although systemic PDT might better meet the needs of a treatment modality to approach an immune disease such as psoriasis two aspects have promoted the concept of topical PDT in general: First, many photosensitizers used in the past caused prolonged and substantial photosensitivity; secondly, immunosuppression as observed in mice treated with systemic PDT is regarded as an inacceptable threat in the eyes of some investigators. In the following sections the tools and protocols for topical PDT of psoriasis developed to avoid these side effects will be reviewed.

16.3 The tools for topical PDT in psoriasis

16.3.1 Photosensitizers

The photosensitizers most frequently used for PDT of psoriasis are porphyrins. These molecules absorb light around the Soret band (400 nm), but they also exhibit four additional absorption bands between 500 and 650 nm thus increasing the tissue penetration from 1.5 to 6 mm.

Historically, hematoporphyrin derivative (HPD) being a complex mixture of different porphyrins has been used systemically. This product caused long lasting and profound photosensitivity and therefore was impractical for application in the treatment of psoriasis.

An endogenous photosensitizer formed during hem biosynthesis is protoporphyrin IX (PpIX). Intracellular accumulation can be obtained by administration of aminolevulinic acid (ALA) which can be metabolized to PpIX by all nuceated cell types. ALA derived PpIX is detectable in the epidermis about 4 h following either systemic [29] or topical application of ALA [30]. Due to its hydrophilic properties, ALA poorly penetrates intact epidermis. Therefore, ALA esters are being developed which might overcome this limitation. Using these novel photosensitizers it has already been demonstrated that similar levels of PpIX accumulation can be reached with 30 to 150-fold less ALA when it is esterified compared to non-esterified ALA [31]. Metabolism of ALA also yields products with absorption peaks beyond 650 nm so that even longer and deeper penetrating wave lengths can be used for PDT [32].

Currently, second generation photosensitizers are explored with absorption bands in the range of 640 to 700 nm. These photosensitizers comprise reduced, hydrophilic porphyrins called chlorins (e.g. benzoporphyrin derivative) and synthetic porphyrins called pthalocyanines.

16.3.2 Light sources

To combine the goal of maximal tissue penetration and absorption of light wavelengths of around 630 nm have primarily been employed for PDT utilising either HPD or ALA. Consequently, gold vapour lasers, copper vapour pumped dye lasers, Nd:YAG lasers, or argon lasers were suitable systems. To reduce the costs and increase convenience diode laser systems are currently being developed as alternative to the lasers mentioned.

From a theoretical point of view laser light is not needed to induce photodynamic reactions and does not have any intrinsic advantage over polychromatic light other than fibre guided application. This could be necessary if the treatment would have to be performed endoscopically. The skin organ, however, can directly be approached. Moreover, if large surface areas are to be treated, lasers are insufficient since the laser beam cannot be widened appropriately. Finally, broad band emissions may activate photoproducts in addition to PpIX [33]. Equivalent or superior effects of polychromatic light sources in comparison to laser systems have indeed been demonstrated [22,34].

Taking advantage of the fact that porphyrins exhibit maximal light absorption around the Soret band several reports document effectiveness of PDT using standard UV irradiation units [35].

16.4 The protocols for topical PDT of psoriasis

Results of the first PDT in psoriasis were published in 1937 when Silver reported marked improvement in seven patients treated with systemic application of HPD and UV light [36]. Almost half a decade later this observation was confirmed by Berns et al. who, while treating a patient for a vulvar neoplasia, also induced clearance of adjacent psoriasis [37].

But it was only in 1989 when Meffert and co-workers first described successful treatment of psoriasis by topical PDT (Table 1). In a placebo-controlled pilot study 12 patients suffering from chronic plaque-stage psoriasis were treated with an ointment containing 1 mg/ml HPD [38]. One hour later plaques were irradiated using a halogen lamp emitting light of 375 to 700 nm (Tungsram, CSSR). The treatment was performed with 3.3 mW/cm^2 and 2, 4 and 6 J/cm^2 at three consecutive days. Using a 4-point scale [39] significant improvement was noted as early as 24h after the first treatment. In a subsequent study by the same group [40] 17 patients with palmoplantar psoriasis were treated using the same light source as in the initial study [38] and HPD at two different concentrations (0.94 and 3.0 mg/ml, respectively) (Table 1). Up to 21 repetitive treatments were performed at 14 mW/cm^2 and 17 J/cm^2. Again, significant improvement was documented. Unfortunately, both reports do not contain detailed informations of individual patients and instead present cumulative data. The latter report does not define the protocol for the actual treatment procedure. Thus, it can only be assumed that the

Table 1. Protocols for topical PDT for psoriasis

Reference	Patients	Diagnosis	Photosensitizer	Light source	Results
[38]	12	plaque-stage psoriasis	HPD (1 mg/ml)	halogen lamp 375–700 nm 3.3 mW/cm^2 2–6 J/cm^2	significant improvement after three treatments at consecutive days
[40]	17	palmoplantar psoriasis	HPD (0.94/3.0 mg/ml)	halogen lamp 375–700 nm 14 mW/cm^2 17 J/cm^2	significant improvement during up to 21 repetitive treatments
[42]	3	plaque-stage psoriasis	10% ALA	Waldmann PDT1200 600–700 nm 70 mW/cm^2 25 J/cm^2	improvement similar or better than dithranol after three treatments at weekly intervals
[44]	22	plaque-stage psoriasis	20% ALA	modified slight projector 400–650 nm 300 mW/cm^2 2–16 J/cm^2	clearing of lesions in 7/22 after a single treatment
[45]	10	plaque-stage psoriasis	20% ALA	modified slight projector 400–650 nm 300 mW/cm^2 2–16 J/cm^2	clearing of lesions in 7/22 during up to 12 treatments every other day

authors have chosen procedures identical to those described in the former report. In one case the therapy had to be stopped prior to completion of the protocol due to "substantial erythema" (not further described). There is no notion of other adverse effects of the treatment.

One year later Kennedy et al. were the first to apply topical PDT using ALA as photosensitizer to dermatology by successfully treating cutaneous tumours this way [41]. It was later confirmed that topically applied ALA also results in accumulation of ALA-derived PpIX in psoriatic plaques [30]. These findings provide the basis for a regimen used by many different groups for the topical PDT of plaque-stage psoriasis: topical application of ALA and subsequent irradiation with visible light. In order to allow treatment of lager surface areas non-coherent light were used instead of laser systems.

Proof of principle for this approach was reported in 1994 [42] (Table 1): Selected plaques of three patients suffering from chronic plaque-stage psoriasis were treated with an ointment containing 10% ALA. 5 h later, the plaques were irradiated using 600–700 nm light emitted from a 1200 W bulb (Waldmann 1200, Waldmann, Germany) at 70 mW/cm^2 and 25 J/cm^2. Three treatments at weekly intervals were performed. All other lesions were treated with dithranol. In all patients, improvement of the PDT treated sites was observed which was at least as substantial as in the case of dithranol treatment. Although the epidermal temperature did not exceed 31°C during irradiation as monitored by a thermo camera, all patients noted moderate burning sensations. No other adverse effects were noted over 6 months of follow-up.

Besides burning sensations during irradiation one has also to be aware of the possibility to actually trigger a Köbner reaction as documented first by a case report in 1996 [43]: A patient was treated with topical ALA-PDT using a cream containing 20% ALA and subsequent irradiation with a 250W slide projector (50 J/cm^2) for actinic keratoses and squamous cell carcinomas. Two days after this therapy has been performed she developed psoriatic lesions at the treatment sites.

In a very carefully designed study Collins et al. investigated the clinical response of 22 patients with chronic plaque-stage psoriasis to topical ALA-PDT [44] (Table 1). An emulsion containing 20% ALA was applied to 2×2 cm squares within areas of plaque-stage psoriasis. 4h later, irradiation was performed using 400–650 nm light emitted by a modified slide projector at doses of 2 to 16 J/cm^2 (300 mW/cm^2). In every patient, the clinical effectiveness of several light doses was evaluated using a scoring system for scaling, erythema, and induration. Seven of the 22 patients showed clearing of psoriasis at some treatment sites. Side effects comprised erosions, hyperpigmentation, and stinging or burning sensations for up to 2 days.

Following up on their initial observations the same group investigated the effectiveness of multiple treatments with topical ALA [45] (Table 1). 10 patients with chronic plaque-stage psoriasis were included and received up to three treatments of ALA-PDT per week, with a maximum of 12 treatments. An oil-in-water emulsion containing 20% ALA was applied to the plaques. Four hours later, irradiation was performed with broad-band visible light at 15 mW/cm^2 and 8 J/cm^2. Eight patients showed clinical responses, but these responses were not predictable. Since discomfort is also associated with topical ALA-PDT, the authors conclude that this approach is unsuitable for the treatment of psoriasis.

16.5 The perspectives of PDT in psoriasis

A variable degree of discomfort ranging from mild and transient sensations to severe and long-lasting burning sensations is a constant feature of topical PDT. Moreover, it is still not possible to predict the clinical outcome of this approach, although proof of priciple has clearly been established. On the other hand, the threat of cancerogenicity of PUVA remains and calls for the development of alternative photochemotherapies. From a theoretical point of view the concept of PDT is still very appealing since many biological PDT-mediated effects may be advantageous for the treatment of psoriasis. To date, investigators have primarily been focussing on the effects on proliferation as well as the immunomodulatory capacity of this approach. In the future, other possible mechanisms may also be worthwhile studying. In this context the antimicrobial effects may be of interest, particularly since bacterial superantigens are considered potential triggers for psoriasis.

Another key issue is whether one should persue the concept of *topical* PDT in psoriasis. Generally, photochemotherapies are considered suitable for treating dermatoses affecting a large portion of the body surface. With second generation photosensitizers becoming available for clinical studies systemic PDT may encounter a renaissance now that phototoxicity is no longer a threat. Indeed, promising observations are already being published [46]. Thus, topical PDT may conquer a place in the treatment of localized forms of psoriasis such as palmoplantar psoriasis, but future studies will have to answer the question of whether systemic PDT may have the potential to be developed to a true alternative for PUVA.

References

1. J.A. Parrish, T.B. Fitzpatrick, L. Tanenbaum, M.A. Pathak (1974). Photochemotherapy of psoriasis with oral methoxsalen and longwave ultraviolet light. *N. Engl. J. Med.*, **291**, 1207–1211.
2. R.S. Stern, L.A. Thibodeau, R.A. Kleinerman et al. (1979). Risk of cutaneous carcinoma in patients treated with oral methoxsalen photochemotherapy for psoriasis. *N. Engl. J. Med.*, **300**, 809–813.
3. R.S. Stern, N. Laird, J. Melski, J.A. Parrish, T.B. Fitzpatrick, H.L. Bleich (1984). Cutaneous squamous-cell carcinoma in patients treated with PUVA. *N. Engl. J. Med.*, **310**, 1156–1161.
4. R.S. Stern, K.T. Nichols, L.H. Väkevä (1997). Malignant melanoma in patients treated for psoriasis with methoxsalen (psoralen) and ultraviolet A radiation (PUVA). *N. Engl. J. Med.*, **336**, 1041–1045.
5. B. Lindelöf, B. Sigurgeirsson, E. Tegner, O. Larkö, A. Johannesson, B. Berne, O.B. Christensen, T. Andersson, M. Törngren, L. Molin, E. Nylander-Lundqvist, L. Emtestam (1991). PUVA and cancer: a large-scale epidemiological study. *Lancet*, **338**, 91–93.
6. K. Wolff (1997). Should PUVA be abandoned? *N. Engl. J. Med.*, **336**, 1090–1091.
7. B. Lindelöf, B. Sigurgeirsson, E. Tegner, O. Larkö, B. Berne (1992). Comparison of the carcinogenic potential of trioxsalen bath PUVA and oral methoxsalen PUVA. *Arch. Dermatol.*, **128**, 1341–1344.
8. P-S. Song, K.J. Tapley (1979). Photochemistry and photobiology of psoralens. *Phtochem. Photobiol.*, **29**, 1179–1197.

9. H. Hönigsmann, T.B. Fitzpatrick, M.A. Pathak, K. Wolff (1993). Oral phtochemotherapy with psoralens and UVA (PUVA): principles and practise. In: T.B. Fitzpatrick, A.Z. Eisen, K. Wolff, I.M. Freedberg, K.F. Austen (Eds), *Dermatology in General Medicine* (pp. 1728–1754). McGraw Hill, New York.

10. C.J. Gomer, N. Rucker, A. Banerjee, W.F. Benedict (1983). Comparison of mutagenicity and induction of sister chromatid exchange in Chinese hamster cells exposed to haematoporphyrin derivate, ionizing radiation or ultraviolet radiation. *Cancer Res.*, **43**, 2622–2627.

11. E. Ben-Hur, T. Fujihara, F. Suzuki, M.M. Elkind (1987). Genetic toxicology of the photosensitization of Chinese hamster cells by phthalocyanines. *Photochem. Photobiol.*, **45**, 227–230.

12. C.J. Gomer, N. Rucker, A.L. Murphee (1988). Transformation and mutagenic potential of porphyrin photodynamic therapy in mammalian cells. *Int. J. Radiat. Biol.*, **53**, 651–659.

13. P. Wolf, R. Fink-Puches, A. Reimann-Weber, H. Kerl (1997). Development of malignant melanoma after repeated topical photodynamic therapy with 5-aminolevulinic acid at the exposure site. *Dermatology*, **194**, 53–54.

14. K. Weishaupt, C.J. Gomer, T. Dougherty (1976). Identification of singlet oxygen as the cytotoxic agent in photoinactivation of a murine tumor. *Cancer Res.*, **36**, 2326–2329.

15. D.J. Granville, C.M. Carthy, H. Jiang, G.C. Shore, B.M. McManus, D.W.C. Hunt (1998). Rapid cytochrome c release, activation of caspases 3, 6, 7 and 8 followed by Bap31 cleavage in HeLa cells treated with photodynamic therapy. FEBS Lett., **437**, 5–10.

16. T. Dougherty, C.J. Gomer, B.W. Henderson, G. Jori, D. Kessel, M. Korbelik, J. Moan, Q. Peng (1998). Photodynamic therapy. *J. Natl. Cancer. Inst.*, **90**, 889–905.

17. S. Gruner, H. Volk, F. Noack, H. Meffert, R. von Baehr (1986). Inhibition of HLA-DR antigen expression and of the mixed leukocyte reaction by photochemical treatment. *Tissue Antigens*, **27**, 147–154.

18. D.E. King, H. Jiang, G.O. Simkin, M.O.K. Obochi, J.G. Levy, D.W.C. Hunt (in press). Photodynamic alteration of the surface receptor expression pattern of murine splenic dendritic cells. *Scand. J. Immunol.*

19. W-H. Boehncke, K. König, R. Kaufmann, W. Scheffold, O. Prümmer, W. Sterry (1994). Photodynamic therapy in psoriasis: Suppression of cytokine production in vitro and recording of fluorescence modification during treatment in vivo. *Arch. Dermatol. Res.*, **286**, 300–303.

20. C.A. Elmets, K.D. Bowen (1986). Immunological suppression in mice treated with hematoporphyrin derivative photoradiation. *Cancer Res.*, **46**, 1608–1611.

21. S Leong, A.H. Chan, J.G. Levy, D.W.C. Hunt (1996) Transcutaneous photodynamic therapy alters the development of an adoptively transferred form of murine experimental autoimmune encephalomyelitis. *Photochem. Photobiol.*, **64**, 751–757.

22. W-H. Boehncke, A. Rück, J. Naumann, W. Sterry, R. Kaufmann (1996). Comparison of sensitivity towards photodynamic therapy of cutaneous resident and infiltrating cell types in vitro. *Lasers Surg. Med.*, **19**, 451–457.

23. M.A. Breider, X. Lu, M. Panjehpour, D.L. Frazier (1993). Cytokine modulation of endothelial cell sensitity to photodynamic therapy. *Lasers Surg. Med.*, **13**, 305–311.

24. E. Schick, R. Kaufmann, Arück, A. Hainzl, W-H. Boehncke (1995). Influence of activation and differentiation of cells on the effectiveness of photodynamic therapy. *Acta. Derm. Venerol.*, **75**, 276–279.

25. E. Hryhorenko, K. Rittenhouse-Diakun, N.S. Harvey, J. Morgan, C.C. Stewart, A.R. Oseroff (1998). Characterization of endogenous protoporphyrin IX induced by δ–amino-levulinic acid in resting and activated peripheral blood lymphocytes by four-coloor flow cytometry. *Photochem. Photobiol.*, **67**, 565–572.

26. N.S. Soukos, L.A. Ximenez-Fyvie, M.R. Hamblin et al. (1998). Target anti-microbial chemotherapy. *Antimicrob. Agents Chemother.*, **42**, 2595–2601.

27. W. Hongcharu, C.R. Taylor, Y. Chang, D. Aghassi, K. Suthamjariya, R.R. Anderson (2000). Topical ALA-photodynamic therapy for the treatment of acne vulgaris. *J. Invest. Dermatol.*, **115**, 183–192.

28. W-H. Boehncke, D. Dressel, T.M. Zollner, R. Kaufmann (1996). Pulling the trigger on psoriasis. *Nature*, **379**, 777.

29. J. Webber, D. Kessel, D. Fromm (1997). Side effects and photosensitization of human tissues after aminolevulinic acid. *J. Surg. Res.*, **68**, 31–37.

30. M.R. Stringer, P. Collins, D.J. Robinson, G.I. Stables, R.A. Sheehan-Dare (1996). The accumulation of protoporphyrin IX in plaque psoriasis after topical application of 5-aminolevulinic acid indicates a potential for superficial photodynamic therapy. *J. Invest. Dermatol.*, **107**, 76–81.

31. Q. Peng, T. Warloe, K. Berg et al. (1997). 5-aminolevulinic acid-based photodynamic therapy – clinical research and future challenges. *Cancer*, **79**, 2282–2308.

32. J-M. Gaullier, K. Berg, Q. Peng, H. Anholt, P.K. Selbo, L-W. Ma, J. Moan (1997). Use of 5-aminolevulinic acid esters to improve photodynamic therapy on cells in culture. *Cancer Res.*, **57**, 1481–1486.

33. R.M. Szeimies, R. Hein, W. Baumler et al. (1994). A possible new incoherent lamp for photodynamic treatment of superficial skin lesions. *Acta Derm. Venerol.*, **74**, 117–119.

34. M.L. de Jode, J.A. Megilligan, M.G. Dilkes (1997). A comparison of novel light sources for photodynamic therapy. *Lasers Med. Sci.*, **12**, 260–268.

35. L. Emtestam, L. Berglund, B. Angelin (1989). Tin-protoporphyrin and long wavelength ultraviolet light treatment of psoriasis. *Lancet*, **1**, 1231–1233.

36. H. Silver (1937). Psoriasis vulgaris treated with hematoporphyrin. *Arch. Derm. Syphilol.*, 1118–1119.

37. M.W. Berns, M. Rettenmaier, J. McCullough et al. (1984). Response of psoriasis to red laser light (630 nm) following systemic injection of hematoporphyrin derivative. *Lasers Surg. Med.*, **4**, 73–77.

38. H. Meffert, H. Pres, W. Diezel, N. Sönnichsen (1989). Antipsoriatische und phototoxische Wirksamkeit von Hämatoporphyrin-derivat nach topischer Applikation und Bestrahlung mit sichtbarem Licht. *Dermatol. Monatsschr.*, **175**, 28–34.

39. G. Weinstein, J. McCullogh, J. Eaglestein, A. Golub, R.C. Cornell, R.B. Stougthon, W. Clendenning, H. Zackheim, H. Maibach, K.R. Kulp, L. King, H.P. Baden, J.S. Taylor, D.D. Deneau (1981). A clinical screening program for topical chemotherapeutic drugs in psoriasis. *Arch. Dermatol.*, **117** 388–393.

40. H. Pres, H. Meffert, N. Sönnichsen (1989). Photodynamische Therapie der Psoriasis palmaris et plantaris mit topisch appliziertem Hämatoporphyrin-Derivat und sichtbarem Licht. *Dermatol. Monatsschr.*, **175**, 745–750.

41. J.C. Kennedy, R Pottier, D.C. Pross (1990). Photodynamic therapy with endogenous protoporphyrin IX: basic principles and present clinical experience. *J. Photochem. Photobiol. B.*, **6**, 143–148.

42. W-H. Boehncke, W. Sterry, R. Kaufmann (1994). Treatment of psoriasis by topical photodynamic therapy with polychromatic light. *Lancet*, **343**, 801.

43. I-M. Stender, H.C. Wulf (1996). Köbner reaction induced by photodynamic therapy using delta-aminolevulinic acid. *Act. Derm. Venerol.*, **76**, 392–393.

44. P Collins, D.J. Robinson, M.R. Stringer, G.I. Stables, R.A. Sheehan-Dare (1997). The variable response of plaque psoriasis after a single treatment with topical 5-aminolevulinic acid photodynamic therapy. *Br. J. Dermatol.*, **137**, 743–749.

45. D.J. Robinson, P. Collins, M.R. Stringer, D.I. Vernon, G.I. Stables, S.B. Brown, R.A. Sheehan-Dare (1999). Improved response to plaque psoriasis after multiple treatments with topical 5-aminolevulinic acid photodynamic therapy. *Acta Derm. Venerol.*, **79**, 451–455.
46. W-H. Boehncke, T. Elshorst-Schmidt, R. Kaufmann (2000). Systemic photodynamic therapy improves psoriasis. *Arch. Dermatol.*, **136**, 271–272.

Photodynamic Therapy and Fluorescence Diagnosis in Dermatology
P.-G. Calzavara-Pinton, R.-M. Szeimies and B. Ortel, editors.

Chapter 17

Photodynamic therapy of pilosebaceous units

Christine C. Dierickx

Table of contents

Abstract

In dermatology, photodynamic therapy (PDT) has been used primarily for treatment of malignant skin tumors using various long-wavelength oxygen-dependent photo-sensitizers such as porphyrins, the porphyrin precursor 5-aminolevulinic acid (ALA), benzoporphyrin derivative (BPD), mono-L-aspartyl chlorin e6 (NPe6), tin ethyl etiopurpurin (SnET$_2$) and chloro-aluminum sulfonated phthalocyanine (CASPc).

It is clear that malignancy is not the only disease process in which PDT will be useful in dermatology. Indeed, "benign" skin lesions may become the most preferred uses for PDT. Hyperproliferative and hypervascular conditions, such as psoriasis and human papilloma virus infections, and benign vascular lesions, such as portwine stains and hemangiomas, can be selectively photosensitized. Selective uptake of certain photo-sentizers in sebaceous glands and hair follicles, implies that PDT might become a potential treatment modality for hair follicle and sebaceous gland disorders such as hypertrichosis, alopecia, and acne.

17.1 Photodynamic therapy and the hair follicle

17.1.1 Photodynamic therapy and alopecia

The use of PDT for alopecia was prompted by the frequent incidence of hypertrichosis in patients with porphyria cutanea tarda. The mechanism for hypertrichosis seen in sun-exposed skin of porphyria cutanea tarda is not known, but probably involves photodynamic action mediated by endogenous porphyrins. Thus, it may be that low-dose, chronic photodynamic action stimulates hair growth. Since PDT is known to cause immunosuppressive effects, it could be a good treatment alternative for therapy-resistant alopecia areata.

Hypertrichosis has been reported to occur following light exposure at the sites of intramuscular hematoporphyrin injections, used for the treatment of depression [1]. Monfrecola et al. have reported hair regrowth in 2 patients with patchy alopecia areata after topical hematoporphyrin and long wave UVA [2]. A halide lamp filtered to produce long-wave UVA was used and the patients were treated 3 times weekly. After 3 to 4 months of therapy, coarse hairs were present on the treatment sites but not the control sites.

However, photodynamic therapy with topical aminolevulinic acid (ALA) was reported to be ineffective in six patients with extensive alopecia areata [3]. Topical ALA lotion at 5%, 10% and 20% as well as the vehicle lotion alone were applied separately to different scalp areas, followed 3 h later by exposure to red light. On fluorescence microscopy, bright red fluorescence was present in the epidermis and sebaceous glands, but not in the inflammatory infiltrate surrounding the hair follicle. No significant hair growth was observed after 20 twice-weekly treatment sessions.

17.1.2 ALA-PDT and hypertrichosis

The biologically important target structures for inactivation of hair growth have not yet been well defined. For many years, the matrix area of the bulb located deep in the dermis, typically 3–7 mm, was considered to be the target. However, recent evidence indicates that follicular stem cells are located in the outer root sheath, in an area called the bulge, near the attachment of the arrector pili muscle, approximately 1.5 mm below the epidermis [4–6] (Fig. 1). Both targets could be important for permanent hair follicle destruction.

The mechanisms by which laser pulses induce selective damage to hair follicles, are based on the principle of selective photothermolysis [7]. Destruction of bulge and/or bulb depends on thermal diffusion from the pigmented hair shaft or matrix respectively. A prerequisite for successful treatment is that the presence of a pigmented hairshaft. Because melanin in the epidermis presents a competing site for absorption, this approach is mainly limited to subjects with dark hair and fair skin.

Photodynamic therapy may therefore be a useful alternative for hair removal. Selective PPIX synthesis in pilosebaceous units is a unique feature of ALA over other photosensitizers and topical application circumvents the photosensitivity that is induced by systemic agents. Subsequent light exposure could lead to selective destruction of hair follicles containing accumulated PPIX [8,9]. Since PPIX tend to localize in the follicular epithelium, photochemical destruction of all hair follicles, no matter what hair

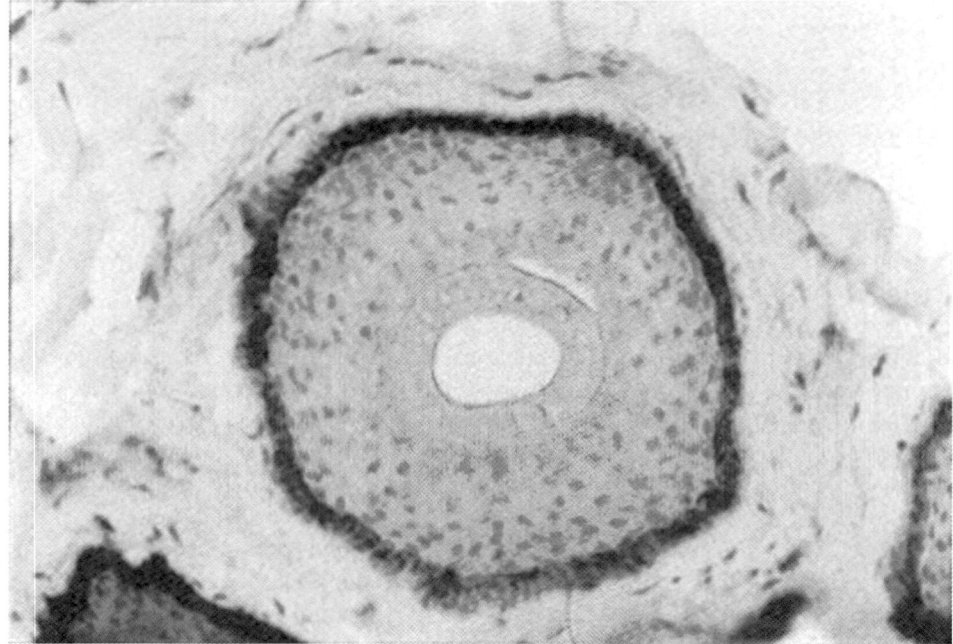

Figure 1. Immunostaining of the human hair stem cells with C8/144B monoclonal antibody. Note the presence of stem cells in the outer rim of the follicular epithelium. (Courtesy of D. Manstein and R. Rox Anderson, Wellman Laboratories of Photomedicine).

Table 1. Main differences between photothermal and photochemical destruction

	LASER	PDT
Chromophore	endogenous melanin	exogenous photosensitizer
Localisation chromomphore	hairshaft and/or hairmatrix	follicular epithelium
Distance target – chromophore	far	close
Damage mode	photothermal	photochemical
Hair color	Only pigmented hairs	All hair colors
Cause of side effect	epidermal melanin	photosensitizer in epidermis
Solution to avoid side effects	cooling	optimization of application conditions

color, could be potentially obtained. The main differences between the photothermal and photochemical destruction are shown in Table 1.

PDT-induced alopecia was reported in a pilot clinical study using ALA topically [10]. In eleven subjects, 10% and 20% topically applied ALA for 3 hours under occlusion, was exposed to 630 nm laser light at doses of 100 and 200 J/cm^2. Treatment was done three hours after ALA was applied to wax epilated skin at all dose combination. Patients were followed up to 6 months following treatment. At the six month follow-up, the control sites and the sites treated at low dose (10% ALA at 100 and 200 J/cm^2 and 20% ALA at 100 J/cm^2) had greater than 90% regrowth. In contrast, the sites treated at the highest doses (20% ALA and 200 J/cm^2) showed a hair reduction of about 40% after a single treatment (Fig. 2). Pre and post treatment H&E and fluorescence biopsies were obtained. Prior to treatment, protoporphyrin IX fluorescence was seen throughout the epidermis and within the follicle (Fig. 3). After treatment, there was epidermal and follicular damage, while sparing adjacent dermis. Mild to moderate pain was experienced during the treatment and transient hyperpigmentation was seen in some patients.

This study established the efficacy of ALA-PDT for hair removal with a growth delay effect at the lower doses and a long term effect at the higher doses of drug and light. However selective PPIX accumulation within the hair follicles was desired to decrease side effects as a consequence of epidermal PPIX accumulation. A second non-invasive pilot study was done to find an application condition where only selective PPIX fluorescence was obtained in the hair follicles. (unpublished observations, Christine C. Dierickx, Suchai Spriprachya-Anunt, William Farinelli and R. Rox Anderson). Ten patients entered this study. Eighteen different application conditions were evaluated on each patient: the hairs were shaved, epilated or not manipulated prior to ALA application, three different concentrations of ALA were used (5%, 10% and 20%) in a vehicle with and without surfactant, and application was done with and without occlusion. Fluorescence imaging was done to monitor PPIX accumulation in epidermis and/or follicles. The fluorescence excitation source was a flashlamp, filtered between 410 and 580 nm. Fluorescence emission was measured at 636 nm. Digital images were taken with a charged coupled device (CCD) camera (Model 4915–2010, Cohu, Japan) and computer frame-capture board (Scanalytics, Inc., Vienna, VA), to record

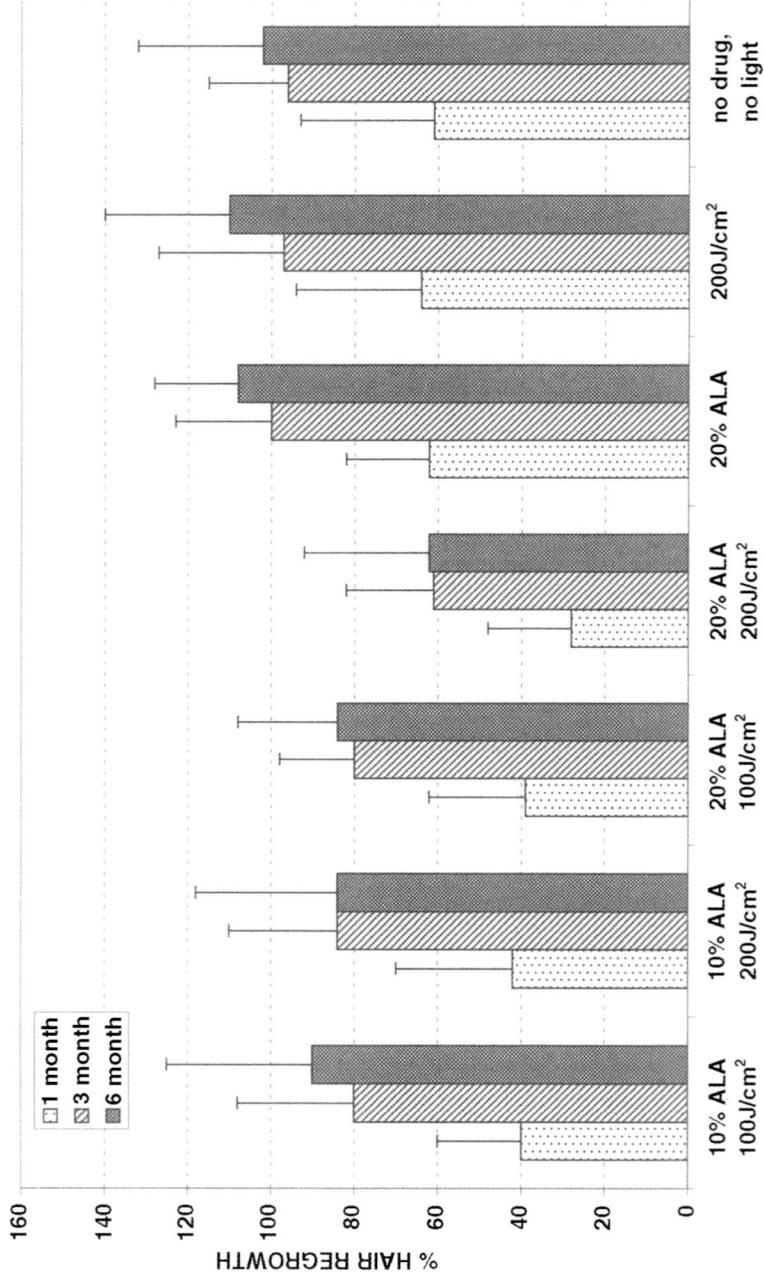

Figure 2. Hair regrowth counts. (Courtesy of M. Grossman and R. Rox Anderson, Wellman Laboratories of Photomedicine).

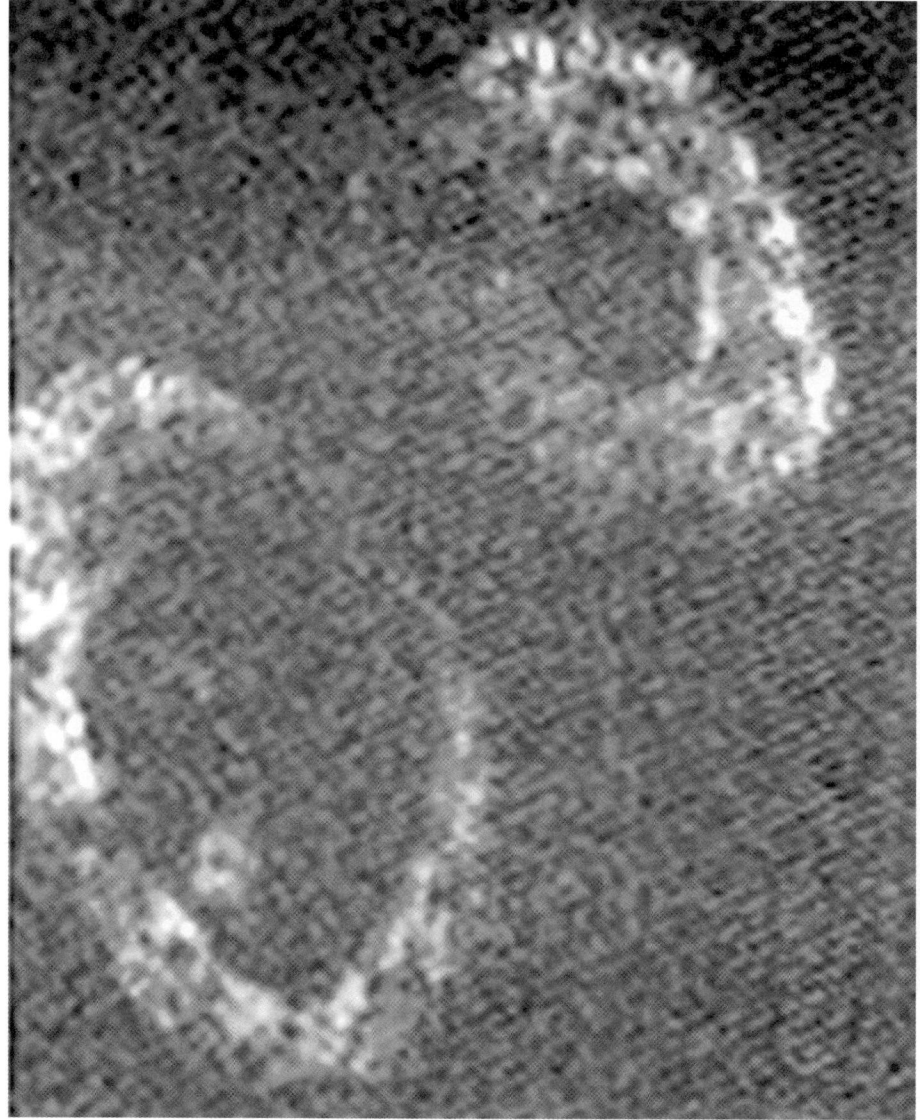

Figure 3. Fluorescence microscopy demonstrates *PpIX* production is mainly located in follicular epithelium of hair follicles (Courtesy of M. Grossman and R. Rox Anderson, Wellman Laboratories of Photomedicine).

fluorescence intensity in each test area. Analysis of the fluorescence intensity was done with image analyzing software (IPLAB,Scion Corporation, Frederick, MD) on a PowerMacintosh computer (model 7500/100). All sites were imaged each hour for the first six hours and at 24 hours. The most selective and most intense PPIX fluorescence signal in hair follicles was seen with 20% ALA in a vehicle with surfactant, after epilation of the hair and application without occlusion (Fig. 4).

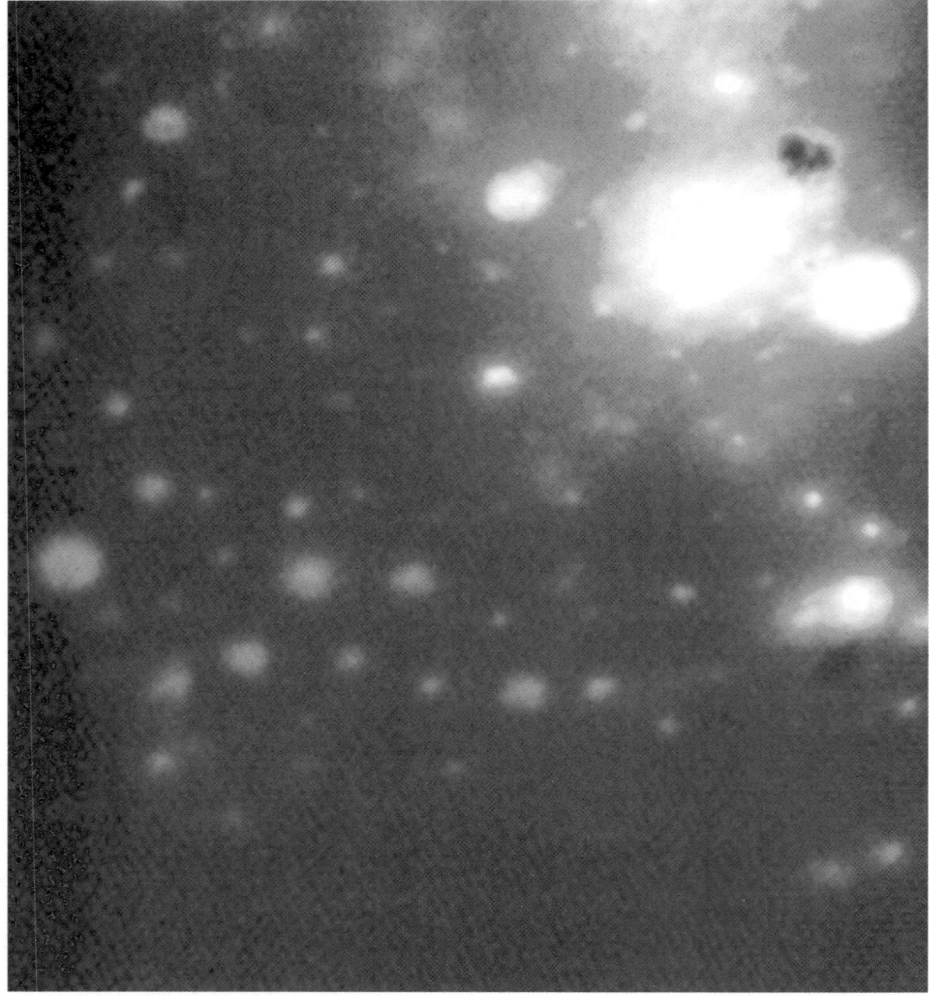

Figure 4. ALA-induced *PpIX* fluorescence is greater in follicles than surrounding tissue.

Preliminary reports from a recent study including 30 patients, examined the ability of Levulan® stick with either a proprietary non-laser light source compared to a laser to remove human hair. The non-laser light plus Levulan® did not result in more significant hair loss than placebo plus light. However, Levulan® plus laser light appears to prevent approximately 30% of the hair from regrowing with one treatment [unpublished results from DUSA pharmaceuticals].

17.2 Photodynamic therapy and the sebaceous gland

As in hair follicles, there is evidence for selective uptake of ALA in sebaceous glands [11]. After systemic administration of 5-aminolevulinic acid, selective phototoxic

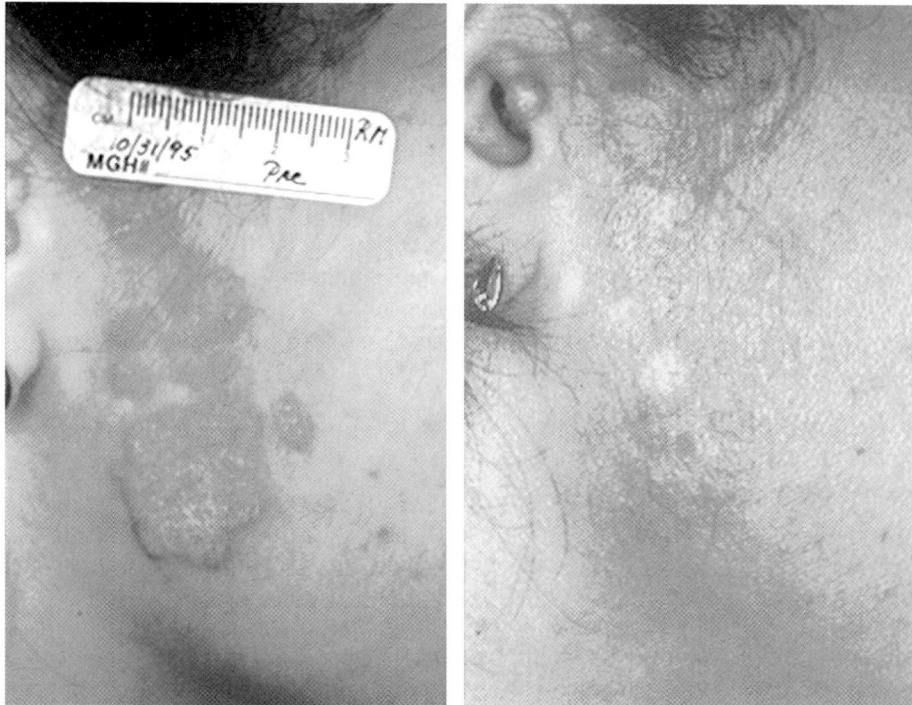

Figure 5. (a) Right pre-auricular area of the patient showing a large tumor lobule of nevus sebaceous 4 weks post the second treatment treatment. Note initial regressing of lesion. (b) Clinical resolution of the tumor 16 month after 13 treatment sessions.

damage occurred to the sebaceous glands and hair follicles of mice. Disorders of the sebaceous glands, like acne, sebaceous hyperplasia and nevus sebaceus, are therefore another potential indication for PDT using ALA.

17.2.1 ALA-PDT and nevus sebaceous

The use of photodynamic therapy (PDT) with topical δ-aminolevulinic acid (ALA) for an extensive nevus sebaceous has been reported [12]. Twenty percent ALA mixed in an oil-in-water emulsion, was applied to the tumor under occlusion for four hours. The area was exposed to 630 nm laser light from an argon pumped tunable dye laser and a total dose of 100 J/cm^2 was given, fractionated in 2 doses of 50 J/cm^2, at an irradiance of 50 mW/cm^2. Similar treatment sessions were repeated every 4–8 weeks with a total of 13 treatment sessions. With each treatment session, the lesion regressed with flattening and decrease in size. Two remaining tumor nodules did not completely resolve and were curetted immediately prior to ALA application at the two last treatment sessions. The final cosmetic result was excellent, with clinical resolution of the NSJ and no textural changes. There has been no recurrence of the lesion in the 16 months following discontinuation of therapy (Figs. 5a and 5b).

A biopsy taken after the final treatment session, from normal-appearing skin at a site of former NSJ, showed normal epidermis. The superficial parts of the sebaceous glands had been replaced by a widened and fibrotic dermis. Large hyperplastic sebaceous lobules were still present in the reticular dermis, presumably corresponding to sites beyond the depth of effective photodynamic therapy.

Although ALA-PDT causes selective destruction of NSJ with excellent cosmetic results, there are several significant limitations. Drug and/or light penetration into skin, may account for the slow clinical clearing of the lesion and the histologic persistence of the deep portion of the lesion. Since the deep component of the lesion was not eliminated, the patient may develop NSJ recurrences as well as the associated benign and malignant tumors. Therefore, long-term, periodic follow-up will be important.

17.2.2 ALA-PDT and acne

Photodynamic therapy with topical ALA was tested for the treatment of acne vulgaris, in an open-label prospective human study [13]. Twenty-two subjects with acne on the back were treated in 4 sites with ALA plus red light (ALA-PDT), ALA alone, light alone, and untreated control. Half of the subjects were treated once; half were treated 4 times. Twenty percent topical ALA was applied with 3hr occlusion, and 150 J/cm^2 broad band light (550–700 nm) was given. ALA-PDT caused selective phototoxicity to sebaceous follicles, prolonged suppression of sebaceous gland function, and apparent decrease in follicular bacteria after PDT. On histology, sebaceous glands showed acute damage and were smaller 20 weeks after PDT. There was clinical and statistically significant clearance of inflammatory acne by ALA-PDT, for at least 20 weeks after multiple ALA-PDT treatments and 10 weeks after a single treatment. Although topical ALA-PDT proved to be an effective treatment of acne vulgaris, there were significant side effects. Each treatment took time, was painful or pruritic, caused acute erythema and edema, occasionally blistering and purpura, caused an acute acneiform eruption, and usually led to hyperpigmentation which faded gradually over weeks to months.

17.3 Conclusion

Potentially, ALA-PDT may be useful for some patients with hypertrichosis and disorders of sebaceous gland. Further optimization of drug and light delivery are needed for more effective treatment and to alleviate the side-effects. Uptake of ALA into hair follilces and/or sebaceous follicles might be enhanced or made more selective by vehicle and application conditions. Addition of dimethyl sulfoxide (DMSO) as a "penetration enhancer", ethylenediaminetetraacetic acid (EDTA) as a chelator, or topical desferrioxamine, an iron chelator, to the ALA vehicle [14–16] could potentially increase ALA penetration and/or PPIX synthesis. ALA can be delivered rapidly into skin by iontophoresis [17]. Curettage immediately before ALA application and PDT could also hasten the resolution of the bulky and resistant tumor areas [12]. Dose-response characteristics for ALA-PDT treatment of acne or hypertrichosis are still unknown. Using a lower light irradiance and fractionated exposures is known to reduce the pain and enhance the efficacy of PDT [18,19]. A series of less-aggressive PDT treatments may therefore be preferable. These and other ideas may be worth pursuing because it is now clear that pilosebaceous units respond to ALA-PDT.

References

1. D. Leroy, P. Deschamps (1986). Long term side effects after intramuscular injection of hematoporphyrin derivative. *Photodermatology*, **3**, 197–199.
2. G. Monfrecola, F. D'Anna, M. Delfino (1987). Topical hematoporphyrin plus UVA for treatment of alopecia areata. *Photodermatology*, **4**, 305–306.
3. R. Bissonnette, J. Shapiro, H. Zeng, D.I. McLean, H. Lui (2000). Topical photodynamic therapy with 5-aminolaevulinic acid does not induce hair regrowth in patients with extensive alopecia areata. *Br. J. Dermatol.*, **143**, 1032–1035.
4. R.M. Lavker, S. Miller, C. Wilson et al. (1993). Hair follicle stem cells: their location, role in hair cycle, and involvement in skin tumor formation. *J. Invest. Dermatol.*, **101**, 16S–26S.
5. T. Sun, G. Cotsarelis, R.M. Lavker (1991). Hair follicular stem cells: the bulge-activation hypothesis. *J. Invest. Dermatol.*, **96**, 77S–78S.
6. S. Lyle, M. Christofidou-Solomidou, Y. Liu, D. Elder, S. Albelda, G. Cotsarelis (1998). The C8/144B monoclonal antibody recognizes cytokeratin 15 and defines the location of human haair follicle stem cells. *J. Cell Science*, **111**, 3179–3188.
7. R.R. Anderson, J.A. (1983). Parrish Selective photothermolysis: precise microsurgery by selective absorption of pulsed radiation. *Science*, **220**, 524–527.
8. J.C. Kennedy, R.H. Pottier, D.C. Pross (1990). Photodynamic therapy with endogenous protoporphyrin IX: Basic principles and present clinical experience. *J. Photochem. Photobiol. B*, **6**, 143–148.
9. J.C. Kennedy, R.H. Pottier (1992). Endogenous protoporphyrin IX, a clinically useful photosensitizer for photodynamic therapy. *J. Photochem.Photobiol. B*, **14**, 275–292.
10. M. Grossman, J. Wimberly, P. Dwyer et al. (1995). PDT for hirsutism. *Lasers Surg. Med. Suppl.*, **7**, 44.
11. D.X.G. Divaris, J.C. Kennedy, R.H. Poittier (1990). Phototoxic damage to sebaceous glands and hair follicles of mice after systemic administration of 5-aminolevulinic acid correlates with localized protoporphyrin IX fluorescence. *Am. J. Pathol.*, **136**, 891–897.
12. C.C. Dierickx, M. Goldenhersch, P. Dwyer, A. Stratigos, M. Mihn, R. Anderson (1999). Photodynamic therpay for nevus sebaceus with topical δ-aminolevulinic acid. *Arch. Dermatol.*, **135**, 637–640.
13. W. Hongcharu, C.R. Taylor, Y. Chang, D. Aghassi, K. Suthamjariya, R.R. Anderson (Aug. 2000). Topical ALA-photodynamic therapy for the treatment of acne vulgaris. *J. Invest. Dermatol.*, **115**, 183–192.
14. T. Warloe, Q. Peng, J. Heyerdahl et al. (1995). Photodynamic therapy with 5-aminolevulinic acid induced porphyrins and DMSO/EDTA for basal cell carcinomas. In 5th International Photodynamic Association Biennial Meeting. *Proc. SPIE*, **2371**, 226–235.
15. S. Fijan, H. Hönigsmann, B. Ortel (1995). Photodynamic therapy of epithelial skin tumors using delta-aminolevulinic acid and desferrioxamine. *Br. J. Dermatol.*, **133**, 282–288.
16. B. Ortel, A. Tanew, H. Hönigsmann (1993). Lethal photosensitization by endogenous porphyrins of PAM cells – Modification by desferrioxamine. *J. Photochem. Photobiol. B*, **17**, 273–278.
17. L.E. Rhodes, M.M. Tsoukas, R.R. Anderson, N. Kollias (1997). Iontophoretic delivery of ALA provides a quantitive model for ALA pharmacokinetics and *PpIX* phototoxicty in human skin. *J. Invest. Dermatol.*, **108**, 87–91.

18. N. van der Veen, K.M. Hebeda, H.S. de Bruijn, W.M. Star (Dec. 1999). Photodynamic effectiveness and vasoconstriction in hairless mouse skin after topical 5-aminolevulinic acid and single- or two-fold illumination. *Photochem. Photobiol.*, **70**, 921–929.

19. H.S. de Bruijn, N. van der Veen, D.J. Robinson, W.M. Star (Feb. 15, 1999). Improvement of systemic 5-aminolevulinic acid-based photodynamic therapy in vivo using light fractionation with a 75-minute interval. *Cancer Res.*, **59**(4), 901–904.

Chapter 18

Topical sensitization – non-oncologic indications – HPV-infections

Claudia Alge and Rolf-Markus Szeimies

Table of contents

Abstract

Although malignant and premalignant lesions are still the major targets, PDT may be used for eradicating non malignant growths such as human papillomavirus (HPV)-induced acanthomas of the skin. HPV-induced skin lesions are very common and can present as foot and hand warts, planar warts and genital warts. Histologically these lesions display an acanthotic epidermis in a hyperproliferative state. The clinical course is characterized by high recurrence rates despite of surgical removal or application of cytotoxic agents. The variety of clinical treatment regimens demonstrates that no single treatment modality to date has been fully satisfactory and additional methods should be sought. HPV-induced skin lesions might provide a possible indication for topical ALA-PDT. The rapidly proliferating cells in viral acanthomas accumulate ALA-induced PPIX selectively when compared to the surrounding non-infected cells [27–28]. Also, more recently ALA-PDT has been shown to have antiviral properties [13] and may be a new approach to destroy cells altered by virus infection by the interaction of absorbed light and the activated phototoxic agent. This chapter shall provide an overview including the different features of HPV-induced skin lesions, epidemiology, recurrence rates and treatment options as well as a brief review comprising the rationale of PDT in treatment of virus acanthomas as well as results of first clinical studies.

18.1 Human papilloma viruses

18.1.1 Background

Human papillomavirus (HPV) infections are widespread in human populations and occur all over the world. HPV infect the skin and mucous membranes and can cause subclinical infection as well as a wide variety of benign and malignant epithelial tumors. Papillomaviruses are highly species-specific and especially in animals they can induce papillomas with increased tendency for malignant transformation.

The HPV is a double stranded DNA virus of the papovavirus class with tropism to squamous epithelia. To present more than 80 HPV-subtypes have been classified which are defined by DNA sequence analysis [1]. HPV types vary in tissue tropism, their association with distinct lesions and their oncogenic potential. Most of them are clearly correlated with specific clinical features. According to DNA sequence homology and tissue tropism they can roughly be grouped in mucosal types, cutaneous types and Epidermodysplasia verruciformis-associated types.

18.1.2 Clinical manifestations

Three clinical manifestations of cutaneous HPV infections occur with high incidence in the general population: common warts, plantar warts, and flat warts. Common warts represent about 70% of all cutaneous warts, occurring in up to 20% of all school-aged children. Plantar warts are common in older children and adults, accounting for 30% of cutaneous warts. Flat/planar warts account for 4% of cutaneous warts [2]. The most common manifestation of mucosal HPV infection is condyloma acuminatum (genital wart), which is the most prevalent sexually transmitted disease in developed countries. Some HPV types play a major role in the pathogenesis of in situ as well as invasive squamous cell carcinoma of the anogenital epithelium. Whereas the majority of condylomata acuminata and low grade CIN are associated with HPV 6 and 11, HPV 16 and 18 are found in approximately 50% of high grade CIN or invasive cancer tissue from various genital sites and are therefore graded as high risk HPV [37]. Also at least further 20 HPV types are at least occasionally found associated with cervical cancer [4].

18.1.3 Transmission and biology of HPV infection

HPV is transmitted by skin contact, sexually or perinatally. Incubation periods for cutaneous warts vary from two weeks to more than a year, those for condyloma acuminata vary from three weeks to eight months.

Infections with subtypes causing skin warts are usually acquired through micro-injuries. Transmission occurs either directly from one person to another or via contaminated objects or surfaces. It is facilitated by the abundance of mature virus

Table 1. Clinical lesions and associated HPV types. Adapted data from [1,85,86] (Rooks, Textbook).

Cutaneous warts	HPV subtype
Deep plantar/palmar warts	1,2,4
Common warts	2,4,26,28,29,63,75–77
Flat warts	3,10,28,41,65
Butchers warts	7
Mucosal warts	
Anogenital warts	6,11,16,18,30,31,35,39,40,42–45,51–59,64,66–68,70,73
CIN	16,18,31,33,35,39,40,45,52,53,56,58,66–68,69
Cervical carcinoma	16,18,31,33,35,39,45,52,53,56,58,66–68
VIN	16,40,61,62
PIN	16,40
Fokal oral epithelial hyperplasia	13
Epidermodysplasia verruciformis	
Macular warts	5,8,9,12,14,15,17,19–25,36–38,47,49,50

Abbreviations: CIN: cervical intraepithelial neoplasia
VIN: vulvar intraepithelial neoplasia
PIN: penile intraepithelial neoplasia

particles from plantar and common warts. Plantar warts for example are commonly acquired from swimming pool or shower room floors. Those rough surfaces abrade moistured keratin from infected feet and help inoculate viruses into the softened skin of others.

Genital HPV infections occur by direct contact. Thus most anogenital warts are sexually transmitted and represent the most common veneral infection. This implies that compared to cutaneous warts the infectivity of anogenital warts is substantially higher. About two thirds of sexual partners of people with condyloma acuminata develop clinical lesions within nine months [5].

HPV viruses enter cells of the basal cell layer of the epidermis via microinjuries. Because transcription of viral genes and replication are linked to an advanced differentiation state HPV are not able to multiply in basal cells. Only with progression of keratinocyte-differentiation in the suprabasal layers expression of viral genes is induced and followed by synthesis of virus-specific proteins. Mature virus particles can finally be found in the granular layer and the stratum corneum. HPV replication is associated with high proliferation rates of several epithelial layers except for the basal cell layer. This results in acanthosis, parakeratosis and hyperkeratosis of the epidermis.

18.1.4 Clinical course and treatment

Treatment of HPV-induced lesions is difficult and often frustrating. No means for elimination of the virus itself is known so far and most treatment modalities aim at the physical destruction of the infected viral cells. Other therapeutic approaches concentrate on immune modulation and also the development of vaccines to either prevent HPV-infections or to boost the hosts immune response to HPV-infected cells is underway [6]. Overall the current failure rate, recurrence rate and side effects make the currently available treatment modalities highly unsatisfactory. Apart from re-infection a main reason might be the presence of latent papillomavirus. HPV has been found in clinically and histologically normal squamous epithelium as far as 1 cm beyond the treatment area. This virus presence in the adjacent skin strongly correlates with the risk of recurrence [7]. In this regard the dual selectivity of ALA-PDT represents an innovative approach to therapy. Selective accumulation of photosensitizing concentrations of ALA-induced PPIX in the infected cells may enable to target altered tissue without harming the adjacent normal skin. In addition subclinical, hyperproliferative cell nests might be attacked as well, thus reducing recurrence rates due to sublinical virus persistence.

18.2 PDT and virus-inactivation

18.2.1 Background

Antiviral photodynamic therapy dates back as early as to the 1930s, when Schultz and Krueger found the ability of photosensitizers to inactivate viruses or bacteriophages. In the past many studies have been performed on enveloped viruses such as the Herpes simplex virus. However despite of therapeutic efficacy in some cases [8,9] PDT has been abandoned for this indication due to the improvement of specific antiviral chemotherapeutic treatments and the potential cancerogenicity of the photosensitizers employed. To date PDT procedures are used to decontaminate blood products for therapeutic use, eg. human plasma is treated with methylene blue in combination with visible light [10,11].

It has been shown that a broad range of enveloped viruses are rapidly inactivated when irradiated in the presence of photoactive dyes. In photodynamic therapy different viral targets, such as the virus-envelope and nucleic acid can be altered to achieve loss of activity. One mechanism might be due to the absorption of the dye to the viruses, especially the DNA, which would improve the local effects of photochemically generated singlet oxygen [12]. Another proposed mechanism is binding of the photosensitizer to a virion's surface and thus reducing infectiosity. As was shown for HSV 1 the amphiphilic aluminium dibenzodisulfophthalocyanine induces changes in a structural protein of the envelope, thus leading to reduced infectiosity [13]. Also in PDT with merocyanine 540 [14], rose bengal [45] or hematoporphyrin derivative [16] this damage results in the inhibition of the early phases of the viral infectious process: the adhesion of the virus to the host cell and its penetration are impaired. In addition to such direct antiviral effects, PDT might exert virus inactivation by rather indirect mechanisms such as destruction of the host-cell. This might especially be important for intracellularly located viruses such as HPV and intracellular HIV. In these cases it is important that the photosensitizer can penetrate the plasma membrane to exert its phototoxic effect. With respect to these considerations the neutral silicon phthalocyanine PC 4 was shown to be taken up by the cells resulting in inactivation of cell-associated HIV via induction of apoptosis of the host cell [17], inactivation of HIV in red blood cell concentrates – apoptosis. In contrast the positively charged aluminum phthalocyanine PC 5 and the negatively charged silicon phthalocyanine were ineffective [18].

18.2.2 Immunological and inflammatory effects

Pronounced clinical and histological inflammatory responses can routinely be observed after PDT. There is convincing evidence that inflammatory cells participate in PDT-induced tumor regression [19,20]. These effects are also being exploited for a number of benign conditions [21]. On the molecular level PDT was shown to increase the production of several pro-inflammatory cytokines [19,22,23]. It is noteworthy that in treatment of HPV infections such PDT-induced immunomodulation might contribute to a boostered antiviral immune response of the host. This may aid elimination of small foci of infected cells, thus reducing the risk of recurrence resulting from virus persistence.

18.2.3 Topical ALA-PDT for HPV-induced skin lesions

In treatment of HPV infections topical application of 5-ALA might provide a novel approach to treat multicentric and overt lesions simultaneously. The mechanisms of

5-ALA-induced virucidal activity on non-enveloped viruses such as HPV are not well studied. It is most likely that ALA-PDT clears HPV by an indirect effect, i.e. exclusively by destruction of the infected cells via PPIX generated photoproducts. Only recently ALA-PDT mediated inactivation of intracellular HIV in vitro and HSV in vivo could be demonstrated. Whereas such virucidal effects of other photosensitizers appear to be mediated by apoptosis of the host cell, this could not be confirmed for 5-ALA [13]. One of the most direct effects of 5-ALA-induced cellular phototoxicity has been shown to be mitochondrial damage [24]. Cell membrane damage may also be involved as PPIX has been found to accumulate there [25].

18.2.4 Selectivity of 5-ALA-induced PPIX formation

When 5-ALA is supplied exogenously on genital warts penetration appears to be enhanced by the abnormal layer of keratin. Cutaneous warts however are covered by massive hyperkeratoses. These thickened horny layers greatly inhibit the penetration of 5-ALA [26]. Impaired 5-ALA penetration results in poor therapeutic efficacy of PDT. In these cases addition of iron chelators or substances to enhance penetration as well as keratolysis or paring of the warts previous to 5-ALA-application might prove useful. In rapidly proliferating tissues such as viral papillomas a higher conversion rate to PPIX can be expected because of higher metabolic activity. Thus enhanced penetration in combination with increased metabolic activity of the infected cells may result in increased PPIX formation within the lesions. Because the adjacent normal skin is less permeable, it is not necessary to restrict topical application of 5-ALA to the lesion itself.

This was demonstrated for genital warts where a peak fluorescence could be localized to the basal cell layer early after application of 5-ALA. These findings suggest that these metabolically active dividing cells definitely convert 5-ALA faster than the more differentiated cells. 3–6 hours after application uniform PPIX fluorescence was found in the whole epithelium. Whereas after 24 hours fluorescence had shifted to the granular and horny layer [27]. Another study reported an increasing arithmetic difference in mean fluorescence of the condylomas and the adjacent normal skin up to 6 hours post 5-ALA application. In contrast the highest ratio between condylomas and surrounding tissue, implicating highest selectivity, was found only 2 hours after topical application of 5-ALA. In this study full thickness fluorescence was noted in all cases. Again, the signal was greatest in the basal cell layer [28]. However the correlation between fluorescence signal and actual photodestruction is unknown. For effective PDT apart from selectivity of PPIX fluorescence also the uniform distribution of the photo-sensitizer within the epithelium might be crucial, since in HPV infections the main virus-load is localized in suprabasal layers.

HPV are known to be present as far as 1 cm from the clinical border [29]. This accounts for the high recurrence rates of HPV infections after removal of the papular component of the warts alone. On the cellular level it is likely that these infected cells do already display an altered metabolic activity. In a clinical trial Smetana et al. noted extension of PPIX fluorescence to apparently normal skin in a patient suffering from cutaneous warts. One likely explanation may be that this area represents virally infected skin without clinical manifestation [13]. This selective uptake of 5-ALA by infected cell nests in subclinical lesions with consequent PPIX formation may help reduce recurrence rates of HPV infections after 5-ALA PDT.

18.3 Clinical trials

18.3.1 Verrucae vulgares

The prevalence of common hand and foot warts in western countries is estimated to be around 22%. For the patient foot and hand warts are a major cosmetic concern and cause social and functional problems. Furthermore keratin shedded off from the lesions represents a potential source of infection for other individuals. Simple treatments such as keratolytic wart paints, paring of the warts, curretage, glutaraldehyde and cryotherapy may cure up to 70% within three months. Warts resistent to these basic treatment modalities are dealt with an assortment of treatments including infrared coagulation, intralesional injection of bleomycin, natural alpha-interferon, topically applied 5-fluor-uracil and the application of immunosensitizing dinitrochlorobenzene. But still, some warts remain resistant. Especially for patients with an impaired immune system or immunosuppressive therapy (e.g. after organ transplantation) new treatment modalities are needed.

18.3.1.1 5-ALA-PDT alone
Encouraged by the good results of topical ALA-PDT in treatment of premalignant and malignant skin lesions Amman et al. in 1995 made an approach to treat refractory verrucae vulgares of the hands with topical 5-ALA -PDT alone. A protocol which had proven successful in the treatment of solar keratoses and superficial skin malignancies was employed. However, only one patient out of six showed a total response within the observation period of two months [30]. Same negative effects were observed by Kennedy et al. in preliminary studies on cutaneous warts [31]. The lack of efficiency in these trials might have been due to an impaired 5-ALA-penetration caused by the hyperkeratoses associated with verrucae vulgares. Optimizing treatment conditions by alteration of treatment parameters such as drug vehicle or preparation (oil in water emulsion, water in oil emulsion or gel), ALA-concentration, incubation time, wavelength for irradiation as well as pretreatment of the warts were documented to achieve satisfactory results:

18.3.1.2 Enhanced drug penetration by addition of DMSO and EDTA
Smetana et al. improved the method by enhancing 5-ALA-penetration by adding 2% EDTA and 2% DMSO to the 20% 5-ALA oil in water emulsion and reported clearance of massive verrucae vulgares in a kidney transplant patient. No recurrence was observed during a two year follow up.

18.3.1.3 5-ALA -PDT combined with regular treatment and irradiation at different wavelengths
Since 5-ALA -PDT has been used primarily for the treatment of malignant lesions highest possible tissue penetration was always matter of main interest. For that purpose

usually laser sources with wavelengths around 635 nm red light are employed. In treatment of warts the deep penetration of light may however not be as important since the virions are detected in the upper parts of the epidermis. PPIX shows various absorption peaks within the visible range. Therefore broadband light, which includes all absorption peaks of PPIX might be more effective than red and blue light. In addition a slide projector may deliver reasonable amounts of hyperthermia, which may have an effect on its own.

Based on these considerations Stender et al. demonstrated that topical 5-ALA-PDT is a promising approach to obtain complete response and reduce recurrence rates when combined to basic treatment regimens such as paring of warts and use of a keratolytic agent [32]. In a study with 30 patients suffering from recalcitrant warts the effectiveness of various wavelengths was compared. Patients were free to use a keratolytic agent at home after paring their warts. Also the hyperkeratoses overlying the warts were pared to the level of visualization of the blood vessels immediately before treatment. Twenty percent 5-ALA cream was applied to the wart and the surrounding normal skin to a distance of 1 cm of the wart margins. To enhance 5-ALA-penetration the treated area was occluded with a hydrocolloid film for 5 hours. Irradiation with a total dose of 40 J/cm^2 was obtained by light from slide projectors at variable wavelengths. Finally significantly more warts responded completely after three times white light irradiation (CR 73%) than after three times irradiation with blue (CR 28%), red light (CR 42%) or cryotherapy (CR 20%). Regression typically started to occur during the first month after treatment. No recurrences were observed in patients cleared with PDT after the 12 month follow-up.

This study was followed by a double blind randomized study on 45 patients. The results confirmed the findings of the pilot study by the same group suggesting that 5-ALA-PDT is effective in reducing the number and area of recalcitrant foot and hand warts when the interventions were combined with regular basic treatments as above. So as to include the 630 and 690 nm peak of PPIX and its photoproducts light was derived from a Waldmann PDT lamp (wavelengths ranging from 590–700 nm) at a fluence rate of 50 mW/cm^2 at a total dose of 70 J/cm^2. These treatments were repeated after 1 and 2 weeks, if warts persisted at week 7, treatment was repeated in another sequence of three weekly intervals. Again, patients were in parallel instructed to pair their warts twice a week and apply a local keratolytic. This time they observed a complete curing rate of 56% in ALA-PDT treated lesions versus 42% in the placebo group. Surprisingly in the placebo group a higher cure rate than expected was observed (42% vs. 30%). As further investigations suggested this might be due to naturally existing PPIX and its photodegradation products within the warts. Taken together these results suggest that 5-ALA-PDT in combination with standard treatment modalities is a promising and safe approach for effective therapy in patients with recalcitrant warts. Compared to most traditional treatment modalities (Table 2) clinical response rates obtained by 5-ALA-PDT appear to be within the same range.

Despite of these promising observations it remains to be remarked that in the authors own studies on six patients with common warts on hands and soles the lesions did not respond to 5-ALA-PDT at all: As described by Stender et al. sufficient keratolysis was done before application of 20% 5-ALA emulsion for 6 hours and subsequent illumination by an incoherent light source (5–150 mW/cm^2, 100 J/cm^2). All patients

Table 2. Summary of selected published data for treatment of cutaneous warts

Treatment	Success Rate	Reference
Cytotoxic Agents		
Salicylic Acid	45–84%	[42]
Glutaraldehyde	70%	[43]
5-FU	64%	[44]
Bleomycin	65–80%	[45–46]
Physically ablative		
Cryotherapy	30–70%	[42–47]
CO_2 Laser	56–75%	[48–49]
Pulsed Dye Laser	66–72%	[50–51]
Infrared Coagulation	70%	[52]
Immunomodulation		
Dinitrochlorobenzene	45–62%	[53,54]
Squaric Acid Dibutylester	69%	[55]
Interferons	31–91%	[56–57]
Imiquimod		

reported unbearable pain but no clinical improvement which could be related to 5-ALA-PDT could be observed. This implies that for critical evaluation of 5-ALA-PDT as an alternative treatment option for common warts further multicentric, randomized, double blind studies should be conducted. The main drawback in PDT with topical 5-ALA however appears to be severe or unbearable pain during illumination. This is a well known side effect of topical 5-ALA-PDT e.g. in treatment of actinic keratoses on the scalp. When treating cutaneous HPV infections the open wounds due to paring of the warts previous to application of 5-ALA might even enhance the susceptibility to pain. Whether intralesional anaesthetics for pain relief would interfere with efficacy of 5-ALA-PDT or not has not been studied yet. Further it remains to be found out whether methylated, more lipophilic 5-ALA derivatives, which were reported to be associated with less pain (Metvix investigators brochure), do improve cure rates and at the same time reduce this main adverse effect.

18.3.2 Genital warts

Visible anogenital condyloma are present 1% of sexually active adults in the USA and at least 15% have subclinical infection [37]. Apart from being highly infectious genital HPV infections are often cosmetically unacceptable may be associated with discomfort and physical or psychosocial dysfunction for the patient. The clinical course is characterized by multiple recurrences despite surgical removal or application of cytotoxic agents. Most destructive modalities target only the papular component of the wart and result in high recurrence rates ranging from 10 to 50%. These recurrences are at least partly due to the presence of HPV as far as 1 cm of the clinical border [29].

Removal can be performed by cryotherapy, electrodissecation or carbon dioxide lasertherapy. Local application of keratolytic or cytotoxic agents such as salicylic acid, trichloroacetic acid, podophyllin, podophyllotoxin or 5-fluoruracil has been reported as treatment option. Further suggestions included immunomodulation by combination of the treatment regimens mentioned above with interferons as well as treatment with topical, systemic or intralesional interferon alone (Table 3). However taken together the results are still not very satisfactory. Within the first three months after treatment recurrence rates within the range of 30 and 40% are to be expected [36a]. Patients with long lasting and recurrent genital warts have even higher recurrence rates [33]: 64% (14/22); 57% (12/21); [34]: 54% (5/11); [35]: 81% (13/16). Only recently a new immunomodulatory drug, imiquimod, has been launched for the treatment of genital warts. Imiquimod represents an immune response modifier, which induces interferon

Table 3. Established therapies for condylomata acuminata. Modified from [36,87]

Treatment	Success Rates	Recurrence Rates at varying timepoints	References
Cytotoxic Agents			
Trichloroacetic acid	64–81%	36%	[58–59]
Podophyllin	38–79%	21–65%	[60–61]
Podofilox	68–88%	16–34%	[62–63]
5-FU	68–97%	0–8%	[64–65a]
Immunomodulation			
IFN-α (natural), topical	23–90%	23–100%	[66–69]
IFN-α (natural), intralesional	48–62%	20–33%	[68,70–71]
IFN-α2b, intralesional	19–60%	17%	[71,73]
IFN-α (natural), systemical	25–69%	5–44%	[74–76a]
rIFN-α2a, systemical	21–71%	0–36%	[77]
IFN in combination to various traditional treatments		9–69%	[34,35,61,72,78]
Imiquimod	37–71%	19%	[36–38]
Physically ablative therapies			
Cryotherapy	70–96%	25–39%	[58–60,79–80]
CO_2-Laser	72–97%	6–49%	[65b-83]
Electrosurgery	72–94%	25–51%	[79,83]
Surgical Excision	89–93%	19–22%	[62,84]

and a number of other cytokines. Several clinical studies reported favorable results, especially as far as recurrence rates are concerned [36–38].

Regarding topical ALA PDT of genital warts only a few reports are available at present although topical ALA-PDT – at least theoretically – might present a promising new approach in treatment of anogenital HPV-infections.

In a pilot study performed by Frank et al. [39] four out of seven patients with perivulvar, perianal, penile and vulvovaginal condyloma respectively, showed a 100% reduction of the lesions. No recurrence was observed within a three month (four weeks in one case) follow-up. In detail 20% ALA cream containing lidocaine hydrochloride was applied under an occlusive dressing for 14 hours. After 14 hours an anaesthetic cream was applied for another two hours. The site then was exposed to irradiation emitted by an argon dye laser at a wavelength of 630 nm to a total dose of 100 J/cm^2. Light intensity was either 150 or 75 mW/cm^2.

Selectivity of PPIX accumulation is crucial for successful ALA-PDT. When evaluating the results of Frank et al. it should be remarked that fluorescence measurements demonstrated highest selectivity early after ALA application [28]. In addition Fehr et al. demonstrated uniform distribution of PPIX fluorescence within the epidermis 3–6 hours after ALA application. At 24 hours fluorescence had shifted to the granular layer alone. For successful treatment however uniform photosensitizer distribution as well as selectivity are desirable conditions. In these studies both was found at earlier timepoints. Thus an incubation period of 14 hours and more might be too long and further studies employing shorter incubation are needed.

Taken together the main objective in treatment of anogenital warts appears to be the reduction of recurrence rates. As it was shown for common warts the combination of ALA-PDT with established, physically ablative treatment modalities might prove useful. Physical destruction of the visible virus acanthomas eg. by carbon-dioxide lasertherapy followed by ALA-PDT to selectively target subclinical lesions may represent a promising approach to increase complete cure rates and reduce recurrence rates.

18.4 Epidermodysplasia verruciformis

Epidermodysplasia verruciformis (EV) represents a worldwide distributed, genetically determined susceptibility to infection with EV-specific HPVs that are harmless for the general population. As a consequence the infection manifests in early childhood as pityriasis versicolor-like macules and flat wart-like skin lesions that undergo malignant transformation in 30–60% [40]. Ultraviolet (UV)-induced immunosuppression may facilitate the development of these skin cancers which are mainly confined to sun-exposed regions. Sun protection is recommended and retinoids or interferon may help reduce the number of viral warts on the skin [41].

In 1998 Karrer et al. published a case report suggesting the use of ALA-PDT in a 65 year old woman with EV who had multiple superficial skin tumors. Topical PDT using 20% 5-ALA ointment applied for 6 hours and irradiating with an incoherent light source at a total dose of 160 J/cm^2, with a fluence rate of 160 mW/cm^2 at wavelengths ranging from 580 to 740 nm resulted in excellent cosmetic results without causing damage to the

Table 4. Advantages and disadvantages of topical ALA-PDT in treatment of HPV-infections

Advantages	Disadvantages
Non-invasive	Pain
Good cosmetic results	Insufficient light delivery in certain sites
Destruction of sublcinical lesions	Adjuvant therapy
Simultaneous treatment of separate lesions	Controversal results
No laser plume	GCP studies needed for true evaluation
Low costs	of effectiveness
No systemic effects	

surrounding skin. Twelve months after PDT a few lesions had recurred on the hands. In situ hybridization revealed that HPV types 5/12/14/19–23/25/36 had disappeared, but HPV 8 was still present in the skin.

Although permanent cure could not be achieved and single lesions continued to reoccur, repeated treatment with PDT, perhaps even just annually or when single lesions reappear, might result in better control of HPV-induced lesions in EV.

18.5. Discussion

In treatment of HPV infections topical ALA-PDT has several advantages over conventional therapies (Table 4). It is non-invasive and produces excellent cosmetic results. The basal membrane remains intact, and therefore no scaring occurs. After topical application of ALA PPIX-induced photosensitization remains restricted to the site of application. Moreover it can be used to treat patients at low costs and on an outpatient basis. Partial selectivity of PPIX formation in metabolically altered infected keratinocytes helps eradication of overt cell nests. Furthermore several separate lesions can be treated simultaneously. Another benefit of PDT in these conditions is the lack of virus containing laser plume. This represents a clear advantage in treatment of patients suffering from HIV or hepatitis C.

The most frequently reported adverse effect of ALA-PDT appears to be severe to unbearable pain during and following illumination. For certain sites such as intraanal and vaginal lesions special light application devices need to be developed so as to guarantee sufficient light delivery.

The pilot studies summarized in this chapter demonstrate that ALA-PDT might be an additional option in treatment of cutaneous and mucosal HPV-infections. The findings also demonstrate that in the future topical ALA-PDT might be viewed as an adjuvant treatment when combined to conservative treatment modalities to reduce recurrence rates. However, to evaluate the overall therapeutic effectiveness and to determine optimal treatment parameters GCP-studies will be needed for both conditions.

References

1a. E.M. De Villiers (1994). Human pathogenic papillomavirus types: an update. *Curr. Top. Microbiol. Immunol.*, **186**, 1–12. Review.

1b. E.M. De Villiers (1994). Human pathogenic papillomavirus types: an update. In: H. zur Hausen (Ed.), *Human Pathogenic Papillomaviruses* (pp. 1–12). Heidelberg: Springer Verlag.

2. T.B. Fitzpatrick, R.A. Johnson, K. Wolff, M.K. Polano, D. Suurmond (1997). *Color Atlas and Synopsis of clinical Dermtology* (3rd ed., pp. 766–771). Mc Graw Hill.

3. L. Koutsky (1997). Epidemiology of genital human papillomavirus infection. *Am. J. Med.*, **102**(5A), 3–8.

4. F.X. Bosch, M.M. Manos, N. Munoz, M. Sherman, A.M. Jansen, J. Peto, M.H. Schiffman, V. Moreno, R. Kurman, K.V. Shah (1995). Prevalence of human papillomavirus in cervical cancer: a worldwide perspective. International biological study on cervical cancer (IBSCC) Study Group. *J. Natl. Cancer Inst.*, **87**(11), 796–802.

5. J.D. Oriel, I.W.Whimster (1971). Carcinoma in situ associated with virus-containing anal warts. *Br. J. Dermatol.*, **84**(1), 71–73.

6. N.D. Christensen, C.A. Reed, N.M. Cladel, R. Han, J.W. Kreider (1996). Immunization with viruslike particles induces long-term protection of rabbits against challenge with cottontail rabbit papillomavirus. *J. Virol.*, **70**(2), 960–965.

7. A. Ferenczy, M. Mitao, N. Nagai, S.J. Silverstein, C.P. Crum (1985). Latent papillomavirus and recurring genital warts. *N. Engl. J. Med.*, **313**(13), 784–788.

8a. E.G. Friedrich Jr. (1973). Relief for Herpes vulvitis. *Obstet. Gynecol.*, **41**(1), 74–77.

8b. E.G. Friedrich Jr. (1974). Editorial: Photodynamic therapy. An editorial comment. *Obstet. Gynecol.*, **43**(2), 304–305.

9. L.E. Bockstahler, T.P. Coohill, K.B. Hellman, C.D. Lytle, J.E. Roberts (1979). Photodynamic therapy for herpes simplex: a critical review. *Pharmacol. Ther.*, **4**(2), 473–949. Review.

10. B. Lambrecht, H. Mohr, J. Knuver-Hopf, H. Schmitt (1991). Photoinactivation of viruses in human fresh plasma by phenothiazine dyes in combination with visible light. *Vox Sang.*, **60**(4), 207–213.

11. H. Mohr, B. Lambrecht, H. Schmitt (1993). Photo-inactivation of viruses in therapeutical plasma. *Dev. Biol. Stand.*, **81**, 177–183.

12. K.G. Specht (1994). The role of DNA damage in PM2 viral inactivation by methylene blue photosensitization. *Photochem. Photobiol.*, **59**(5), 506–514.

13. Z. Smetana, Z. Malik, A. Orenstein, E. Mendelson, E. Ben-Hur (1997). Treatment of viral infections with 5-aminolevulinic acid and light. *Lasers Surg. Med.*, **21**(4), 351–358.

14. J.M. O'Brien, D.K. Gaffney, T.P. Wang, F. Sieber (1992). Merocyanine 540-sensitized photoinactivation of enveloped viruses in blood products: site and mechanism of phototoxicity. *Blood*, **80**(1), 277–285.

15. J. Lenard, R. Vanderoef (1993). Photoinactivation of influenza virus fusion and infectivity by rose bengal. *Photochem. Photobiol.*, **58**(4), 527–31.

16. L.J. Kadish, D.B. Fisher, A.B. Pardee (1967). Photodynamic inactivation of free and vegetative bacteriophage T4. *Biochem. Biophys. Acta.*, **138**(1), 57–65.

17. E. Ben-Hur, J. Oetjen, B. Horowitz (1997). Silicon phthalocyanine Pc 4 and red light causes apoptosis in HIV-infected cells. *Photochem Photobiol.*, **65**(3), 456–460.

18. N.L. Oleinick, A.R. Antunez, M.E. Clay, B.D. Rihter, M.E. Kenney (1993). New phthalocyanine photosensitizers for photodynamic therapy. *Photochem. Photobiol.*, **57**, 242–247.

19. W.J. De Vree, M.C. Essers, J.F. Koster, W. Sluiter (1997). Role of interleukin 1 and granulocyte colony-stimulating factor in photofrin-based photodynamic therapy of rat rhabdomyosarcoma tumors. *Cancer Res.*, **57**(13), 2555–2558.

20. M. Korbelik (1996). Induction of tumor immunity by photodynamic therapy. *J. Clin. Laser Med. Surg.*, **14**(5), 329–334. Review.

21. T.J. Dougherty, C.J.Gomer, B.W. Henderson, G. Jori, D. Kessel, M. Korbelik, J. Moan, Q, Peng (1998). Photodynamic therapy. *J. Natl. Cancer Inst.*, **90**(12), 889–905.
22. U.O. Nseyo, R.K. Whalen, M.R. Duncan, B. Berman, S.L. Lundahl (1990). Urinary cytokines following photodynamic therapy for bladder cancer. A preliminary report. *Urology*, **36**(2), 167–171.
23. S.O. Gollnick, X. Liu, B. Owczarczak, D.A. Musser, B.W. Henderson (1997). Altered expression of interleukin 6 and interleukin 10 as a result of photodynamic therapy in vivo. *Cancer Res.*, **57**(18), 3904–3409.
24. S. Iinuma, S.S. Farshi, B. Ortel, T. Hasan (1994). A mechanistic study of cellular photodestruction with 5-aminolaevulinic acid-induced porphyrin. *Br. J. Cancer*, **70**(1), 21–28.
25. P. Steinbach, H. Weingandt, R. Baumgartner, M. Kriegmair, F. Hofstadter, R. Knuchel (1995). Cellular fluorescence of the endogenous photosensitizer protoporphyrin IX following exposure to 5-aminolevulinic acid. *Photochem. Photobiol.*, **62**(5), 887–895.
26. J.C. Kennedy, R.H. Pottier (1992). Endogenous protoporphyrin IX, a clinically useful photosensitizer for photodynamic therapy. *J. Photochem. Photobiol. B*, **14**(4), 275–292. Review.
27. M.K. Fehr, C.F. Chapman, T. Krasieva, B.J. Tromberg, J.L. McCullough, M.W. Berns, Y. Tadir (1996). Selective photosensitizer distribution in vulvar condyloma acuminatum aftertopical application of 5-aminolevulinic acid. *Am. J. Obstet. Gynecol.*, **174**(3), 951–957.
28. E.V. Ross, R. Romero, N. Kollias, N. Crum, R.R. Anderson (1997). Selectivity of protoporphyrin IX fluorescence for condylomata after topical application of 5-aminolaevulinic acid: implications for photodynamic treatment. *Br. J. Dermatol.*, **137**(5), 736–742.
29. A. Ferenczy (1991). Laser treatment of genital human papillomavirus infections in the male patient. *Obstet. Gynecol. Clin. North Am.*, **18**(3), 525–535. Review.
30. R. Ammann, T. Hunziker, L.R. Braathen (1995). Topical photodynamic therapy in verrucae. A pilot study. *Dermatology*, **191**(4), 346–347.
31. J.C. Kennedy, R.H. Pottier, D.C. Pross (1990). Photodynamic therapy with endogenous protoporphyrin IX: basic principles and present clinical experience. *J. Photochem. Photobiol. B*, **6**(1–2), 143–148.
32a. I.M. Stender, H.C. Wulf, (1996). Treatment of recalcitrant verrucae by photodynamic therapy with topical application of delta-aminolaevulinic acid. *Clin. Exp. Dermatol.*, **21**(5), 390.
32b. I.M. Stender, J. Lock-Andersen, H.C. Wulf (1999). Recalcitrant hand and foot warts successfully treated with photodynamic therapy with topical 5-aminolaevulinic acid: a pilot study. *Clin. Exp. Dermatol.*, **24**(3), 154–159.
32c. I.M. Stender, H.C. Wulf (1999). Photodynamic therapy of recalcitrant warts with 5-aminolevulinic acid: a retrospective analysis. *Acta. Derm. Venereol.*, **79**(5), 400–401
32d. I.M. Stender, R. Na, H. Fogh, C. Gluud, H.C. Wulf (2000). Photodynamic therapy with 5-aminolaevulinic acid or placebo for recalcitrant foot and hand warts: randomised double-blind trial. *Lancet*, **355**(9208), 963–966.
33. B.R. Duus, T. Philipsen, J.D. Christensen, F. Lundvall, J. Sondergaard (1985). Refractory condylomata acuminata: a controlled clinical trial of carbon dioxide laser versus conventional surgical treatment. *Genitourin Med.*, **61**(1), 59–61.
34. K.H. Tiedemann, T.M. Ernst (1988). Combination therapy of recurrent condylomata acuminata with electrocautery and alpha-2-interferon. *AKT Derm.*, **14**, 200–204.
35. U. Hohenleutner, M. Landthaler, O. Braun-Falco (1990). Postoperative adjuvant therapy with interferon alfa-2B following laser surgery of condylomata acuminata. *Hautarzt.*, **41**(10), 545–548.

36a. K.R. Beutner, A. Ferenczy (1997). Therapeutic approaches to genital warts. *Am. J. Med.*, **102**(5A), 28–37.

36b. K.R. Beutner, S.L.Spruance, A.J. Hougham, T.L. Fox, M.L. Owens, J.M. Douglas Jr. (1998). Treatment of genital warts with an immune-response modifier (imiquimod). *J. Am. Acad. Dermatol.*, **38**(2 Pt 1), 230–239.

37. L. Edwards, A. Ferenczy, L. Eron, D. Baker, M.L. Owens, T.L. Fox, A.J. Hougham, K.A. Schmitt (1998). Self-administered topical 5% imiquimod cream for external anogenital warts. HPV Study Group. Human Papilloma Virus. *Arch. Dermatol.*, **134**(1), 25–30.

38. C.M. Perry, H.M. Lamb (1999). Topical imiquimod: a review of its use in genital warts. *Drugs*, **58**(2), 375–390.

40. S. Majewski, S. Jablonska (1995). Epidermodysplasia verruciformis as a model of human papillomavirus-induced genetic cancer of the skin. *Arch. Dermatol.*, **131**(11), 1312–1318. Review.

41. C. Blanchet-Bardon, M.A. Lutzner (1985). Interferon and retinoid treatment of warts. *Clin. Dermatol.*, **3**(4), 195–199.

42. M.H. Bunney, M.W. Nolan, D.A. Williams (1976). An assessment of methods of treating viral warts by comparative treatment trials based on a standard design. *Br. J. Dermatol.*, **94**(6), 667–679.

43. R. Hirose, M. Hori, T. Shukuwa, M. Udono, M. Yamada, T. Koide, H. Yoshida (1994). Topical treatment of resistant warts with glutaraldehyde. *J. Dermatol.*, **21**(4), 248–253.

44. C.L. Halasz (1998). Treatment of warts with topical pyruvic acid: with and without added 5-fluorouracil. *Cutis.*, **62**(6), 283–285.

45. M. Amer, N. Diab, A. Ramadan, A. Galal, A. Salem (1988). Therapeutic evaluation for intralesional injection of bleomycin sulfate in 143 resistant warts. *J. Am. Acad. Dermatol.*, **18**(6), 1313–1316.

46. M.H. Bunney, M.W. Nolan, P.K. Buxton, S.M. Going, R.J. Prescott (1984). The treatment of resistant warts with intralesional bleomycin: a controlled clinical trial. *Br. J. Dermatol.*, **111**(2), 197–207.

47. J.F. Bourke, J. Berth-Jones, P.E. Hutchinson (1995). Cryotherapy of common viral warts at intervals of 1, 2 and 3 weeks. *Br. J. Dermatol.*, **132**(3), 433–436.

48. J.E. Mancuso, S.P. Abramow, B.R. Dimichino, M.J. Landsman (1991). Carbon dioxide laser management of plantar verruca: a 6-year follow-up survey. *J. Foot Surg.*, **30**(3), 238–243.

49. R.A. Logan, C.B. Zachary (1989). Outcome of carbon dioxide laser therapy for persistent cutaneous viral warts. *Br. J. Dermatol.*, **121**(1), 99–105.

50. O.T. Tan, R.M. Hurwitz, T.J. Stafford (1993). Pulsed dye laser treatment of recalcitrant verrucae: a preliminary report. *Lasers Surg. Med.*, **13**(1), 127–137.

51. K.J. Robson, N.M. Cunningham, K.L. Kruzan, D.S. Patel, C.D. Kreiter, M.J. O'Donnell, C.J.Arpey (2000). Pulsed-dye laser versus conventional therapy in the treatment of warts: a prospective randomized trial. *J. Am. Acad. Dermatol.*, **43**(2 Pt 1), 275–280. Review.

52. C.L. Halasz (1994). Treatment of common warts using the infrared coagulator. *J. Dermatol. Surg. Oncol.*, **20**(4), 252–256.

53. G. Orecchia, H. Douville, L. Santagostino, G. Rabbiosi (1988). Treatment of multiple relapsing warts with diphenciprone. *Dermatologica*, **177**(4), 225–231.

54. M.F. Naylor, K.H. Neldner, G.K. Yarbrough, T.J. Rosio, M. Iriondo, J. Yeary (1988) Contact immunotherapy of resistant warts. *J. Am. Acad. Dermatol.*, **19**(4), 679–683.

55. A.N. Lee, S.B. Mallory (1999). Contact immunotherapy with squaric acid dibutylester for the treatment of recalcitrant warts. *J. Am. Acad. Dermatol.*, **41**(4), 595–599.

56. J.R. Gibson, S.G. Harvey, D. Kemmett, J. Salisbury, P. Marks (1986). Treatment of common and plantar viral warts with human lymphoblastoid interferon-alpha-pilot studies with

intralesional, intramuscular and dermojet injections. *Br. J. Dermatol.*, **115**, (Suppl. 31), 76–79.

57. R.T. Brodell, D.L. Bredle (1995). The treatment of palmar and plantar warts using natural alpha interferon and a needleless injector. *Dermatol. Surg.*, **21**(3), 213–218.

58. A.N. Abdullah, M. Walzman, A. Wade (1993). Treatment of external genital warts comparing cryotherapy (liquid nitrogen) and trichloroacetic acid. *Sex. Transm. Dis.*, **20**(6), 344–345.

59. M.J. Godley, C.S. Bradbeer, M. Gellan, R.N. Thin (1987). Cryotherapy compared with trichloroacetic acid in treating genital warts. *Genitourin. Med.*, **63**(6), 390–392.

60. S.A. Bashi (1985). Cryotherapy versus podophyllin in the treatment of genital warts. *Int. J. Dermatol.*, **24**(8), 535–536.

61. J.M. Douglas Jr., L.J. Eron, F.N. Judson, M. Rogers, M.B. Alder, E. Taylor, D. Tanner, E. Peets (1990). A randomized trial of combination therapy with intralesional interferon alpha 2b and podophyllin versus podophyllin alone for the therapy of anogenital warts. *J. Infect. Dis.*, **162**(1), 52–59.

62. S.L. Jensen (1985). Comparison of podophyllin application with simple surgical excision in clearance and recurrence of perianal condylomata acuminata. *Lancet*, **2**(8465), 1146–1148.

63. A. Edwards, A. Atma-Ram, R.N. Thin (1988). Podophyllotoxin 0.5% v podophyllin 20% to treat penile warts. *Genitourin. Med.*, **64**(4), 263–265.

64. G.L. Pride (1990). Treatment of large lower genital tract condylomata acuminata with topical 5-fluorouracil. *J. Reprod. Med.*, **35**(4), 384–387.

65a. A. Ferenczy (1984). Comparison of 5-fluorouracil and CO_2 laser for treatment of vaginal condylomata. *Obstet. Gynecol.*, **64**(6), 773–778.

65b. A. Ferenczy (1984). Laser therapy of genital condylomata acuminata. *Obstet. Gynecol.*, **63**(5), 703–707.

66. D. Ikic, M. Orescanin, J. Krusic, Z. Cestar (1985). Preliminary study of the effect of human leukocyte interferon on condyloma acuminata in women. In: *Proceedings of the Symposium on Clinical Use of Interferon* (pp. 223–225). Zagreb, Yugoslavian Academy of Science and Arts.

67. M.A. Byrne, B.R. Moller, D. Taylor-Robinson, J.R. Harris, C. Wickenden, A.D. Malcolm, M.C. Anderson, D.C. Coleman (1986). The effect of interferon on human papillomaviruses associated with cervical intraepithelial neoplasia. *Br. J. Obstet. Gynaecol.*, **93**(11), 1136–1144.

68. A.E. Friedman-Kien, L.J. Eron, M. Conant, W. Growdon, H. Badiak, P.W. Bradstreet, D. Fedorczyk, J.R. Trout, T.F. Plasse (1988). Natural interferon alfa for treatment of condylomata acuminata. *JAMA.*, **259**(4), 533–538.

69. E. Vesterinen, B. Meyer, E. Purola, K. Cantell (1984). Treatment of vaginal flat condyloma with interferon cream. *Lancet*, **1**(8369), 157.

70. J.R. Geffen, R.J. Klein, A.E. Friedman-Kien, (1984). Intralesional administration of large doses of human leukocyte interferon for the treatment of condylomata acuminata. *J. Infect. Dis.*, **150**(4), 612–615.

71. J.C. Vance, B.J. Bart, R.C. Hansen, R.C. Reichman, C. McEwen, K.D. Hatch, B. Berman, D.J. Tanner (1986). Intralesional recombinant alpha-2 interferon for the treatment of patients with condyloma acuminatum or verruca plantaris. *Arch. Dermatol.*, **122**(3), 272–277.

72. J.C. Vance, D. Davis (1990) Interferon alpha-2b injections used as an adjuvant therapy to carbon dioxide laser vaporization of recalcitrant ano-genital condylomata acuminata. *J. Invest. Dermatol.*, **95**(6 Suppl), 146S-148S.

73. J.M Boot, F.B. Blog, E. Stolz (1989). Intralesional interferon alfa-2b treatment of condylomata acuminata previously resistant to podophyllum resin application. *Genitourin Med.*, **65**(1), 50–53.

74. S.A. Gall, C.E. Hughes, K. Trofatter (1985). Interferon for the therapy of condyloma acuminatum. *Am. J. Obstet. Gynecol.*, **153**(2), 157–163.

75. E.A. Olsen, K.R. Trofatter, S.A. Gall, J.R. Medoff, C.E. Hughes, M.S. Weiner, F.F. Kelly. Human lyphoblastoid alpha-interferon in the treatment of refractory condyloma acuminata. *Clin. Res.*, **33**, 673A

76a. R.C. Reichman, D. Oakes, W. Bonnez, C. Greisberger, S. Tyring, L. Miller, R. Whitley, H. Carveth, M. Weidner, G. Krueger et al. (1988). Treatment of condyloma acuminatum with three different interferons administered intralesionally. A double-blind, placebo-controlled trial. *Ann. Intern. Med.*, **108**(5), 675–679.

76b. R.C. Reichman, J.P. Micha, P.K. Weck, W. Bonnez, D. Wold, J.K. Whisnant, P. Mounts, K.F. Trofatter, P. Kucera, S.A. Gall (1988). Interferon alpha-n1 (Wellferon) for refractory genital warts: efficacy and tolerance of low dose systemic therapy. *Antiviral Res.*, **10**(1–3), 41–57.

77. G. Gross, H. Ikenberg, A. Roussaki, N. Drees, E. Schopf (1986). Systemic treatment of condylomata acuminata with recombinant interferon-alpha-2a: low-dose superior to the high-dose regimen. *Chemotherapy*, **32**(6), 537–541.

78. L.J. Eron, M.B. Alder, J.M. O'Rourke, K. Rittweger, J. DePamphilis, D.J. Pizzuti (1993). Recurrence of condylomata acuminata following cryotherapy is not prevented by systemically administered interferon. *Genitourin Med.*, **69**(2), 91–93.

79. K.M. Stone, T.M. Becker, A. Hadgu, S.J. Kraus (1990). Treatment of external genital warts: a randomised clinical trial comparing podophyllin, cryotherapy, and electrodesiccation. *Genitourin Med.*, **66**(1), 16–19.

80. R.J. Damstra, W.A. van Vloten (1991). Cryotherapy in the treatment of condylomata acuminata: a controlled study of 64 patients. *J. Dermatol. Surg. Oncol.*, **17**(3), 273–276.

81. M.S. Baggish (1980). Carbon dioxide laser treatment for condylomata acuminata venereal infections. *Obstet. Gynecol.*, **55**(6), 711–715.

82. J.W. Calkins, B.J. Masterson, J.F. Magrina, C.V. Capen (1982). Management of condylomata acuminata with the carbon dioxide laser. *Obstet. Gynecol.*, **59**(1), 105–108.

83. A. Ferenczy, Y. Behelak, G. Haber, T.C. Wright Jr., R.M. Richart (1995). Treating vaginal and external anogenital condylomas with electrosurgery vs. CO_2 laser ablation. *J. Gynecol. Surg.*, **11**(1), 41–50.

84. H.T. Khawaja (1989). Podophyllin versus scissor excision in the treatment of perianal condylomata acuminata: a prospective study. *Br. J. Surg.*, **76**(10), 1067–1068.

85. M.W. Cobb (1990) Human papillomavirus infection. *J. Am. Acad. Dermatol.*, **22**(4), 547–566. Review.

86. G. Meyers, H.-U. Bernhard, H. Delius et al. (1995). *Human papillomaviruses*. Los Alamos, New Mexico: Los Alamos National Laboratory.

87. S. Karrer, R.M. Szeimies, C. Abels, U. Wlotzke, W. Stolz, M. Landthaler (1999). Epidermodysplasia verruciformis treated using topical 5-aminolaevulinic acid photodynamic therapy. *Br. J. Dermatol.*, **140**(5), 935–938.

88. Metvix[R] 160mg/g cream. *Investigator Brochure* (6th ed.), 27.06.2000.

89. Condylomata international Collaborative Study Group (1990). Recurrent condylomata acuminata treated with recombinant interferon alfa-2a. *JAMA.*, **265**, 2684–2687.

Photodynamic Therapy and Fluorescence Diagnosis in Dermatology
P.-G. Calzavara-Pinton, R.-M. Szeimies and B. Ortel, editors.

Chapter 19

Topical sensitization – non-oncologic indications – others (scleroderma, sarcoidosis, eczema)

Sigrid Karrer

Table of contents

19.1 Introduction

Besides the above discussed chronic inflammatory indications for PDT, e.g. psoriasis, for which already some experience has been reported, also other promising indications appear for which only little clinical experience is available yet. Similar to PDT of psoriasis, one single treatment would not be sufficient to achieve the desired clinical effect in such chronic skin disorders.

19.2 Localised scleroderma and lichen sclerosus et atrophicans

Localised scleroderma is characterised by collagen accumulation and excessive sclerosis of the skin. The major complaints are tightness and itching and the disease is often complicated by contractures and cosmetic disfigurement. Numerous treatments have been tried, such as local corticosteroids, D-penicillamine, intravenous penicillin, intralesional interferon gamma, plasmapheresis, oral PUVA therapy or PUVA-bath photochemotherapy, UVA$_1$ phototherapy and in patients with severe disease immuno-suppressive and vasoactive drugs [2–4]. However, treatments have only limited success or considerable side effects. To assess if ALA-PDT could be useful in localised scleroderma, five patients with progressive disease, in whom conventional therapies had failed were treated by application of a gel containing 3% 5-aminolevulinic acid followed by irradiation with an incoherent lamp (PDT 1200, Waldmann, Germany, light dose 40 mW/cm^2, power density 10 J/cm^2) [1]. The treatment was performed once or twice weekly for three to six months. Sclerotic plaques were assessed by a clinical skin score [5] and a durometer score [6]. The clinical skin score assesses the degree of thickening and induration by palpation of the skin on an analogue scale graded from 10 (severe sclerosis, no folding, wooden hard) to 0 (normal skin folding) [5]. The hardness of the lesion was examined by means of a hand-held type durometer provided with a calibrated gauge that registers linearly divided units present on a scale from 0 to 100 [6].

ALA-PDT was well tolerated, there was only a slight stinging by the end of irradiation. In all patients sclerosis greatly regressed, skin score markedly decreased and durometer measurements objectively showed a reduction of skin hardness. The only side effect was a transient hyperpigmentation in the treated area which resolved within some weeks after cessation of ALA-PDT in all patients. Recurrence or worsening of the disease was not observed within a follow-up of up to two years after treatment. This preliminary clinical trial documents for the first time the beneficial effect of ALA-PDT in patients with localised scleroderma in whom other therapies like PUVA-bath photochemotherapy and intravenous penicillin had failed.

An advantage of PDT as compared to PUVA or UVA$_1$ phototherapy [6,7], is that carcinogenic effects are not being discussed for PDT. However, prospective, double blind placebo controlled trials with larger number of patients are essential to confirm the value of ALA-PDT for the treatment of localised scleroderma.

The investigation of PDT-induced immunological effects, as also known for other light therapies e.g. PUVA-therapy, is an important aim of current research and will help to define new indications for PDT. In scleroderma fibroblasts type I collagen synthesis is elevated, whereas collagenase activity and production is remarkably reduced [9], resulting in excessive accumulation of dermal collagen. Photosensitization of human dermal fibroblasts with uroporphyrin followed by long-wave ultraviolet irradiation resulted in an increase of collagenases (matrix metalloproteinases: MMP-1 and MMP-3) with singlet oxygen being the major intermediate in the upregulation of MMPs [10]. We could demonstrate that scleroderma fibroblasts incubated with ALA (100 µmol/l) and irradiated with red light (24 J/cm^2, 40 mW/cm^2) showed an increase of MMP-1 and MMP-3 protein levels as measured by ELISA in cell supernatants, whereas synthesis of the counteracting TIMP 1 was reduced. MMP-2 and TIMP-2 protein levels remained unaltered after PDT.

Induction of collagenase could also be confirmed in vivo. Biopsies taken before and at the end of PDT in three patients with localized scleroderma were stained immunohistochemically for MMP-1. In all patients a marked increase of the signal for MMP-1 in the dermis of the lesional skin could be seen after PDT. These results suggest that PDT-induced reduction of skin hardness, as observed in 5 patients with localized scleroderma treated repeatedly by ALA-PDT, is at least in part mediated by the induction of MMP-1 and MMP-3.

Another disorder characterized by sclerotic changes of the skin is lichen sclerosus et atrophicans. Hillemanns and coworkers [13] recently reported on the successful treatment of lichen sclerosus of the vulva by topical ALA-PDT. Twelve women with biopsy proven vulvar lichen sclerosus and pronounced pruritus were treated with a 20% ALA solution applied topically to the vulva for 4–5 hours followed by illumination with an argon ion-pumped dye laser that emitted light with a 635 nm wavelength (80 J/cm^2, 70 mW/cm^2). Patients with persistent pruritus were offered a second treatment of PDT after 1–3 weeks. Treatment was tolerated very well with 8 patients not requiring any analgesia. Six to 8 weeks after PDT, pruritus significantly improved in 10 of the 12 women. The authors suggest that topical ALA-PDT is a novel alternative treatment for vulvar lichen sclerosus. Local cytotoxic effects were minimal and there is no generalized cutaneous photosensitization. Furthermore, PDT is not destructive and it preserves the integrity of the vulvar skin.

These observations offer promise that also other disorders characterized by sclerosis of the skin, such as systemic scleroderma, sclerodermatous graft-versus host disease or keloid formation may profit from ALA-PDT.

19.3 Cutaneous sarcoidosis

Sarcoidosis is a multi-organ system disease which is characterized by the formation of non-caseating epithelioid cell tubercles in all or several organs. The lung is by far the most common involved organ in more than 90% of all cases [11]. In contrast, skin involvement occurs only in about 25% of cases, but may be the only manifestation of the disease. The etiology of sarcoidosis remains unclear [12]. Several therapeutic approaches are tried for cutaneous sarcoidosis, including topical and systemic corticosteroids, cryotherapy, cytostatic drugs, allopurinol, thalidomide or antimalarials. However, up to now no specific treatment of sarcoidosis exists.

We treated a 67-year-old female patient who has been suffering from disseminated cutaneous sarcoidosis (small nodular type) for more than 17 years with ALA-PDT. The patient had not responded to various conventional and alternative therapies, e.g. cryotherapy, topical or systemic corticosteroids, clofazimine, allopurinol as well as bath-PUVA therapy. After topical application of a gel containing 3% ALA to the red-brown papular lesions on legs and arms for 6 h under occlusion, irradiation was performed using an incoherent light source ($\lambda = 580$–740 nm, 40 mW/cm^2, 10–20 J/cm^2). Within 3 months of therapy (1–2 times weekly) the lesions resolved completely leaving a slight transient hyperpigmentation. A skin biopsy taken 4 months after the last PDT showed normal skin. Within a follow up of 18 months after PDT the patient remained free of recurrence.

PDT has not been described yet for the treatment of granulomatous skin disorders. Since sarcoidosis is a disease with an abnormal immune regulation, ALA-PDT may be an effective agent for this disease. Topically applied ALA sensitizes primarily epidermal keratinocytes. Upon irradiation with red light choosing parameters which probably do not induce cytotoxicity cytokines are released, possibly resulting in a change of the cytokine environment and finally leading to a disruption of the granuloma formation process. However, the exact mechanism by which ALA-PDT exerts its anti-inflammatory effect in sarcoidosis has to be elucidated.

Due to this excellent clinical response in cutaneous sarcoidosis we tried ALA-PDT also in another granulomatous disease in two patients with recalcitrant, extensive disseminated granuloma anulare persisting for more than two years. The same, above mentioned treatment parameters were used. Within some weeks of treatment in one patient the lesions resolved, in the second patient there was a marked improvement of the skin lesions.

These case reports suggest that ALA-PDT might be effective in the treatment of granulomatous skin disorders. However, controlled clinical trials are necessary involving a larger number of patients to prove these preliminary results.

19.4 Eczema

Another potential indication for topical PDT in dermatology might be eczema. We have treated four patients (2 male, 2 female) aged between 59 and 72 years with recalcitrant chronic hyperkeratotic eczema of the hands in whom conventional treatments and bath-PUVA therapy had failed. Patients were repeatedly treated with a 3% ALA gel applied topically to the hands for 6 hours followed by light exposure (Waldmann PDT 1200; 10 J/cm^2 and 40 mW/cm^2), treatment was performed once or twice weekly for several weeks. Already after a few treatments pruritus improved, after several weeks of treatment eczema markedly improved in two patients and completely resolved in one patient and the disease remained stable within a follow up of at least 6 months. One patient did not respond to PDT. There were no side effects of therapy besides some itching during and immediately following irradiation.

19.5 Perspectives

Up to now only studies including a small numbers of patients or case reports demonstrate the potential efficacy of PDT in the treatment of non-oncologic inflammatory diseases of the skin, such as scleroderma or sarcoidosis. In contrast to the treatment of skin tumors, destruction of the tissue is not desired when treating inflammatory dermatoses. Depending on the light dose and the drug concentration applied PDT involves either cytotoxic effects or immunomodulatory effects, which exact mechanisms are not known yet. However, the knowledge of these mechanisms is substantial to define new indications in which PDT could potentially work.

References

1. S. Karrer, C. Abels, M. Landthaler, R.M. Szeimies (2000). Topical photodynamic therapy for localized scleroderma. *Acta. Dermatol. Venereol.*, **80**, 26–27.
2. N. Hunzelmann, K. Scharffetter-Kochanek, C. Hager, T. Krieg (1998). Management of localised scleroderma. *Sem. Cutan. Med. Surg.*, **17**, 34–40.
3. H. Stege, M. Berneburg, S. Humke, M. Klammer, M. Grewe, S. Grether-Beck et al. (1997). High-dose UVA1 radiation therapy for localised scleroderma. *J. Am. Acad. Dermatol.*, **36**, 938–944.
4. W.L. Morison (1997). Psoralen UVA therapy for linear and generalized morphea. *J. Am. Acad. Dermatol.*, **37**, 657–659.
5. A.H. Rook, B. Freundlich, B.V. Jegasothy, M.I. Perez, W.G. Barr, Jimenes (1992). Treatment of systemic sclerosis with extracorporal photochemotherapy. *Arch. Dermatol.*, **128**, 337–346.
6. M.M.B. Seyger, F.H.J. van den Hoogen, T. De Boo, E.M.G.J. De Jong (1997). Reliability of two methods to assess morphea: skin scoring and the use of a durometer. *J. Am. Acad. Dermatol.*, **37**, 793–796.
7. R.S. Stern, E.J. Lunder (1998). Risk of squamous cell carcinoma and methoxsalen (psoralen) and UV-A radiation (PUVA). A meta-analysis. *Arch. Dermatol.*, **134**, 1582–1585.
8. T.M. Rünger (1995). Genotoxizität, Mutagenität und Karzinogenität von UVA und UVB. *Z. Hautkr*, **70**, 877–881.
9. K. Takeda, A. Hatamochi, H. Ueki, M. Nakata, Y. Oishi (1994). Decreased collagenase expression in cultured systemic sclerosis fibroblasts. *J. Invest Dermatol.*, **103**, 359–363.
10. G. Herrmann, M. Wlaschek, K. Bolsen, K. Prenzel, G. Goerz, D. Scharffetter-Kochanek (1996). Photosensitization of uroporphyrin augments the ultraviolet A-induced synthesis of matrix metalloproteinases in human dermal fibroblasts. *J. Invest. Dermatol.*, **107**, 398–403.
11. L.S. Newman, C.S. Rose, L.A. Maier (1997). Medical progress: sarcoidosis. *N. Engl. J. Med.*, **336**, 1224–1234.
12. R.E. Jones, W.W. Chatham (1999). Update on sarcoidosis. *Curr. Opin. Rheumatol.*, **11**, 83–87.
13. P. Hillemanns, M. Untch, F. Pröve, R. Baumgartner, M. Hillemanns, M. Korell (1999). Photodynamic therapy of vulvar lichen sclerosus with 5-aminolevulinic acid. *Obstet. Gynecol.*, **93**, 71–74.

Part VI:

Concluding Remarks

Chapter 20

Practical guidelines

Piergiacomo Calzavara-Pinton

Table of contents

Abstract

In the past 25 years, a growing number of scientific and clinical reports have showed the efficacy of photodynamic therapy (PDT). However, the great number of photosensitisers, light sources, clinical indications and treatment protocols may confuse clinicians without a specific training in PDT. In addition, for many indications, controlled randomised clinical trials are lacking. This chapter aims at providing practical information and guidelines to dermatologists by suggesting those techniques that are most suitable for each drug and light combination. In addition, efficacy and safety of PDT treatments are viewed in comparison to the features of conventional therapies. Finally, perspectives of anticipated developments are outlined.

20.1 Introduction

The basic principle of photodynamic therapy (PDT) is very intriguing. Three elements, photosensitiser (PS), light and oxygen, which are individually non-toxic at those concentrations present in PDT, are photochemically combined to obtain a selective destruction of cancer cells without damage to normal surrounding tissues and without significant systemic toxicity.

Although PDT effects have been studied on almost all body organs, skin remains a preferred area of investigation because it is readily accessible to light and biological effects can be easily evaluated. In addition, skin can be sensitised topically enhancing selectivity and avoiding the general skin photosensitivity associated with the parenteral delivery of the drugs.

PDT has been studied for the treatment of actinic keratoses (AK), non-melanoma skin cancers (NMSC), i.e. basal cell carcinomas (BCC), squamous cell carcinomas (SCC) and Bowen's disease, and skin metastasis since the 1970s but only in recent years very promising results have aroused a growing awareness of PDT among dermatologists. However, the high number of PS, widely differing treatment protocols and the variety of clinical indications can puzzle the physicians without specific training or experience.

In addition, the assessment of the efficacy and convenience of PDT in comparison to well-established conventional treatment modalities is made more difficult by the absence of controlled randomised clinical trials and long-term follow-up in many indications, and the lack of cost-effectiveness analysis. This chapter aims at providing practical guidelines in order to give orientation to clinicians in the field of PDT.

20.2 Light sources

Emission spectrum, fluence rate and geometry of the light beam are the main physical characteristics of a light source for PDT.

The emission spectrum must match as much the absorption spectrum of the PS as much as possible. Protoporphyrin IX (PpIX) has its absorption maximum in the Soret band and additional smaller peaks in the green and red regions. Most PS are chosen to have one or more peaks in the red and near-infrared ranges in order to optimise the depth of light penetration into the tissue, which is directly proportional to the wavelength of the incident light.

Light intensity is another crucial point. The power density of the incident light is usually kept between 50 and 150 mW/cm^2. If the fluence rate of broadband light is greater than 200 mW/cm^2, a hyperthermic action may overlap the photochemical effects [1] and, if it is lower than 50 mW/cm^2, exposures may become excessively long.

Light sources for PDT belong to two groups: non-coherent lamps and lasers.

ALA-PDT of skin diseases was initially performed using slide projectors. These emit a polychromatic white light that can be filtered using long-pass filters [2]. In addition, most projectors lamps feature dichroic reflectors that reduce the emission of the infrared portion. On the other hand, slide projectors can not focus the beam on the lesions and the fluence rate may be insufficient to keep light exposure times within practical limits, specifically if large areas need to be irradiated. Therefore, slide projectors are being largely replaced by incoherent light sources that have been constructed especially for specific applications of PDT. A few devices have a single halogen lamp and a flexible liquid light guide that facilitates the delivery to the target area. Another light source contains several lamps that allow a broad, albeit non-uniform [3], filtered light beam with high-power density. Finally, fluorescent lamps with emission in the blue range are used for ALA PDT with Levulan Kerastick® (see below).

Unlike broad-band lamps, lasers have a rather homogeneous light beam and the emission wavelength may be chosen to exactly match an absorption peak of the PS. Continuous wave argon laser-pumped dye tuned at 630 nm and pulsed lasers, such as the gold vapour laser with emission at 628 nm and the copper vapour laser with peaks at 510 nm and 578 nm, have been widely used with HPD and Photofrin® without apparent differences in cure rates [4,5].

Tunable solid state lasers, such as the neodymium: YAG laser cover the range of 690 to 1100 nm and are particularly useful for PDT mediated by second-generation photosensitisers with absorption peaks in the far-red/near-infrared spectrum [6]. Unfortunately, high costs, variable reliability and thus frequent need of repairs limit the use of lasers. Large dimensions and lack of mobility are further disadvantages.

Through the development of semiconductor diode lasers most of these disadvantages can be avoided. These small units emit narrow spectral bands within the range from 770 to 850 nm. Diode lasers are very small solid state devices with no moving parts, minimal energy requirements, limited cooling needs and a very long expected lifetime. The crystals are made with gallium arsenide that is doped with one of several elements such as aluminium and silicon. Recently a diode laser emitting at 635 nm has become available for ALA-PDT.

20.3 Topical photodynamic therapy

ALA-PDT is a simple technique: within a few hours after the application of an ALA containing cream, skin tumours are irradiated with visible light. A great amount of clinical information suggests a high efficacy for AK and NMSC. The published data also show, that clinical results may depend on small differences in the complex interplay of treatment parameters. ALA-PDT involves the variable steps of ALA delivery (dosage, formulation, time of application, delay between application and irradiation) and photoactivation (wavelength, fluence rate, total light dose).

In December 1999, FDA approved a patented ALA formulation, Levulan Kerastick® (DUSA Inc., USA), for the treatment of AK of the face and scalp. Patients are treated as follows. Fourteen to 18 hours after application (without occlusion) of Levulan alcoholic solution containing 20% ALA, patients receive 10 J/cm^2 of blue fluorescent light at a fluence of 10 mW/cm^2.

However, this protocol is not generally accepted as a gold standard and is only approved for the treatment of AK. Further evaluation of the different methods that have been employed in the past is needed for the optimisation of the therapeutic technique. ALA is hydrophilic and may be mixed with various proprietary O/W emulsions without apparent differences in the sensitising activity. Emulsions conventionally contain a fixed concentration (20%) of ALA although 10% or 30% ALA emulsions were found equally effective [7]. This may indicate that under all these conditions, ALA concentration is not a limiting factor for PpIX formation.

Immediately after the application, the emulsion is covered by an occlusive dressing with a dark plastic foil that enhances penetration of the drug into the tissue and also prevents photodegradation of PpIX by visible light.

The application time usually ranges from 4 to 8 hours. Application times of less than 6 hours are not always followed by a homogeneous sensitisation of tumour cells in the dermis [8] whereas it is unclear if applications longer than 8 hours can enhance the cellular accumulation of Pp IX.

ALA- induced PpIX has maximal absorption at 407 nm and smaller peaks at 510, 545, 580, and 630 nm. Corresponding to this spectrum, wavelengths ranging from the blue absorbing Soret band to the red can be used for activation. In a comparison of white light with four filtered radiation bands (>515; >530; >570; >610 nm), full spectrum visible light was more effective for ALA-PDT of AK [9] although it was more painful [10]. However, for the treatment of BCC and SCC, red light encompassing wavelengths around 630 nm seems preferable because it penetrates deeper into the skin.

Light doses are principally related to the spectrum of the light source and the spectral response of the dosimeter. Typical fluences are 60–250 J/cm^2 if a laser is used and 30 up to 540 J/cm^2 with incoherent lamps.

Immediately following irradiation, the skin becomes edematous and pale, with collateral erythema surrounding the lesion. Over the next few days, treated tumours undergo necrosis with crusts and, rarely, eschar formation. If ulceration occurs, it generally heals by secondary intention in a few days (Fig. 1).

Side effects are limited to localised pain of variable degree, which starts almost immediately with irradiation and reaches a plateau after 2–10 minutes. Fractionated irradiation and local anaesthesia before the irradiation may reduce or abrogate pain.

The optimal targets of ALA-PDT are non-hyperkeratotic AK and superficial NMSC. In contrast, thick, nodular NMSC and hyperkeratotic AK [2,11–13] are rather resistant although surgical debulking of nodules [14] and curettage of scales as well as the repetition of treatments on consecutive days [11] may improve clearing rates. Morpheiform and pigmented BCC are the only NMSC that are almost always resistant to the treatment.

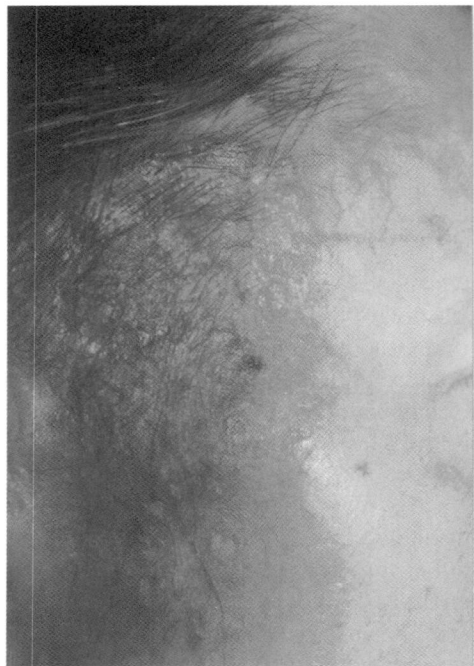

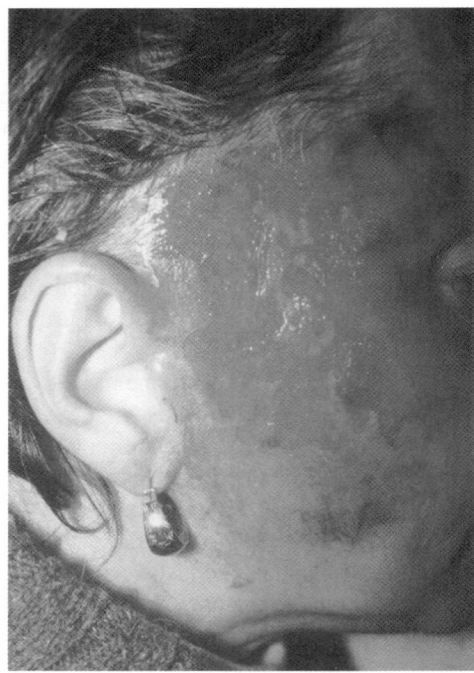

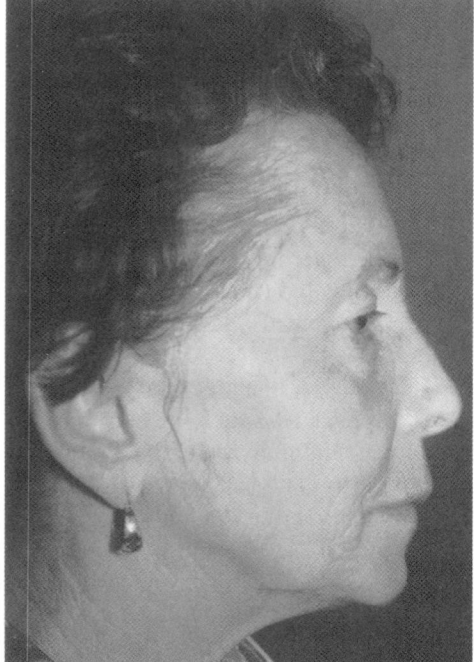

Figure 1. Superficial squamous cell carcinoma of the face before (a), soon after (b) and 4 weeks after topical ALA-PDT.

Results of large controlled and randomised clinical trials are still lacking for most indications. However, the response rates to ALA-PDT of non-hyperkeratotic AK and superficial NMSC seem to rival those reported for conventional treatments, such as 5-fluorouracil cream and locally destructive techniques, including cryotherapy, curettage, electro-desiccation, radiation therapy and excisional surgery. Topical ALA-PDT may offer several advantages: lack of invasiveness, lack of significant side effects, good tolerability and acceptance by patients, wound healing and excellent cosmetic and functional outcomes. It can be applied to patients who refuse surgery, have pacemakers or impaired coagulation, ALA-PDT may be suitable for the treatment of lesions in "surgically difficult" body locations, such as oral mucous membrane or the uro-genital area. In addition, ALA-PDT is not affected by previous therapies that the patient may have received, including radiation therapy. If recurrences occur, PDT may be repeated until complete clinical remission is reached without loss of effectiveness or cumulative toxicity with subsequent treatments. Finally, PDT does not prevent the patient from receiving other therapies subsequently.

On the other side, technical and organisational drawbacks include the delay between ALA application and the irradiation as well as the excessive duration of treatment sessions, if several lesions are present.

The efficacy of ALA-based PDT may be enhanced by the use of additive compounds, i.e. desferrioxamine [15], DMSO, and EDTA [16] that enhance penetration or interfere with certain enzymes of the heme synthesis pathway. All these additives are used to increase intracellular concentration of PpIX.

The application of more lipophilic ALA derivatives is a more recent strategy for inducing skin sensitisation. These compounds cause stronger porphyrin fluorescence and better tumour selectivity, most likely because of better penetration through cellular membranes when compared with the hydrophilic ALA [17].

In a controlled, randomised trial, topical PDT with the methyl ester of ALA was as effective as cryotherapy but showed a substantial reduction of pain with an improved compliance to the treatment by the patients [18].

In the past few years, the interest in possible non-oncological applications of PDT have remarkably increased. At present, efficacy as well as possible advantages over standard therapies remain to be established.

The application of ALA leads to accumulation of PpIX in hair follicles and sebaceous glands, suggesting the potential use for disorders of skin appendages [19].

Acne is considered a possible target because single treatments with 20% ALA plus 150 J/cm^2 broad band light (550–700 nm) can strongly reduce sebum excretion for several weeks. However, its use is limited by relevant side effects, namely severe post-treatment inflammation and transient hyperpigmentation [20]. Hair reduction of 40% was seen 6 months after a single treatment with 20% ALA cream topically applied for 3 hours under occlusion and 200 J/cm^2 of 630 nm laser light [19]. These studies showed that ALA-PDT might represent a potential treatment modality for diseases of skin appendages. Future developments will come from studies using less aggressive protocols that encompass more treatments with smaller PS and light doses.

Warts (including plantar lesions) were treated with three treatments with 20% ALA cream applied with a hydrocolloid occlusion for 4–5 h and irradiation with 45 J/cm^2 of white light. Before treatment, warts were pared down with a scalpel to visualisation of the blood vessels. Severe pain was reported and the cure rate was rather low (58%) but a few advantages over standard treatments were seen: wound or inflammation were absent, lesion care was not required and scarring was not observed [21]. In another study, a quite similar protocol, without prior debulking of lesions, was effective only for 1 out of 6 patients [22].

Results in the treatment of plaque-type psoriasis are encouraging but still experimental and far from the possibility of a clinical use. A single treatment with a 20% ALA emulsion plus 2–16 J/cm^2 of light in the spectral range 400–650 nm resulted in variable clinical responses in 22 patients [23]. In another study enrolling 3 patients, a partial regression was obtained with alternate-day treatments [24].

Unlike the prodrug ALA and its esters, most exogenous dyes are not used for topical sensitisation because they have a low solubility in aqueous media, low penetration into the skin and poor access to the interior of cells. Topical tetra-sodium-meso-tetraphenylporphine-sulfonate (TPPS4) plus exposures to light at 630 nm was highly effective (complete response rate of 93.5% and a two year recurrence rate of 18–20%) for skin tumours [25]. However, its use is limited by its potential neurotoxicity [26]. Several second generation sensitisers (verteporfin, porphycenes etc.) are currently under evaluation for topical use with phase II/III clinical trials but results are not available so far.

20.4 PDT with systemically delivered sensitisers

The first reports of PDT of skin cancers with intravenous delivery of the sensitiser date back to the 1970s. In the past years, the most commonly applied PS were HPD and porphimer sodium (Photofrin®, QLT Phototerapeutics Inc.). Both are complex mixtures of porphyrins that are derived from hematoporphyrin and contain variable amounts (>80% in the case of Photofrin®) of the active porphyrin dimers/oligomers. HPD preparations are not officially approved for clinical use in Dermatology but Photofrin® is now approved for several types of internal cancers, namely lung and oesophageal cancers in USA, Canada, Japan and several European countries.

In several experimental trials, NMSC [27,28], Kaposi's sarcoma [29], and metastatic breast carcinoma [30, 31] have been treated successfully with parenterally administered 2.5–5.0 mg HPD/kg of body weight or 1.5–3.0 mg Photofrin®/kg of body weight and 25–200 J/cm^2 of laser light [27]. Irradiation takes place 24 to 48 hours after the injection, once the maximal tumour to normal tissue is achieved.

Discomfort in the form of burning, stinging, pruritus, or pain may be experienced during light irradiation and likely reflects direct nerve stimulation and/or damage by the reactive singlet oxygen and released mediators. Premedication of patients with benzodiazepines and narcotic analgesics may help to prevent some of this discomfort.

The main disadvantage of this form of PDT is that blood distribution and retention of HPD and Photofrin® over the entire skin surface induce long-lasting skin photo-sensitivity. This requires careful photoprotective measures, i.e. clothing and physical sunscreens, during at least 4 to 6 weeks after the treatment. Although very popular, chemical sunscreens are of no value because they do not protect the patient against visible light and, indeed, their use is potentially dangerous because they may leave patients with a false sense of protection.

Despite the lack of formal clinical trials, PDT seems as effective as surgery or radiation therapy for the treatment of NMSC. In comparison to topical ALA-PDT, the clearing rate is similar but the durability of the complete remissions (recurrences ~ 4% ± 1% at 3 years) seems superior [28].

HPD or Photofrin® is generally used only for patients with Gorlin's Syndrome, or large infiltrating NMSC, particularly if in "surgically difficult", body areas, where surgery and radiotherapy are not applicable and topical application of ALA can be difficult and time-consuming. It may be considered also for Kaposi's sarcoma [29] and metastatic breast cancers [30,31] that are unresponsive to conventional treatments. The aim of PDT in these settings is palliation because distant metastases eventually develop in most patients.

A single PDT treatment showed some efficacy also in the treatment of plaque-type psoriasis [32] although prolonged photosensitivity and skin necrosis [32] were unacceptable adverse effects. A protocol encompassing several irradiations with UVA following a single injection of tin-protoporphyrin (2 μmol/kg) was tolerated much better, but efficacy was rather poor [33]. However, this study was important because it showed that PDT with low light and drug doses can be used to treat non-oncological skin diseases through the modulation of inflammatory mechanisms without necrotic processes triggered by high PDT doses.

This observation as well as the availability of "second generation" PS gave a renewed impulse to the investigation of clinical effects of systemic PDT. The new dyes display features close to the "ideal" ones, i.e. chemical purity, lack of dark toxicity, high quantum yield, selectivity, short time interval between the administration of the drug and its maximal accumulation in the tumour as well as short half life and rapid clearance from normal tissues. In addition they can be activated at wavelengths with optimal tissue penetration (around 650 nm or longer) allowing the treatment of large volumes of tissue and the use of diode lasers.

Several second generation PS are currently under clinical evaluation. In uncontrolled pilot studies, chloroaluminim phtalocyanine tetrasulfonate (AlPcTS) [34], benzoporphyrin derivative-monoacid ring A (BPD-MA or verteporfin) (QLT Phototherapeutics Inc., Vancouver, BC, Canada) [35] tin etiopurpurin (SnET2) (Miravant, St. Barbara, CA) [36] and lutetium texaphyrin (Lu-Tex) (Pharmacyclics Inc., USA) [37] have demonstrated a very high efficacy for NMSC as well as Kaposi' Sarcoma and skin metastasis of breast cancer.

Treatment protocols can be obtained from PDT companies but, since health authorities have not approved these techniques so far, they must be amended according to the legal and procedural operations of each country and institution.

In a preliminary study, verteporfin improved psoriatic plaques of 20 patients but, in 5 cases, the treatment was stopped due to side effects.

Therefore, verteporfin-PDT can therefore currently not be considered a therapeutic option for psoriasis although the reported results suggest a potential for further development as an alternative to PUVA and UVB phototherapy [38].

20.5 Conclusions

The present domain of topical PDT with ALA or ALA esters in Dermatology consists of non- hyperkeratotic AK and superficial NMSC. In contrast, treatment of nodular, thick NMSC and hyperkeratotic AK does not seem advantageous over treatments already available because of the need to debulk lesions prior to the treatment or to deliver several treatments. In contrast, topical ALA-PDT of non-oncological skin diseases is still experimental and does currently not seem to offer substantial advantages over standard treatments.

PDT with systemic delivery of HPD or Photofrin® represents a useful alternative for those patients with widespread lesions of NMSC, large NMSC or lesions in anatomically difficult areas, where topical PDT and conventional invasive methods are less applicable.

These statements will be probably modified in the future, as results of clinical trials with second generation PS will become available. These drugs will represent the driving force behind PDT in the 21st century.

References

1. L.O. Svasaand (1984). Thermal and optical dosimetry for photoradiation therapy of malignant tumours. In: A. Andreoni, R. Cubeddu (Eds), *Porphyrins in Tumour Phototherapy* (pp. 261–279). Plenum Press, New York.

2. J.C. Kennedy, R.H. Pottier, D.C. Pross (1990). Photodynamic therapy with endogenous protoporphyrin IX: basic principles and present clinical experience. *J. Photochem. Photobiol. B. Biol.*, **6**, 143–148.

3. S. Karrer, S.M. Szeimies, U. Hohenleutner, A. Heine, M. Landthaler (1995). Unilateral localized basaliomatosis: treatment with topical photodynamic therapy after application of 5 aminolevulinic acid. *Dermatology*, **190**, 218–222.

4. A.L. McKenzie, J.A.S. Carruth (1986). A comparison of gold vapour and dye lasers for photodynamic therapy. *Lasers Med. Sci.*, **1**, 117–120.

5. M.J. Shikowitz (1992). Comparison of pulsed and continuous wave light in photodynamic therapy of papillomas: an experimental study. *Laryngoscope*, **102**, 300–310.

6. A.M.R. Fischer, A.L. Murphree, C.J. Gomer (1995). Clinical and preclinical photodynamic therapy. *Lasers Surg. Med.*, **17**, 2–31.

7. E.W. Jeffes, J.L. McCullough, G.D. Weinstein, P.E. Fergin, J.S. Nelson, T.F. Shull, K.R. Simpson, L.M. Bukaty, W.L. Hoffman, N.L. Fong (1997). Photodynamic therapy of actinic keratosis with topical 5-aminolevulinic acid. *Arch. Dermatol.*, **133**, 727–732.

8. R.M. Szeimies, R. Hein, W. Baumler, A. Heine, M. Landthaler (1994). A possible new incoherent lamp for photodynamic treatment of superficial skin lesions. *Acta. Derm. Venereol.*, **74**, 117–119.

9. R. Fink Puches, A. Hofer, J. Smolle, H. Kerl, P. Wolf (1997). Primary clinical response and long-term follow-up of solar keratoses treated with topically applied 5- aminolevulinic acid and irradiation by different wavebands of light. *J. Photochem. Photobiol. B. Biol.*, **41**, 145–153.

10. P. Wolf, E. Rieger, H. Kerl (1993). Topical photodynamic therapy with endogenous porphyrins after application of 5-aminolevulinic acid: an alternative treatment modality for solar keratoses, superficial squamous cell carcinomas, and basal cell carcinomas? *J. Am. Acad. Dermatol.*, **28**, 17–21.

11. P.G. Calzavara-Pinton (1995). Repetitive photodynamic therapy with topical δ-aminolevulinic acid as an appropriate approach to the routine treatment of superficial non-melanoma skin cancers. *J. Photochem. Photobiol. B. Biol.*, **29**, 53–57.

12. R.M. Szeimies, S. Karrer, A. Sauerwald, M. Landthaler (1996). Photodynamic therapy with topical application of 5-aminolevulinic acid in the treatment of actinic keratoses: an initial clinical study. *Dermatology*, **192**, 246–251.

13. K. Svanberg, T. Anderson, D. Killander, I. Wang, U. Stenram, S. Andersson-Engels, R. Berg, J. Johansson, S. Svanberg (1994). Photodynamic therapy of non-melanoma malignant tumours of the skin using topical δ-aminolevulinic acid sensitisation and laser irradiation. *Br. J. Dermatol.*, **130**, 73–76.

14. M.R.M Thissen, C.A. Schroter, H.A.M. Neumann (2000). Photodynamic therapy with delta aminolevulinic acid for nodular basal cell carcinomas using a prior debulking technique. *Br. J. Dermatol.*, **142**, 338–339.

15. B. Ortel, A. Tanew, H. Honigsmann (1993). Lethal photosensitisation by endogenous porphyrins of PAM cells: modification by desferoxiamine. *J. Photochem. Photobiol. B. Biol.*, **17**, 273–278.

16. Z. Malik, G. Kostenich, L. Roitman, B. Ehrenberg, A. Orenstein (1995). Topical application of 5-aminolevulinic, DMSO and EDTA: protoporphyrin IX accumulation in skin and tumours of mice. *J. Photochem. Photobiol. B. Biol.*, **28**, 213–218.

17. Q. Peng, T. Warloe, C. Berg, J Moan, M. Kongshaug, K.E. Giercksky (1997). 5-aminolevulinic acid-based photodynamic therapy: clinical research and future challenges. *Cancer*, **79**, 2282–2308.

18. R.M. Szeimies, S. Karrer, A. Tanew, S. Radakovic, P.G. Calzavara-Pinton, C. Zane, A. Sidoroff, M. Hempel, J. Ulrich, T. Proebstle, H. Meffert, M. Mulder, D. Salomon, H. Dittmar, J.W. Bauer, K. Kernl.L. Braathen (In press). Photodynamic therapy using topical methyl-ALA (Metvix®). is as efficacious as cryotherapy in actinic keratosis, but with superior cosmetic results and high patient satisfaction. A prospective, randomised study. *J. Am. Acad. Dermatol.*

19. M. Grossman, J. Wimberly, P. Dwyer, T. Flotte, R.R. Anderson (1995). PDT for hirsutism. *Laser Surg. Med.*, **7**, 44.

20. W. Hongcharu, C.R. Taylor, Y. Chang, D. Aghassi, K. Suthamjariya, R.R. Anderson (2000). Topical ALA-photodynamic therapy for the treatment of acne vulgaris. *J. Invest. Dermatol.*, **115**, 183–192.

21. I.M. Stender, H.C. (1999). Wulf Photodynamic therapy of recalcitrant warts with 5-aminolevulinic acid: a retrospective analysis. *Acta Derm. Venereol.*, **79**, 400–401.

22. P. Amman, T. Hunziker, L.R. Braathen (1995). Topical photodynamic therapy in verrucae. *Dermatology*, **191**, 346–347.

23. P. Collins, D.J. Robinson, M.R. Stringer, G.I. Stables, R.A. Sheehan-Dare (1997). The variable response of plaque psoriasis after a single treatment with topical 5-aminolaevulinic acid photodynamic therapy. *Br. J. Dermatol.*, **137**, 743–749.

24. W.H. Boehnke, W. Sterry, R. Kaufmann (1994). Successful treatment of psoriasis by topical photodynamic therapy using polychromatic light. *Lancet*, **343**, 80.

25. O. Santoro, G. Bandieramonte, E. Melloni, R. Marchesini, F. Zunino, P. Lepera (1990). Photodynamic therapy by topical mesotetraphenylporphinesulfonate tetrasodium salt administration in superficial basal cell carcinomas. *Cancer Res.*, **50**, 4501–4503.

26. J.W. Winkelmann, G.H. Collins (1987). Neurotoxicity of tetraphenylporphinesulfonate TPPS4 and its relation to photodynamic therapy. *Photochem. Photobiol.*, **46**, 801–807.

27. P.G. Calzavara-Pinton, B. Ortel, R.M. Szeimies, C. Zane (1996). Photodynamic therapy with systemic administration of photosensitisers in Dermatology. *J. Photochem. Photobiol. B. Biol.*, **36**, 225–231.

28. A.R. Oseroff, A.S. Dee, N.P. Frawley, J.C. Parsons, B.W. Henderson, L. Blumensol (2000). Recent advances in ALA PDT in the skin, *Abstract Book 13th International Congress on*

Photobiology and 28th annual Meeting American Society for Photobiology (p. 148). San Francisco, CA.

29. F. Calzavara, L. Tomio (1991). Photodynamic therapy: clinical experience at the Department of Radiotherapy at Padova General Hospital. *J. Photochem. Photobiol. B. Biol.*, **11**, 91–95.

30. R.B. Buchanan, J.A.S. Carruth, A.L. McKenzie, S. Rhys Williams (1989). Photodynamic therapy in the treatment of malignant tumours of the skin of head and neck, *Eur. J. Surg. Oncol.*, **15**, 400–406.

31. P.W. Sperduto, T.F. DeLaney, G. Thomas, P. Smith, L.J. Dachowsky, A. Russo, R. Bonner, E. Glatstein (1991). Photodynamic therapy for chest wall recurrence in breast carcinoma. *Int. J. Rad. Oncol. Biol. Phys.*, **21**, 441–446.

32. M.W. Berns, M. Rottenmaier, J. Mc Cullough (1984). Response of psoriasis to red laser light (630 nm). following systemic injection of hematoporphyrin derivative. *Lasers Surg. Med.*, **4**, 73–77.

33. L. Emtestam, L. Berglud, B. Angelin, A. Kappas (1989). Tin-protoporphyrin and long wavelength ultraviolet light in treatment of psoriasis. *Lancet*, **1**, 1231–1233.

34. V.V. Sokolov, V.I. Chissov, R.I. Yakubovskya, E.I. Aristarkhova, E.V. Filonenko, T.A. Belous (1996). Photodynamic therapy (PDT). of malignant tumours by photosensitiser Photosens: results of 45 clinical cases. In: B. Ehrenberg, G. Jori, J. Moan (Eds), *Photochemotherapy: Photodynamic Therapy and Other Modalities* (pp. 281–287). Proc. SPIE 2625.

35. M. Ochsner (1997). Photodynamic therapy: the clinical perspective. *Arzneim Forsch. Drug Res.*, **47**, 1185–1194.

36. B.D. Wilson, T. Mang (1995). Photodynamic therapy for cutaneous malignancies. *Clin. Dermatol.*, **13**, 91–96.

37. M.F. Renschler, A. Yuen. T.J. Panella, T.J. Wieman, C. Julius, M. Panjehpur (1997). Photodynamic trials with lutetium texaphyrin PCI-0123 (Lu-Tex). *Photochem. Photobiol.*, **65s**, 47.

38. W.H. Boehncke, T. Elshort-Schmidt, R. Kaufmann (2000). Systemic photodynamic therapy is a safe and effective treatment for psoriasis. *Arch. Dermatol.*, **136**, 271–272.

Chapter 21

Cutaneous photosensitivity and photoprotection for photodynamic therapy patients

Manju Trehan and Charles R. Taylor

Table of contents

Abstract

Photodynamic therapy (PDT) is an exciting new field of photomedicine, but one of its major limitations is the associated skin photosensitivity, especially if prolonged. There are many new PDT compounds on the horizon and as they leave the laboratory and reach the forum for clinical human studies, one important issue to address will be precise characterization of their photosensitivity properties. Certain precautions are strongly recommended during the time period that the PDT patient remains photosensitive. Physical barriers, particularly protective clothing, can provide some protection against ultraviolet and visible light, but by far, the optimum method of protection in these patients is sun avoidance. The period of time during which these measures have to be maximized depends specifically on the individual photosensitizer. This chapter comprehensively reviews the various PDT photosensitizers, their associated lengths of photosensitivity, and above all, photoprotection measures of importance for the practitioner of this new therapy.

21.1 Introduction

One of the major drawbacks of this exciting modality known as PDT has been the issue of unwanted cutaneous photosensitivity, which can sometimes be quite prolonged. Newer compounds are under investigation in an effort to improve selectivity and minimize cutaneous phototoxicity. Most of these compounds are tetrapyrrole derived and therefore, have the strongest absorption peak at the Soret band (360–430 nm). Consequently, maximal photosensitivity lies in the UVA and visible blue light range. Factors that need to be taken into careful consideration when evaluating PDT compounds include absolute doses, drug pharmacokinetics, skin localization properties and the specific absorption spectrum of the photosensitizer. Topically administered photosensitizers such as aminolevulinic acid (ALA) represent one solution towards reducing generalized photosensitivity; however, most PDT agents, especially systemically administered ones, suffer some degree of photosensitivity as a side effect. Therefore, PDT patients must be educated on the best ways to protect themselves from this adverse event.

   This chapter will address cutaneous photosensitivity and phototesting as it relates to those compounds that are in the more advanced stages of development. We will also be discussing photosensitizers that are not used for the treatment of skin disease because cutaneous photosensitivity is an important issue with all forms of PDT. Pertinent issues of photoprotection for the PDT patient will also be reviewed.

21.2 Photosensitivity

21.2.1 Porphyrins

21.2.1.1 Hematoporphyrin derivative (HPD) and porfimer sodium (PF)
Porphyrin PDT sensitizers include HPD in its crude form, as well as the more purified forms which comprise dihematoporphyrin ether (DHE) and porfimer sodium, the FDA

approved purified form of HPD. HPD was the first purified derivative of hematoporphyrin to be studied as a photosensitizer for PDT. This first generation PDT sensitizer
is synthesized by extracting hematoporphyrin from bovine blood followed by treatment
with sulfuric acid and acetic acid and base hydrolysis. HPD is composed of a mixture
of porphyrin oligomers with ether and or ester linkages [1]. Besides the Soret
absorption, the clinically relevant absorption peak lies at 630 nm and is associated with
a generalized cutaneous photosensitivity that can last for up 6–8 weeks [2]. For skin
photosensitivity, the Soret absorption is most significant, whereas for the therapeutic
application, the 630 nm peak is used. DHE is one of the components of HPD and is felt
to be one of the more active fractions. PDT with DHE for bladder cancer at doses of
2 mg/kg IV and dye laser light at 630 nm resulted in phototoxicity in 6 out 7 patients.
In 4 patients the reactions consisted of erythema and edema of the hands and face. In
2 patients there were second degree burns requiring topical therapy [3]. Clinical
photosensitivity reactions occured within 10 days after administration of DHE, but the
overall duration of photosensitivity was not determined.

PF (Photofrin™) is the FDA approved form of purified HPD and has provided the
basis for most of our pre-clinical and clinical experience in PDT. As a porphyrin based
photosensitizer, PF exhibits a maximum absorption spectrum in the Soret band followed
by 4 smaller peaks between 500 and 635 nm. In order to allow for maximum tissue
penetration, PF is usually activated with sources of 630 nm, thus allowing for
approximately 3–5 mm depth in most tissues. By 1995, PF had received regulatory
approval in several countries, including the USA, but one of its major limitations has
been its prolonged cutaneous phototoxicity on exposure to visible light. This
photosensitivity can last for several weeks and even months when used at typical doses
of 1–2 mg/kg [4,5]. PF takes about 48 hours after injection to reach a selective
concentration in tumor tissue [6]. While clearance from a variety of tissues occurs over
40–72 hours, slow elimination from the skin accounts for the prolonged cutaneous
photosensitivity. Interestingly, Bellnier reported measurable amounts of PF in human
serum for up to 1 year after injection [7]. Based on a triexponential, 3-compartment
pharmacokinetic model, the alpha, beta, and gamma half-lives of PF were noted to be
16 h, 7.53 days and 155.56 days, respectively. There was no fluorescence detected in
serum 560 days after the PF injection. An earlier study by the same group showed a
direct correlation between cutaneous photosensitivity and the serum concentration of PF
[8]. One review of PF in the treatment of psoriasis also demonstrated the
photosensitivity of PF at 630 nm to be mild and dose-related with a duration of 6–8
weeks [9]. In practical terms, most authorities consider the peak photosensitivity to last
from 4–6 weeks and to a lesser degree up to 3 months, the periods for which maximum
photoprotective measures are geared. Clearly, lower doses of PF reduce both the severity
and the incidence of photosensitivity [10].

21.2.1.2 Delta-aminolevulinic acid (δALA) induced protoporphyrin IX (PpIX)
δALA is the prodrug or precursor to protoporphyrin IX (PpIX). Although δALA itself
is not photoreactive, PpIX is a photosensitizer. PpIX can be excited with blue light
(Soret band) and red light (630 nm absorption peak) [11]. Fortunately, PpIX and its
precursors are rapidly eliminated from the body, usually within 48 hours and therefore,
the risk of photosensitization reactions from systemically administered δALA are

limited to the first few days only [12]. Topical δALA has the added advantage of the absence of generalized cutaneous photosensitivity. Whether given topically or systemically, δALA predominantly results in an accumulation of PpIX in the epidermis and skin appendages, while sparing the cells of the dermis and the vascular endothelium. A study involving iontophoresis of δALA by Rhodes et al. demonstrated that phototoxicity could be predicted by the δALA dose, the irradiation dose, and the photobleaching of PpIX [13]. While PpIX is an efficient photosensitizer, it is also very easily photobleached in a 2 step process, thus minimizing photosensitivity because it can decay by both light-independent conversion to heme and light-dependent conversion to "bleached PpIX". Photo-protoporphyrin IX is the first product of the photobleaching pathway and is a potent photosensitizer in itself, which is then further photobleached to an inert compound. Unbleached PpIX can either be excreted by the hepatobiliary pathway or can be metabolically converted into heme. Excretion of PpIX and its conversion to heme and, potentially, photobleaching, limit clinical photosensitivity after systemic δALA.

21.2.1.3 Porphines
The porphines are another class of synthetic porphyrins of which tetra-sodium-meso-tetraphenylporphine-sulfonate (TPPS$_4$) has been studied for its application in skin disorders. TPPS$_4$ is a hydrophilic compound which is an effective PDT agent. Because it is most often locally administered via intratumoral injection, prolonged photo-sensitivity is less of an issue. It also has an absorption peak at 630 nm. In the past, the application of this compound in PDT-medicine was tainted by reports of potential neurotoxicity [14]. More recently, intratumoral injection of a purified form of TPPS$_4$ was evaluated in 9 patients with locally recurrent breast cancer and was found to have no photosensitivity of the skin and negligible neurotoxicity [15]. As more data becomes available, this compound may resurface as a viable PDT agent.

21.2.2 Chlorins

The chlorin-derived agents are a heterogenous group of compounds containing a tetrapyrrole nucleus with one double bond reduced. The red absorption peaks are found at wavelengths greater than 650 nm.

21.2.2.1 Benzoporphyrin derivative monoacid ring-A (BPD-MA)
BPD-MA (Verteporfin™) is a semisynthetic potent lipophilic chlorin, classified as a second-generation photosensitizer. This drug has been approved by the FDA for its application in "age related macular degeneration". Besides the Soret band, the major absorption peak of BPD-MA is at 690 nm and this peak is four times greater than PF absorption at 630 nm [16,17]. Currently, BPD-MA is formulated in a liposomal preparation and is administered intravenously. Pharmacokinetic studies demonstrate a higher concentration of this drug in blood than tissue during the first 8 hours after infusion. The study by Lin et al. demonstrated that skin phototoxicity induced by BPD-MA was biphasic and closely correlated with the plasma pharmacokinetics rather than tissue drug levels [18]. Skin phototoxicity was maximum immediately after drug

administration and then decreased rapidly. The second phase of phototoxicity started about 1 hour after the injection and had an approximate duration of 5–6 days [19]. The skin phototoxicity decreases rapidly as a function of time after BPD-MA injection at all drug doses. This correlates with the finding that clinical skin photosensitivity due to BPD-MA occurs maximally during the first 72 hours and then rapidly disappears.

21.2.2.2 5,-10,-15,-20-tetra(m-hydroxyphenyl) chlorin (mTHPC)

mTHPC (temoporfin or Foscan™) is also a second generation photosensitizer that appears to be superior to PF and HPD in many ways [20,21]. First, this drug has a single component purity of 98% and an absorption coefficient in the red at 652 nm, which is greater than that of PF at 630 nm by one order of magnitude. The Soret absorption maximum is located at 420 nm. Secondly, mTHPC is also much more phototoxic than PF. Significant photosensitivity was observed during the first week in 12 of 66 subjects who experienced either redness, edema and or occasional blistering in sunlight after undergoing treatment with mTHPC-PDT [22]. A skin photosensitivity study demonstrated that the duration of cutaneous photosensitivity was proportional to the mTHPC drug dose, when the fluence rate was kept constant at 100 mW/cm^2 [21]. Peak skin photosensitivity was observed between 2 and 6 days after injection and persisted for about 6 weeks for an injected dose of 0.15 mg/kg. There were no photoxicity related adverse events that were recorded after the first week, indicating a more rapid clearance from the skin in comparison to PF. In general, the maximum critical period for strict sun avoidance was said to be up to 2 weeks after the administration of mTHPC at 0.15 mg/kg. Hence, the authors suggest that doses of less than 0.3 mg/kg or intermediate doses of about 0.15 mg/kg may be optimal for clinical use of mTHPC-PDT in order to prevent unacceptable skin phototoxicity. In addition, Wagnieres's study also highlighted the use of fluorescence spectroscopy as a useful predictor of the degree of photosensitivity produced by mTHPC [21]. The intensity of the mTHPC fluorescence correlated highly with the skin reaction and therefore, could be predictive of the critical time period during which sunlight had to be strictly avoided.

Interestingly, one subject who received both an intravenous dose of PF and mTHPC three years apart, showed more rapid clearing of the mTHPC from the skin [21]. The duration of photosensitization was three times shorter than that induced by PF, but since this was a single case, inter-subject variability with respect to pharmacokinetics has to be taken into account. While this report emphasizes that mTHPC has a shorter time period of associated phototoxicity, its stronger phototoxicity during the 2 week period after administration must be remembered. In particular, the strong phototoxicity of mTHPC also has to be considered very carefully in a patient undergoing surgery. Prolonged exposure to the operating room lights during a surgical procedure could result in tissue necrosis as documented by Savary et al. [23].

More recently, Ettiaratchy et al., reported partial thickness skin burns in 6 of 14 volunteers (43%) who received injections of mTHPC [24]. The subjects received a single dose of 0.1–0.129 mg/kg and then 2 weeks later underwent phototesting with sunlight. Within 48 hours, 6 subjects demonstrated a phototoxic response. Prior clinical studies have reported a much lower incidence, on the order of 15 cases of burns or photosensitivity reactions in 931 patients (1.6%). On further evaluation, it was found that all 14 volunteers experienced a photosensitivity reaction localized to the site of the infusion arm [25]. The high incidence of cutaneous reactions in this particular study has been attributed to the extravasation of the photosensitizing drug into the adjacent tissue with a subsequently delayed photosensitivity reaction.

21.2.2.3 Mono-L-aspartyl chlorin e_6 (MACE or Npe6)

PDT with MACE (Npe6) is a more recent recruit to the systemic PDT armamentarium. It is a relatively hydrophilic compound, largely bound to albumin and high density lipoproteins (HDL) and only 1–2% is bound to low density lipoproteins (LDL) [26,27]. The main therapeutic absorption peak for Npe6 is at 664 nm. It is rapidly cleared from tissues resulting in a reduced duration of cutaneous photosensitivity. Early animal studies utilizing this compound found absolutely no skin phototoxicity 24 hours after the MACE injection, correlating with the rapid clearance from plasma and tissues as demonstrated by other groups [28–30]. In contrast, a phase I clinical study in human subjects evaluated the role of this compound in conjunction with light at 664 nm for various carcinomas [31]. The protocol included a single IV injection of MACE at 0.5–3.5 mg/kg, with irradiation 4hrs later at 25–100 J/cm². The only significant toxicity noted with this modality was a temporary generalized skin photosensitivity, lasting a maximum of 4 weeks. More specifically, phototesting was performed in the 11 study participants using a solar simulator at set time intervals for up to 6 weeks. Subjects treated at doses of up to 3.5 mg/kg demonstrated minimal or mild erythema at the test site within the first 96 hours. There was no significant skin photosensitivity noted 4 days after the initial injection. Two of the 11 patients, however, did exhibit persistent mild erythema on phototesting at 4 weeks. Both subjects had received doses of 3.5 mg/kg, but since there were so few patients, no direct correlation could be established with respect to degree of skin photosensitivity and dosage.

21.2.2.4 Tin etiopurpurin (SnET_2)

SnET_2 (PhotoPoint drug) is a synthetic chlorin analogue with an absorption peak at 660 nm. This systemic sensitizer is associated with minimal cutaneous phototoxicity since it is usually cleared from the skin within one day although reactions after one or more months have been reported [32]. No clinical photosensitivity was observed in 8 patients treated with a single dose of intravenous SnET_2-PDT for recurrent cutaneous metastatic breast cancer [33]. SnET_2-PDT is currently the focus of a large clinical study for age-related macular degeneration (AMD) which is in its final stages of completion, but there is no data available at this time with respect to cutaneous photosensitivity.

21.2.3 Other tetrapyrrole compounds

21.2.3.1 Porphycenes

The porphycenes are a new class of photosensitizers that are lipophilic in nature and chemically pure with a high quantum yield of singlet oxygen on activation. They show a strong absorption peak at 640 nm. 9-Acetoxy-2,7,12,17-tetrakis-(Beta-methoxyethyl)-porphycene (ATMPn) is one such promising agent. It is characterized by better cell penetration properties and more rapid metabolism [34]. It can be applied topically or administered intravenously. Fickweiler et al. demonstrated via co-staining experiments that ATMPn tends to localize in lysosomes [34]. Ex-vivo studies with human skin suggest rapid penetration of ATMPn into tumor tissue after topical application, with poor penetration into surrounding tissue [35]. Animal studies with the porphycenes indicate the absence of cutaneous phototoxicity [36]. Systemic application of ATMPn has not been studied in humans.

21.2.3.2 Phthalocyanines

Phthalocyanines (PCs) are another group of porphyrin-like photosensitizers that may contain a variety of different metal ions such aluminum, zinc, silicon or gallium and have a strong absorption in the red region. One of the earliest studied PCs is chloroaluminum sulfonated phthalocyanine (CASPc), which is currently being evaluated for its role in patients with basal cell carcinoma, Kaposi's sarcoma and early lung cancer [32]. This water soluble substance has an absorption peak at 675 nm [37,38]. Low accumulation levels in normal skin and rapid drug elimination result in minimal cutaneous photosensitivity [38,39].

A zinc based highly substituted PC (ZnODPc) was found in animal studies to have a high affinity for tumor and serum low-density lipoproteins (LDLs). Skin photo-sensitivity was short lived (less than 24 hours) and seemed to correlate with plasma levels of the dye [40]. In mice studies, silicon PC 4 intravenously administered at a dose of 2 mg/kg resulted in a phototoxic response that could be demonstrated for up to 10 days, with return to baseline by day 14 [41,42]. The cutaneous response was elicited by exposure to solar-simulated radiation and was noted to resolve within 24 hours. At doses of 15 J/cm^2, the PC-4 induced photosensitivity was 90% inhibited when the animals were pretreated with dexamethasone 0.6 mg/kg or cyproheptadine 7.5 mg/kg, 2–6 hours prior to illumination. A 67% reduction in photosensitivity was observed in those animals that received 50 mg/kg pentoxifylline prior to solar irradiation. When PC-4-PDT was performed with an argon-pumped dye laser at 135 J/cm^2, 24 hours after administration of PC-4, tumor regression was found to be unaffected despite pretreatment with certain inhibitors of the photosensitivity response. The absorption at longer wavelengths allowing for greater depth of penetration of light, along with reduced cutaneous photosensitivity and their high purity, makes the PCs' very promising PDT agents.

21.2.4 Benzophenothiazinium dyes

The benzophenothiazinium dyes belong to the class of cationic photosensitizers. Unlike sensitizers that target tumor vasculature, the preferential intracellular localization of these dyes results in minimal damage to the surrounding vasculature and overlying skin. Cincotta et al. demonstrated high cure rates with the cationic photosensitizers in two animal models of sarcoma with minimal damage to surrounding skin tissue [43,44]. As more data accumulates with regard to animal and human studies, the cutaneous photosensitivity issues will become more clearly defined.

Methylene Blue Derivative (MBD) is a phenothiazinium dye that is highly lipophilic and also localizes in the mitochondria. Topical delivery of MBD by iontophoresis at a concentration of 0.05% resulted in a peak in photosensitivity at 24 hours with a maximum duration of 7 days. (personal communication; Dr. Mabet Alora, Wellman Laboratories, Boston, MA)

21.2.5 Lutetium texaphyrin (Lu-Tex)

Lu-Tex (Antrin™ and Lutrin™) is a synthetic, water soluble, photosensitizer with a therapeutically relevant absorption peak at 732 nm and low inherent dark toxicity. This drug has been evaluated for its potential role as a systemic PDT agent in metastatic tumors to the skin and for chest wall recurrences in breast cancer [45]. Antrin is the Lu-Tex drug formulation being evaluated for photoangioplasty and the treatment of peripheral arterial disease. In human studies so far, Lu-Tex is proving to be a highly selective systemic agent which is rapidly cleared within hours and is therefore characterized by a lack of significant cutaneous photosensitivity [46].

21.3 Photoprotection

21.3.1 Avoidance

Lessons learned from managing photosensitivity diseases including lupus erythematosus (LE), xeroderma pigmentosum, erythropoietic protoporphyria (EPP), and the other porphyrias as well as from managing patients with skin phototype I have taught us that sometimes simple *sun avoidance is the best and most adequate photoprotective measure*. Like the EPP patient who is forced to avoid sunlight, PDT patients may simply need to avoid light exposure in order to minimize phototoxicity. Unlike in EPP, PDT associated photosensitivity eventually subsides and the period of avoidance is usually limited by the clearance of the photosensitizing drug. In practice, PDT patients should avoid going out during peak sun periods, and preferably not at all during daylight hours. For ALA or BPD-MA that only cause a limited period of photosensitivity, such a measure may not be so difficult. On the other hand, for longer acting agents such as PF, avoidance alone may be problematic.

During the photosensitive period, we recommend PDT patients avoid exposure of their eyes and skin to sunlight or direct light through window glass and light from photocopiers. They should avoid brightly focused indoor lights such as dental lamps, halogen lamps, operating room lamps, skylights, etc. They should also absolutely refrain from going to tanning salons because such units emit not only UV, but also substantial amounts of visible light. Watching television or movies is safe. Windows at home should have curtains or shades to block out direct sunlight. Total darkness for PDT patients is not warranted since the presence of ambient indoor lighting may allow a gradual photobleaching reaction with subsequent inactivation of the remaining drug.

Therefore, indoor lighting is not only acceptable, but also actually beneficial since it promotes the gradual inactivation of the photosensitizer in the skin. Subjects should probably also avoid bonnet style hair dryers during the photosensitivity period, since concentrated heat has been known to trigger a photosensitivity-like reaction with resultant erythema and induration in some PDT patients (Photofrin brochure – Axcan Scandipharm Inc., Birmingham, AL 35242).

21.3.2 Physical barriers

Ideally, as mentioned above, sun avoidance during the critical periods is strongly recommended, but if PDT patients absolutely have to go out, specially designed highly protective fabrics such as "Solumbra™" or "Sun Precautions™" can provide additional protection against UVA and visible light between 360–800 nm, although the degree of protection afforded against UV was found to be greater than that for visible light [11]. Special tightly-woven, light-colored gloves may be used for the hands, wide-brimmed and hats may help protect the scalp and face. Wrap-around dark glasses are essential. Nevertheless, such protective articles cannot cover the whole body surface. Naturally, PDT physicians have looked towards sunscreens as part of the solution. However, broad spectrum chemical sunscreens do not sufficiently absorb in the visible range. Therefore, they play a minimal role in photoprotection of the PDT patient, and their use is mostly limited to those PDT agents with some absorption component in the UV range. Physical sunblocks containing zinc oxide or titanium dioxide are protective in the visible range, but may not be cosmetically acceptable to some people. Another agent, dihydrox-yacetone (DHA), the active agent in sunless tanning products, has been shown to provide some photoprotection in the UVA and visible blue range [47]. Therefore, DHA may provide some protection in the Soret region. Ingestion of agents such as beta-carotene and polypodium leucotomos may provide additional avenues yet to be systemically explored for patients undergoing PDT [48,49].

21.3.3 Special precautions for PDT patients undergoing surgery

A study by Webber published in 1996 assessed the side effects associated with δALA ingestion for 42 patients undergoing abdominal surgery [50]. Subjects received either 30 or 60 mg/kg δALA as a single bolus orally 3–6 hours prior to the scheduled time for surgery, and no cutaneous photosensitivity was observed. Since δALA phototoxicity reactions are usually limited to the first 48 hours, a strict protocol of precautions similar to those used in EPP patients, was devised to minimize any potentially adverse reactions [51–53]. From the moment of the ingestion of δALA, the patients were kept in a semi-darkened room for 48 hours. The surgical procedures were performed under lights filtered with Kodak Wratten No. 21 filters, which excluded wavelengths less than 550 nm in order to minimize photoactivation of PpIX. In the operating room, all exposed skin surfaces were covered during the operation. Except for the anesthesia equipment, no other extraneous sources of light in the operating room were allowed. In addition, the patients wore sunglasses throughout their hospitalization. The only side effects noted in the 42 patients were nausea (15%) and vomiting (5%). An analogous protocol could probably be adapted to minimize potential adverse events for any individual who requires a surgical procedure and has in the recent past received a PDT treatment with a photosensitizer associated with prolonged phototoxicity.

21.4 Additional strategies to reduce photosensitivity

In order to minimize the photosensitivity further, efforts at reducing the drug dose, the drug distribution, or drug reactivity can be entertained. Quenchers, alternate drug delivery systems, and combination therapies are some of the avenues that have been pursued.

21.4.1 Quenchers

In photosensitization reactions, reactive oxygen species (ROS) such as singlet O_2, superoxide radical O_2^-, hydroxyl radical OH and hydrogen peroxide H_2O_2 are formed. These molecules play a significant role in tissue damage as well as in cutaneous photosensitivity. Consequently, agents that quench reactive oxygen species conceptually have a role in minimizing unwanted photosensitivity.

In EPP patients, therapeutic options such as beta-carotene and N-acetyl-cysteine (NAC) have been tried, but the beneficial effects of these agents is questionable. There are several reports in the literature that attest to the photoprotective properties of beta-carotene and its ability to increase tolerance to light, but a single controlled study of 200 patients showed no significant improvement [54]. It has been argued that the lack of benefit observed in the study by Corbett et al. was as a result of lower doses. A similar study with NAC also found minimal photoprotective benefit when evaluated in 6 patients [55]. The true efficacy of these compounds with respect to photoprotection may become more clear as additional controlled studies are done. PDT patients and their associated limited photosensitivity may provide a unique forum where the application of some of these quenching agents can be put to the final test.

Consider the following observations. One study evaluated the effect of applying chemical quenchers together with psoralen on guinea pig skin prior to UVA irradiation [56]. Topical application of sodium azide (NaN$_3$) at 100ug/cm^2 resulted in almost complete inhibition of erythema and edema as did applications of DABCO (1,4-diazabicyclo{2.2.2}octane) and beta-carotene, with NaN$_3$ being the most effective and beta-carotene least. Glutathione also provided some degree of protection, but less than the other three quenchers whereas ascorbic acid had no effect. This study focused on psoralen phototoxicity and demonstrated some potential photoprotective benefits. Previous studies in albino mice, however, have demonstrated insufficient photoprotection from UV light despite oral doses of beta-carotene [57].

Dimethyl thiourea, a hydroxyl radical scavenger, demonstrated a cutaneous photoprotective effect in animals that were given 10 mg/kg intraperitoneal injections of photosensitizer dihematoporphyrin ether (DHE). McLear et al. demonstrated a protective effect of singlet oxygen scavengers 1,3 diphenylisobenzofuran (DPIBF) and L-tryptophan on the skin, in association with DHE-PDT [58]. The protection was not complete indicating the presence of other mechanisms that mediate continued phototoxicity. Other studies by the same group have shown photoprotection in 67% of animals given DPIBF [59]. These studies all seem to support the role of reactive oxygen species in the mechanism of photosensitization. The application of topical quenchers may in the future allow for the development of a therapeutic regimen that reduces cutaneous phototoxicity; however, at present, no such strategies are being used as part of PDT regimens.

21.4.2 Alternate drug delivery systems for enhanced PDT

An animal model study evaluating the use of liposome carriers for porphyrin (HPD, DHE and Polyhematoporhyrin esters-PHE) compounds in PDT was associated with a mild decrease in skin phototoxicity [4]. Ear-skin phototoxicity in mice was found to be lower with liposome delivery of HPD and DHE than with free drug. Although an increase in skin photosensitivity was observed with PHE liposome delivery, this value was still lower at 24 hours than for either of the other compounds. In addition, the uptake of DHE by tumor was enhanced by the liposomes selectivity. Combinations of photosensitizers and drug delivery strategies along with improved light sources may allow for more efficient target specific destruction and reduced photosensitivity.

21.4.3 Combination drug therapy

In conventional chemotherapy, side effects are reduced by combining drugs with different spectrums of unwanted side effects. A similar approach was used in a pilot study of four patients with skin metastases secondary to breast carcinoma. The addition of mitomycin C (MMC) enabled the use of a lower dose of PF and lower light doses, thereby producing a reduced skin phototoxic response [60]. By reducing the PF dose, there is an automatic reduction in the duration of skin phototoxicity. Lower PF concentration in the target tissue is easily compensated by increasing the irradiation time, but prolonged illumination times may become onerous for patients with multiple lesions or large target areas. The combination of MMC and PDT takes advantage of the chronic tumor hypoxia already produced by PDT. By adding a hypoxic toxin such as MMC, which specifically targets cells at low oxygen tensions, the result is an enhanced tumoricidal effect, while allowing for a reduction in either the light dose or photosensitizer dose. In terms of skin reactions, subjects demonstrated an increase in sensitivity immediately after the PF injection (0.75 mg/kg) which lasted for 2–3 weeks. Overall this finding represents a dramatic improvement since PF is often used at 2.0 mg/kg and the duration of associated photosensitivity is significantly longer.

21.5 Phototesting

There is no standardized way to phototest compounds employed in PDT. A firm knowledge of the pharmacokinetics and pharmacodynamics of the drug is valuable. The action spectrum of the photosensitizer is also important. Phototesting for PDT compounds follows a procedure similar to that used for determining a patient's minimal erythema dose (MED) before broad-band or narrow-band UVB phototherapy. In general, a grid template is used and varying amounts of solar energy, delivered by a solar simulator, are given to each unit window in the grid. At our institution, we use an adhesive-backed aluminum template through which eight 1×1 cm square holes are cut. Each square gets a different dose of light. Only the skin showing through the template holes is exposed during each phototest and careful attention is given towards protecting all other skin. This procedure is carried out before the photosensitizer's administration in order to establish the patient's baseline sensitivity. Once the PDT drug is given, similar phototests are conducted using the visible portion of a solar simulator. Typically, such phototesting might be done immediately after the drug is given and at 4–12 hours, 1, 2, 3, 7, 14, 30 days after, etc. depending on the drug's pharmacokinetics, and skin distribution. Ordinarily this phototesting procedure is repeated until the results return to baseline. Variables recorded at each light dose and at each time point include erythema, edema, vesiculation, purpura and necrosis as well subjective complaints such as burning, tingling or pain. Adequate phototesting should be part of any new photosensitizer's development in order to prospectively assess potential side effects.

Individual PDT subjects can perform their own phototest by exposing a small area of skin to sunlight for 10 minutes (for example, by placing hand into a paper bag with a 2-inch hole, followed by 10 minutes of timed sun exposure). If there is no evidence of erythema, edema or blistering within 24 hours, the subject can gradually increase exposure to sunlight with caution. If a positive reaction occurs, then continued photoprotective precautions are advised. Repeat self-phototesting is recommended bimonthly until the subject demonstrates no skin reactivity on exposure to sunlight. Since the phototoxicity can last for months, each individual has to adhere to a self-testing schedule until the sensitivity subsides.

References

1. H. Lui, R.R. Anderson (1993). Photodynamic therapy in dermatology: recent developments. *Dermatol. Clin.*, **11**, 1–13.
2. B. Ortel, P.G. Calzavara-Pinton, R.M. Szeimies, T. Hasan (1996). Perspectives in cutaneous photodynamic sensitization. *J. Photochem. Photobiol. B*, **36**, 209–211.
3. J.I. Harty, M. Amin, T.J. Wieman, M.T. Tseng, D. Ackerman, W. Broghamer (1989). Complications of whole bladder dihematoporphyrin ether photodynamic therapy. *J. Urol.*, **141**, 1341–1346.
4. R.K. Davis, R. Straight, Z. Kereszti (1990). Comparison of photosensitizers in saline and liposomes for tumor photodynamic therapy and skin phototoxicity. *Laryngoscope*, **100**, 682–686.
5. H. Mukhtar, R. Agarwal, M. Athar, R.L. Lewen, C.A. Elmets, D.R. Bickers (1991). Photodynamic therapy of murine skin tumors using Photofrin-II. *PhotoDermatol. Photo-immunol. Photomed.*, **8**, 169–175.
6. J.G. Levy (1994). Photosensitizers in photodynamic therapy. *Semin. Oncol.*, **21**, 4–10.
7. D.A. Bellnier, T.J. Dougherty (1996). A preliminary pharmacokinetic study of intravenous Photofrin in patients. *J. Clin. Laser Med. Surg.*, **14**, 311–314.
8. D.A. Bellnier, T.J. Dougherty (1989). The time course of cutaneous porphyrin photo-sensitization in the murine ear. *Photochem. Photobiol.*, **49**, 369–372.
9. P.G. Calzavara-Pinton, R.M. Szeimies, B. Ortel, C. Zane (1996). Photodynamic therapy with systemic administration of photosensitizers in dermatology. *J. Photochem. Photobiol. B.*, **36**, 225–231.

10. J.J. Schuitmaker, P. Baas, H.L. van Leengoed, F.W. van der Meulen, W.M. Star, N. van Zandwijk (1996). Photodynamic therapy: a promising new modality for the treatment of cancer. *J. Photochem. Photobiol. B.*, **34**, 3–12.

11. J.M. Menter, T.D. Hollins, R.M. Sayre, A.A. Etemadi, I. Willis, S.N. Hughes (1998). Protection against photodynamic therapy (PDT)-induced photosensitivity by fabric materials. *Photodermatol. Photoimmunol. Photomed.*, **14**, 154–159.

12. P. Mlkvy, H. Messmann, J. Regula, M. Conio, M. Pauer, C.E. Millson, A.J. MacRobert, S.G. Bown (1995). Sensitization and photodynamic therapy (PDT) of gastrointestinal tumors with 5-aminolaevulinic acid (ALA) induced protoporphyrin IX (PPIX). A pilot study. *Neoplasma*, **42**, 109–113.

13. L.E. Rhodes, M.M. Tsoukas, R.R. Anderson, N. Kollias (1997). Iontophoretic delivery of ALA provides a quantitative model for ALA pharmacokinetics and PpIX phototoxicity in human skin. *J. Invest. Dermatol.*, **108**, 87–91.

14. V. Sacchini, E. Melloni, R. Marchesini, T. Fabrizio, N. Cascinelli, O. Santoro, F. Zunino, S. Andreola, G. Bandieramonte (1987). Topical administration of tetrasodium-meso-tetra-phenylporphinesulfonate (TPPS) and red light irradiation for the treatment of superficial neoplastic lesions. *Tumori*, **73**, 19–23.

15. M. Lapes, J. Petera, M. Jirsa (1996). Photodynamic therapy of cutaneous metastases of breast cancer after local application of meso-tetra-(para-sulphophenyl)-porphin (TPPS4). *J. Photochem. Photobiol. B.*, **36**, 205–207.

16. M.M. Tsoukas, G.C. Lin, M.S. Lee, R.R. Anderson, N. Kollias (1997). Predictive dosimetry for threshold phototoxicity in photodynamic therapy on normal skin: red wavelengths produce more extensive damage than blue at equal threshold doses. *J. Invest. Dermatol.*, **108**, 501–505

17. M.M. Tsoukas, S. Gonzalez, T.J. Flotte, R.R. Anderson, M.E. Sherwood, N. Kollias (2000). Wavelength and fluence effect on vascular damage with photodynamic therapy on skin. *J. Invest. Dermatol.*, **114**, 303–308.

18. G.C. Lin, M.L. Tsoukas, M.S. Lee, S. Gonzalez, C. Vibhagool, R.R. Anderson, N. Kollias (1998). Skin necrosis due to photodynamic action of benzoporphyrin depends on circulating rather than tissue drug levels: implications for control of photodynamic therapy. *Photochem. Photobiol.*, **68**, 575–583.

19. H. Lui (1994). Photodynamic therapy in dermatology with porfimer sodium and benzoporphyrin derivative: an update. *Semin. Oncol.*, **21**, 11–14.

20. H.B. Ris, T. Krueger, A. Giger, C.K. Lim, J.C. Stewart, U. Althaus, H.J. Altermatt (1999). Photodynamic therapy with mTHPC and polyethylene glycol-derived mTHPC: a comparative study on human tumour xenografts. *Br. J. Cancer*, **79**, 1061–1066.

21. G. Wagnieres, C. Hadjur, P. Grosjean, D. Braichotte, J.F. Savary, P. Monnier, H. van den Bergh (1998). Clinical evaluation of the cutaneous phototoxicity of 5,10,15,20-tetra(m-hydroxyphenyl)chlorin. *Photochem. Photobiol.*, **68**, 382–387.

22. P. Grosjean, J.F. Savary, J. Mizeret, G. Wagnieres, A. Woodtli, J.F. Theumann, C. Fontolliet, H. Van den Bergh, P. Monnier (1996). Photodynamic therapy for cancer of the upper aerodigestive tract using tetra(m-hydroxyphenyl)chlorin. *J. Clin. Laser Med. Surg.*, **14**, 281–287.

23. J.F. Savary, P. Monnier, C. Fontolliet, J. Mizeret, G. Wagnieres, D. Braichotte, H. van den Bergh (1997). Photodynamic therapy for early squamous cell carcinomas of the esophagus, bronchi, and mouth with m-tetra (hydroxyphenyl) chlorin. *Arch. Otolaryngol Head Neck Surg.*, **123**, 162–168.

24. S. Hettiaratchy, J. Clarke, J. Taubel, C. Besa (2000). Burns after photodynamic therapy. *Br. Med. J.*, **320**, 1245.

25. R. Bryce (2000). Burns after photodynamic therapy. Drug point gives misleading impression of incidence of burns with temoporfin (Foscan). *Br. Med. J.*, **320**, 1731–1732.

26. D. Kessel (1989). Determinants of photosensitization by mono-L-aspartyl chlorin e6 [published erratum appears in *Photochem. Photobiol.*, **50**(6), 1 (Dec. 1989)]. *Photochem. Photobiol.*, **49**, 447–452.

27. D. Kessel, K.L. Whitcomb, V. Schulz (1992). Lipoprotein-mediated distribution of N-aspartyl chlorin-E6 in the mouse. *Photochem. Photobiol.*, **56**, 51–56.

28. A. Ferrario, D. Kessel, C.J. Gomer (1992). Metabolic properties and photosensitizing responsiveness of mono-L-aspartyl chlorin e6 in a mouse tumor model. *Cancer Res.*, **52**, 2890–2893.

29. C.J. Gomer, A. Ferrario (1990). Tissue distribution and photosensitizing properties of mono-L-aspartyl chlorin e6 in a mouse tumor model. *Cancer Res.*, **50**, 3985–3990.

30. W.G. Roberts, K.M. Smith, J.L. McCullough, M.W. Berns (1989). Skin photosensitivity and photodestruction of several potential photodynamic sensitizers. *Photochem. Photobiol.*, **49**, 431–438.

31. S.W. Taber, V.H. Fingar, C.T. Coots, T.J. Wieman (1998). Photodynamic therapy using mono-L-aspartyl chlorin e6 (Npe6) for the treatment of cutaneous disease: a Phase I clinical study. *Clin. Cancer Res.*, **4**, 2741–2746.

32. K. Kalka, H. Merk, H. Mukhtar (2000). Photodynamic therapy in dermatology. *J. Am. Acad. Dermatol.*, **42**, 389–413.

33. T.S. Mang, R. Allison, G. Hewson, W. Snider, R. Moskowitz (1998). A phase II/III clinical study of tin ethyl etiopurpurin (Purlytin)-induced photodynamic therapy for the treatment of recurrent cutaneous metastatic breast cancer. *Cancer J. Sci. Am.*, **4**, 378–384.

34. S. Fickweiler, C. Abels, S. Karrer, W. Baumler, M. Landthaler, F. Hofstadter, R.M. Szeimies (1999). Photosensitization of human skin cell lines by ATMPn (9-acetoxy-2,7,12,17-tetrakis-(beta-methoxyethyl)-porphycene) in vitro: mechanism of action. *J. Photochem. Photobiol. B*, **48**, 27–35.

35. S. Karrer, C. Abels, R.M. Szeimies, W. Baumler, M. Dellian, U. Hohenleutner, A.E. Goetz, M. Landthaler (1997). Topical application of a first porphycene dye for photodynamic therapy – penetration studies in human perilesional skin and basal cell carcinoma. *Arch. Dermatol. Res.*, **289**, 132–137.

36. C. Abels, R.M. Szeimies, P. Steinbach, C. Richert, A.E. Goetz (1997). Targeting of the tumor microcirculation by photodynamic therapy with a synthetic porphycene. *J. Photochem. Photobiol. B*, **40**, 305–312.

37. N.L. Oleinick, A.R. Antunez, M.E. Clay, B.D. Rihter, M.E. Kenney (1993). New phthalocyanine photosensitizers for photodynamic therapy. *Photochem. Photobiol.*, **57**, 242–247.

38. C.J. Tralau, H. Barr, D.R. Sandeman, T. Barton, M.R. Lewin, S.G. Bown (1987). Aluminum sulfonated phthalocyanine distribution in rodent tumors of the colon, brain and pancreas. *Photochem. Photobiol.*, **46**, 777–781.

39. H.I. Pass (1993). Photodynamic therapy in oncology: mechanisms and clinical use. *J. Natl. Cancer Inst.*, **85**, 443–456.

40. C. Ometto, C. Fabris, C. Milanesi, G. Jori, M.J. Cook, D.A. Russell (1996). Tumour-localising and -photosensitising properties of a novel zinc(II) octadecylphthalocyanine. *Br. J. Cancer*, **74**, 1891–1899.

41. C. Anderson, S. Hrabovsky, Y. McKinley, K. Tubesing, H.P. Tang, R. Dunbar, H. Mukhtar, C.A. Elmets (1997). Phthalocyanine photodynamic therapy: disparate effects of pharmacologic inhibitors on cutaneous photosensitivity and on tumor regression. *Photochem. Photobiol.*, **65**, 895–901.

42. C.Y. Anderson, K. Freye, K.A. Tubesing, Y.S. Li, M.E. Kenney, H. Mukhtar, C.A. Elmets (1998). A comparative analysis of silicon phthalocyanine photosensitizers for in vivo photodynamic therapy of RIF-1 tumors in C3H mice. *Photochem. Photobiol.*, **67**, 332–336.

43. L. Cincotta, J.W. Foley, T. MacEachern, E. Lampros, A.H. Cincotta (1994). Novel photodynamic effects of a benzophenothiazine on two different murine sarcomas. *Cancer Res.*, **54**, 1249–1258.

44. L. Cincotta, D. Szeto, E. Lampros, T. Hasan, A.H. Cincotta (1996). Benzophenothiazine and benzoporphyrin derivative combination phototherapy effectively eradicates large murine sarcomas. *Photochem. Photobiol.*, **63**, 229–237.

45. J.L. Sessler, R.A. Miller (2000). Texaphyrins: new drugs with diverse clinical applications in radiation and photodynamic therapy. *Biochem. Pharmacol.*, **59**, 733–739.

46. S.W. Young, K.W. Woodburn, M. Wright, T.D. Mody, Q. Fan, J.L. Sessler, W.C. Dow, R.A. Miller (1996). Lutetium texaphyrin (PCI-0123): a near-infrared, water-soluble photo-sensitizer. *Photochem. Photobiol.*, **63**, 892–897.

47. C.R. Taylor, C. Kwangsukstith, J. Wimberly, N. Kollias, R.R. Anderson (1999). Turbo-PUVA: dihydroxyacetone-enhanced photochemotherapy for psoriasis: a pilot study. *Arch. Dermatol.*, **135**, 540–544.

48. W.P. Raab, H. Tronnier, A. Wiskemann (1985). Photoprotection and skin coloring by oral carotenoids. *Dermatologica*, **171**, 371–373.

49. M.V. Alcaraz, M.A. Pathak, F. Rius, N. Kollias, S. Gonzalez (1999). An extract of Polypodium leucotomos appears to minimize certain photoaging changes in a hairless albino mouse animal model. A pilot study. *Photodermatol. Photoimmunol. Photomed.*, **15**, 120–126.

50. J. Webber, D. Kessel, D. Fromm (1997). Side effects and photosensitization of human tissues after aminolevulinic acid. *J. Surg. Res.*, **68**, 31–37.

51. G.M. Murphy (1998). Evaluation of porphyria. *Photodermatol. Photoimmunol. Photomed.*, **14**, 58–63.

52. D.A. Paslin (1992). The porphyrias. *Int. J. Dermatol.*, **31**, 527–539.

53. D.J. Todd (1994). Erythropoietic protoporphyria. *Br. J. Dermatol.*, **131**, 751–766.

54. M.F. Corbett, A. Herxheimer, I.A. Magnus, C.A. Ramsay, A. Kobza-Black (1977). The long term treatment with beta-carotene in erythropoietic protoporphyria: a controlled trial. *Br. J. Dermatol.*, **97**, 655–662.

55. J.C. Bijlmer-Iest, H. Baart de la Faille, B.S. van Asbeck, J. van Hattum, H. van Weelden, J.J. Marx, J.C. Koningsberger (1992). Protoporphyrin photosensitivity cannot be attenuated by oral N-acetylcysteine. *Photodermatol. Photoimmunol. Photomed.*, **9**, 245–249.

56. R.S. Ray, P.C. Joshi (1995). Protection of UV-radiation induced skin toxicity by reactive oxygen scavengers. *Indian J. Exp. Biol.*, **33**, 383–386.

57. R.M. Sayre, H.S. Black (1992). Beta-carotene does not act as an optical filter in skin. *J. Photochem. Photobiol. B*, **12**, 83–90.

58. P.W. McLear, R.E. Hayden (1989). Prevention of cutaneous phototoxicity in photodynamic therapy, *Am. J. Otolaryngol.*, **10**, 92–98.

59. N.D. Hogikyan, R.E. Hayden, P.W. McLear (1991). Cutaneous photoprotection using a hydroxyl radical scavenger in photodynamic therapy. *Am. J. Otolaryngol.*, **12**, 1–5.

60. P. Baas, I.P. van Geel, H. Oppelaar, M. Meyer, J.H. Beynen, N. van Zandwijk, F.A. Stewart (1996). Enhancement of photodynamic therapy by mitomycin C: a preclinical and clinical study. *Br. J. Cancer*, **73**, 945–951.

Appendix

Table 1. PDT compounds and their associated Photosensitivity Periods

Photosensitizer	Operational Absorption peak	Duration of Photosensitivity	Precautions/ Recommendations
Photofrin (PF)	630 nm	Up to 6 months	Strict avoidance of sunlight & bright lights for 30 days, protective clothing. Self phototesting every 2 weeks.
Protoporphyrin IX (δALA-prodrug)	630–633 nm	48 hours	Strict avoidance for 48 hours.
BPD-MA	690 nm	Up to 96 hours	Strict avoidance of sunlight during the first 4–7 days.
mTHPC	652 nm	Peaks at 2–6 days Duration is up to 6 weeks	Strict avoidance for 2 weeks Self phototesting every week thereafter.
MACE/Npe6	664 nm	Peaks within 96 hours Duration 1–4 weeks	Strict avoidance for 4–5 days and then self phototesting weekly till reaction subsides.
Tin etiopurpurin ($SnET_2$)	660 nm	Minimal clinical photosensitivity	Strict avoidance for 48 hours and then self phototest.
Phthalocyanines	675 nm	Minimal clinical photosensitivity	Strict avoidance for 48 hours and then self phototest.
Lu-Tex	732 nm	Minimal clinical photosensitivity	Strict avoidance for 24 hours and then self phototest.

Photodynamic Therapy and Fluorescence Diagnosis in Dermatology
P.-G. Calzavara-Pinton, R.-M. Szeimies and B. Ortel, editors.

Chapter 22

Future directions – photosensitizer targeting and new disease indications

Michael R. Hamblin and Bernhard Ortel

Table of contents

Abstract

This chapter will focus on some aspects of PDT that may be more speculative in nature and attempt to identify some applications that may become useful in the future both in dermatology and in wider clinical practice. We will attempt to outline areas of research underway in PS development and targeting strategies that have the aim of enhancing selectivity for specific cells in tumors or other tissues, This will include the use of non-covalent complexes and covalent conjugates between PS and macromolecular vehicles, where the object is to allow the macromolecular vehicle to control the targeting and pharmacokinetics, and to choose the PS based only on its photophysical and photochemical properties. Although these targeted PS will usually be systemically injected, the possibility of topical or intralesional administration should not be overlooked. The development of methods to enhance the penetration of topical PS or PS precursors through the skin's permeability barrier, and the development of new light sources and dosimetry methods will be briefly covered. Finally some new possible indications for PDT in dermatology and related fields will be mentioned, many of which are still under pre-clinical investigation. These will be grouped under oncological, non-oncologic and infectious disease headings.

22.1 Introduction

Previous chapters in this volume have concentrated on fundamental priciples and real-life clinical applications of PDT in dermatology. This chapter, however, has the relative luxury of taking a more speculative outlook, and attempting to look forward into the future and identify those areas of PDT research that are presently being conducted, and which may some day lead to clinical applications for various diseases, especially in skin. These areas will undoubtedly reflect the authors' interests which include targeting strategies for enhancing the selectivity of PDT, the use of differentiation therapy in combination with ALA-PDT, and PDT for eradicating localized infections and for modulating wound healing. The skin is an organ ideally suited to PDT with the comparative ease of light delivery and dosimetry complemented with the ability to deliver PS with topical application. The ease of making detailed inspections and accessibility for taking biopsies for follow-up evaluation will preserve the skin organ as a preferred target for preclinical experimentation and clinical applications.

22.2 Photosensitisers

22.2.1 ALA

The use of topical ALA and oral ALA and topical application of ALA esters will continue to play a major role in cutaneous and dermatological application of PDT. Advances will consist in combinations of ALA application and treatments designed to enhance the selective accumulation of ALA-induced PPIX in target lesions.

22.2.1.1 Differentiation therapy and ALA-PDT

Neoplasia involves the loss of normal mechanisms that are associated with the differentiated state. The structure and function of a given tissue is determined by a genetic program that balances cellular proliferation, differentiation, and apoptosis. Understanding the regulatory mechanisms that maintain this balance can be crucial in the rational design of new therapies. Among a host of strategies to deal with neoplasia, differentiation therapy takes advantage of the fact that a "normal" state of the cancer cells may be restored by differentiating agents [1]. The best-known indication for differentiation therapy is acute promyelocytic leukemia. For that particular malignancy, pharmacological levels of retinoic acid induce pathologically immature promyelocytic cells to terminally differentiate toward mature neutrophils [2]. A newer example of differentiation therapy is the use of antidiabetic drugs (troglitazone), ligands for the nuclear transcription factor PPARγ, to stimulate terminal differentiation of malignant breast epithelial cells [3]. Our approach differs from conventional differentiation therapy in the sense that we alter the state of the tumor cells only temporarily, to make them more susceptible to PDT [4].

In skin, the differentiation program has been well characterized, making it a premier model for the study of cellular differentiation. The majority of events in this program, such as the altered expression of keratins and other differentiation markers, and growth arrest, are under tight transcriptional control. The importance of this program is underscored by the fact that most skin diseases, including skin cancer and its precursors, involve a disturbance in the normal balance between proliferation, differentiation, and apoptosis of epidermal keratinocytes, leading to altered anatomy and function.

Our approach is based on the finding that certain differentiated cells accumulate much higher levels of ALA-induced PpIX than their proliferating, undifferentiated counterparts [4]. Consequently, photosensitivity is increased in differentiated cells. In mouse keratinocytes, differentiation is induced in vitro by increasing calcium levels in the medium. This results in a time-dependent differentiation program. Very similarly, the layers of the epidermis in vivo represent different stages of the terminal differentiation process. If skin is exposed to systemic ALA, more differentiated, superficial layers show more PpIX fluorescence than the more basal layers. These findings may help to develop combined therapeutic regimens that use differentiating agents such as retinoids to increase PpIX formation and photosensitization. Strategies combining induction of differentiation with ALA-PDT have potential applications in skin cancers as well as in benign disorders that are associated with increased proliferation, including psoriasis, where differentiating agents have shown efficacy on their own.

22.2.2 Photosensitizer targeting

22.2.2.1 Tumor-localizing PS

Since the original observation of tumor localizing ability of HPD [5] many workers have investigated the mechanism of this tendency of PS to preferentially localize in tumors and other specific organs and anatomical sites [6–8]. There has been much effort made to determine which factors in the chemical structures of the PS are optimal for maximizing the selectivity for the tumor over normal tissue and organs. This has proved quite complicated because the pharmacokinetics can vary dramatically. For instance, one PS can have its best tumor to normal tissue ratio at a relatively early time point. After administration such as 3 hours, while with another this can be at 7 days after injection. One of the particular properties of these tetrapyrrole compounds relevant to their tumor localizing ability, is their tendency to bind strongly to serum proteins and to each other. This means that most PS when injected into the bloodstream behave as macromolecules either because they are more or less firmly bound to large protein molecules or because they have formed intermolecular aggregates of similar size. Many workers have reported [9–11] on the distribution of PS between the various classes of serum proteins when mixed with serum in vitro. These are usually divided into albumin and other heavy proteins, high-density lipoprotein (HDL), low-density lipoprotein (LDL), and very low-density lipoprotein (VLDL). However even this study has been complicated by the fact that the most lipophilic PS are insoluble in aqueous media and need to be delivered in a solvent mixture which may alter the serum protein distribution of the PS. It has been found that the more lipophilic compounds generally bind preferentially to LDL and VLDL, while those of moderate lipophilicity bind to HDL, and those of more hydrophilic character bind to albumin and other heavy proteins (globulins). It has been argued that PS that preferentially bind to LDL are better tumor localizers (see below) [8], but this is by no means always the case [12].

It is useful to make a distinction between selective accumulation and selective retention. The tumor localizing ability of the PS with the faster pharmacokinetics is probably due to selective accumulation in the tumor, while the localization of PS with slower acting pharmacokinetics is more likely due to selective retention. In the selective accumulation model it is though that the increased vascular permeability to macromolecules typical of tumor neovasculature is chiefly responsible for the preferential extravasation of the PS. These quick acting PS frequently bind to albumin which is of ideal size and Stokes radius to pass through the "pores" in the endothelium of the tumor microvessels [13]. The selective retention of PS in tumors has been the subject of much speculation. As mentioned above, a popular theory maintains that the binding of the PS to LDL is of major importance [8]. In this theory, it is proposed that cancer cells overexpress the LDL (apoB/E) receptor. Upregulation of the expression of LDL receptors is one way that rapidly growing malignant cells gain cholesterol needed for the biosynthesis of lipids needed for the rapid turnover of cellular membranes. There is experimental evidence both for and against this theory [12,14]. Other theories have been proposed to account for the selective retention of PS in tumor tissue. One is that tumors have poorly developed lymphatic drainage, and that macromolecules, which extravasate from the hyperpermeable tumor neovasculature, are retained in the extravascular space [15]. Another involves the macrophages, which infiltrate solid tumors to varying extents [7]. These tumor-associated macrophages have been shown to accumulate up to thirteen times the amount of some PS compared to cancer cells [16]. The explanation for this has been proposed to be either the phagocytosis of aggregates of PS [17], or the preferential uptake by macrophages of lipoproteins which have been altered by the binding of porphyrins [7]. Another theory proposes that the low pH commonly found in tumors has the effect of trapping some of the anionic PS, which are ionized at normal physiological pH. These PS then become neutrally charged and hence more lipophilic, when they encounter the lowered pH in the tumor environment [18].

22.2.2.2 Non-covalent complexes

As mentioned above, many of the most effective PS are too hydrophobic to dissolve in aqueous solvents, and this necessitates the use of a delivery vehicle to keep the molecules in a sufficiently disaggregated state to be able to travel in the blood vessels and to extravasate into tumors. It has been found that the choice of delivery vehicle can influence the tumor selectivity of the PS [19]. The castor oil derivative Cremophor EL has been used as a delivery vehicle, but Kessel et al. have shown that its use actually changes the lipoprotein profile [20] and can affect the biodistribution of PS and the efficacy of the tumor treatment [21]. An alternative method of PS delivery is encapsulation in liposomes. It has been suggested that when liposomal PS are administered to animals, the PS is more efficiently transferred to LDL than an aqueous formulation [22]. In vivo liposomal delivery has been shown to give advantages in either biodistribution or tumor destruction compared to non-liposomal delivery for Photofrin [23], benzoporphyrin derivative (BPD) [24] and Zn phthalocyanine[25]. Other workers have used polyethylene-glycol-coated poly(lactic acid) nanoparticles [26], or albumin microspheres [27] to deliver PS.

Some workers have investigated the pre-complexing with various serum proteins. The protein most frequently investigated has been LDL for two reasons. Firstly the hydrophobic core of the LDL particle can act as a solubilizing medium for hydrophobic PS in a similar way to liposomes or Cremophor, and secondly it was proposed to increase tumor targeting by taking advantage of receptor-mediated endocytosis of the complex by the LDL receptor that has been reported to be overexpressed on cancer cells (see above). In a study of PDT of ocular melanoma in a rabbit model, LDL complexed to BPD was used. Despite the use of LDL as a carrier, early damage to the vasculature was demonstrated by light and electron microscopy [28]. Larroque et al. [29] showed that the insoluble Zn-PC could be formulated for i.v. administration as a complex with serum albumin, after which it redistributed primarily to HDL.

22.2.2.3 Conjugates with passive macromolecular tumor targeting vehicles

In recent years much progress has been made in the use of conjugates between cytotoxic drugs and polymeric carriers to increase the therapeutic ratio of tumor treatment i.e. to increase tumor selectivity while simultaneously reducing toxicity to normal organs [30]. Similar arguments have been made in favor of the use of polymeric PS conjugates to increase the targeting ability of PS. Many of the synthetic schemes for preparing polymer-PS conjugates have arisen from work on conjugates between Mabs and PS (see later). These polymers may be either natural polymers such as dextran [30] and polyamino-acids [31,32], or synthetic polymers such as N-(2-hydroxypropyl)-methacrylamide [33] and polyvinyl alcohol [34]. The preparation of conjugates between tetrapyrrole PS and polymers poses special challenges. They may aggregate to a greater or lesser extent due to the amphiphilic nature of hydrophobic tetrapyrroles joined to hydrophilic polymer chains. Water-soluble PS-polymer conjugates must be relatively hydrophilic, but may bear net cationic, anionic or neutral charges. Soukos et al. [31] showed that the cationic conjugate had the highest cellular uptake, but that the neutral conjugate was considerably more phototoxic. Conjugates may also be prepared with varying size distributions according to the molecular weight range of the original polymer. A recent study by Hamblin et al. [35] showed that poly-l-lysine chlorin e6 conjugates that had been succinylated to give them an overall anionic charge, were able to extravasate from tumor vasculature faster than conjugates with a cationic charge, and that the molecular weight of the polymers also had a remarkable effect.

Ris and coworkers have studied the effects of covalent attachment of polyethylene glycol (PEGylation) on the PS *meta*-tetrahydroxyphenylchlorin (mTHPC) [36,37]. The PS molecule was modified by the covalent attachment of 4 PEG chains in order to make it water-soluble and improve its tumor targeting and PDT properties. It was found that the pegylated molecule was less phototoxic in vitro (even on an equimolar basis) than the native molecule, but that in vivo this may have benefits in reducing phototoxicity to normal tissue. Recently, there has been a report [38] on both PEG and polyvinylalcohol being axially attached to aluminum phthalocyanine. Improved water solubility and longer serum half lives in mice were found, and some indication of improved tumor control after illumination. Bachor et al. [39] prepared conjugates between chlorin e6 and 1 μm diameter microspheres and found they gave a distinct increase in phototoxicity compared to unconjugated PS. They showed that some cancer cells take up several times more PS than non-malignant cells, due to increased rates of phagocytosis.

22.2.2.4 Conjugates with active macromolecular tumor targeting vehicles: monoclonal antibodies

The emergence in recent years of monoclonal antibodies (Mabs) which recognize tumor associated antibodies and are available in reasonable quantities, has led to their being investigated for targeted delivery of PS, in addition to being well known targeting vehicles for toxins and radioisotopes. The advantages claimed for this approach include those specific to Mabs (high specificity and affinity for their target antigens) and those specific for PS (non-toxic to normal organs which do not receive light). Because relatively large quantities of immunoconjugates can accumulate in organs such as liver and kidneys, there can be toxicity to normal organs when the cytotoxic moiety is a radioisotope or a protein toxin, but this is avoided if these organs do not receive light. It was thought that in contrast to immunoconjugates with toxin molecules and cytotoxic drugs, photoimmunoconjugates (PICs) could be active if they bound to the plasma membranes of tumor cells, since the reactive species generated upon illumination could diffuse into the cells and produce fatal damage. However it is now thought that PICs will kill cells more efficiently if they are internalized [40]. In contrast to Mab-toxin or Mab-radionuclide conjugates, photoimmunotargeting requires conjugates with high photosensitizer-to-Mab ratios, which makes the synthesis and purification complicated. The goal of any such synthesis should be to retain features essential for both photosensitizer and antibody activities and at the same time allow maximal photosensitizer incorporation.

Two basic approaches for the synthesis of antibody-photosensitizer conjugates have been used: (a) PS are linked chemically to Mabs directly, and (b) PS are linked to Mabs via polymers. The PS is bound to polymeric carriers in the first step, and the carriers are attached to the Mab in a second step. This method allows for a high PS: Mab ratio with only a small number of attachment sites on the Mab itself and, therefore, in principle, minimal losses in the immunoreactivity of the Mab. A variety of PS-carrying polymers have been used. These include dextrans [41], polyglutamic acid (PGA) [42], polyvinyl alcohols (PVA) [43], and poly [N-(2-hydroxypropyl) methacrylamide] [44] and poly-Llysines [45]. Since the antigen binding capabilities of antibodies largely reside in the Fab portion of the antibodies, conjugation at sites removed from these antigen recognition sites are most desirable, and such site specific syntheses have recently been developed [42,43,45]. In the first study of Mab-photosensitizer conjugates [46], the photosensitizer hematoporphyrin (Hp) was coupled directly to a Mab directed against the DBA/2J myosarcoma, M-1. Modestly increased photosensitized inhibition of tumor growth in mice treated with these conjugates and light was demonstrated, compared with controls treated with Hp, Mab, or light alone.

A different approach to photoimmunotargeting was exemplified in a study by Steele et al. [47] in which immune stimulation was demonstrated by targeting T-suppressor cells using a Mab (B16G)-Hp conjugate directed against an epitope on T-suppressor cells in DBA/2J mice. Photosensitized tumor regression, reported in 10 to 40% of the mice, was correlated with an increase in the killing activity of specific cytotoxic T lymphocytes against the target tumor cells. Hasan and collaborators have developed this approach as a potential application for treating the disseminated intraperitoneal spread of ovarian cancer [48–50]. It is hoped that the intraperitoneal administration of the conjugate will avoid difficulties posed by liver uptake, and restricted transport of large macromolecules from the capillaries to the tumor [51]. Light can be administered laparoscopically via a fiber optic into the peritoneal cavity where a diffusing liquid will allow fairly uniform illumination of all the surfaces [52]. Schmidt et al. [53,54] have used a similar approach to target ovarian cancer and in addition to showing selective phototoxicity to target cells in vitro, and in vivo in a tumor bearing nude rat model, they treated 3 patients with advanced ovarian cancer by i.p. administration of 1 mg Mab-phthalocyanine conjugate. At laparotomy 72 h later, after removal of gross tumor the peritoneum was irradiated with 50 J/cm^2 670 nm light and histological evidence of tumor cell death was obtained.

Hamblin and co-workers have shown [45] that the overall charge conferred on the immunoconjugate by the polymer-PS moiety can produce major differences in both the phototoxicity towards cells, and also in the biodistribution in tumor-bearing mice [51]. Recently such a positively charged conjugate targeted against human ovarian cancer was shown to combine well with cisplatin in ex vivo primary cultures, and it was found that pre-treatment with photoimmunotherapy had the potential to reverse platinum resistance which frequently presents a major problem in ovarian cancer therapy [55]. A recent report [40] describes the preparation of immunoconjugates using mTHPC and an anti-head and neck cancer Mab. Biodistribution studies showed greater tumor selectivity than unconjugated PS in a nude mouse xenograft model. A slightly different approach [44] uses polymeric linkers that bear both a PS and a cytotoxic drug (adriamycin) attached to anti ovarian cancer Mabs. These conjugates showed dark toxicity, which could be significantly increased after illumination, and which could also be competed by unconjugated Mab.

22.2.2.5 Conjugates with active macromolecular tumor targeting vehicles: other cellular receptor ligands

Malignant cells have both faster growth rates and faster metabolic rates than normal cells, and consequently have a much higher demand for many of the nutrients and building blocks that are delivered to normal cells by well regulated pathways. Cancer cells satisfy this increased demand by upregulating the expression of many of the membrane receptors that are responsible for supplying cells with these necessary molecules by receptor mediated endocytosis. There are two approaches for taking advantage of this selective overexpression for tumor targeting of PS. The approach of attaching PS to Mabs which recognize the extracellular portion of these membrane receptors has been discussed above; an alternative approach however, is to attach the PS to the natural ligand of these receptors, which may have advantages in that the maximum cellular uptake may be higher, and the intracellular location may be more sensitive due to normal intracellular processing.

As discussed above LDL may be a tumor-targeting delivery vehicle for PS and the formation of covalent conjugates with the apolipoprotein may serve to obviate the possibility of the PS exchanging between LDL and cellular membranes. Schmidt-Erfurth et al. [56] showed that retinoblastoma cells took up four times as much PS when delivered as a covalent conjugate with LDL compared to a complex with LDL, and the uptake could be partly competed by added native LDL, while Hamblin and Newman [57] reported that a LDL hematoporphyrin conjugate, although recognized by the LDL receptor on colon cancer cells and fibroblasts, was also recognized by a scavenger receptor on macrophages.

Transferrin receptors are also overexpressed on many types of cancer cell that can have an increased demand for iron, and have been used as targets to deliver various cytotoxic moieties by both Mab- and transferrin-conjugates. Hamblin and Newman [58] found that a conjugate between holotransferrin and hematoporphyrin was recognized by transferrin receptors on a colorectal cancer cell line, as demonstrated by the observation that both uptake and phototoxicity could be competed by addition of native transferrin. Gijsins and de Witte [59] used epidermal growth factor (EGF) attached to a polymer-PS carrier to target EGF receptors that are known to be frequently upregulated in malignant cells. A small effect of the EGF ligand was observed together with a larger effect depending on the nature of the carrier. Akhlyina and co-workers have used insulin-photosensitizer conjugates to produce receptor-mediated endocytosis via insulin receptors [60], and in addition have shown that a nuclear localization signal from SV40 large T antigen increases phototoxicity by directing the PS to the cell nucleus [61].

22.3 Modes of delivery

The accessibility of skin lesions to topical delivery of PS has resulted in many attempts to enhance penetration of PS through the skin permeability. This effort is not peculiar to the field of PDT as transdermal drug delivery is a major focus of research for many pharmaceuticals and biologically active substances. The major obstacle of percutaneous drug delivery is the stratum corneum and topically applied lipophilic PS may remain largely in the uppermost skin layer and the relatively high concentrations of chromophores can form an optical barrier to light delivery into deeper layers.

It has been known that certain physical procedures can enhance the permeability of the skin and two of these have been applied to increase the amount of ALA that penetrates the skin and consequently increases the amount of PPIX formed. Rhodes et al. [62] used iontophoresis which involves using an electric current to drive ionically charged molecules into the skin. ALA was iontophoresed from a 2% solution into upper inner arm skin of healthy volunteers and the time course of fluorescence was measured. Five hours post-iontophoresis, sites were irradiated with broad-band yellow-red light and resultant erythema was measured by reflectance spectroscopy. The time course of PpIX fluorescence was iontophoresis-dose-dependent. With low values of total charge PpIX fluorescence peaked at 3 h and returned to zero at 9–10 h, whereas high charges had a sustained peak at 5–10 h, falling to zero by 24 h. Both PpIX fluorescence intensity pre-irradiation and fall in PpIX fluorescence post-irradiation correlated with erythema. Another method is the use of laser-generated shock waves that have been shown to create temporary pores in the stratum corneum and can allow the passage of normally impermeant molecules [63]. This was demonstrated by Lee et al. [64] using photomechanical waves generated with a 23 nsec Q-switched ruby laser to deliver ALA into human skin. In vivo fluorescence spectroscopy was used as a non-invasive assay of transport following the application of a single photomechanical wave, which caused no pain. The dose delivered across the stratum corneum depended on the peak pressure and had a threshold at approximately 350 bar. A 30% increase in peak pressure, produced a 680% increase in the amount delivered. Two further physical methods that have been used for increasing the penetration of topically applied pharmacologically molecules, but not as yet for PS or ALA are electroporation and sonophoresis. In the former high-voltage pulses lead to increased transdermal transport as measured by systemic blood uptake and/or pharmacological response, and has been demonstrated for calcein, a fluorescent tracer, fentanyl, a potent analgesic and flurbiprofen, an antiinflammatory drug [65]. In sonophoresis, application of low-frequency ultrasound enhances skin permeability; a short application of ultrasound is used to permeabilize skin for a prolonged period of time. During this period, ultrasonically permeabilized skin may be utilized for drug delivery [66].

22.4 Light sources and dosimetry

The human skin is readily accessible to light exposure using a variety of sources. Fluorescent light sources deliver a defined spectral emission to large surface areas and may thus be the best choice for PDT of indications such as psoriasis. Incandescent lamps, halogen lamps, metal halide lamps and xenon lamps can be used with appropriate filtering to exclude possible contamination with ultraviolet radiation. Diode lasers and light emitting diodes (LEDs) offer increasing choices of wavelengths and make monochromatic irradiations less costly and more handy [67]. PDT is a rapidly developing treatment modality and can be improved by taking advantage of the new technologies available for light generation, delivery and dosimetry. There has been an enormous shift in laser technology over the past 10 years away from unstable gas-dye lasers towards stable solid-state laser technology. In concert with this, the tools for fiber optic delivery of light and monitoring of light in vivo have changed as well and the pricing of fiber optics has had a major shift within the past 5 years as commercial companies coalesce to corner specific markets. Because of the changing marketplace in this technology, there is always a gap between what is used in the medical research world and what is commercially readily available. Sometimes the research drives industry to provide new technology, but all too often fundamental medical research must make use of what presently exists in the commercial market place. One such system under development is a uniquely stable fiber optic probe for multiple site measurements in vivo [68]. These multi-tasking fibers use the spectral signature of several dyes embedded in a dosimetry fiber to simultaneously monitor the light fluence rate along a single implanted fiber. By embedding this fiber at the base of a tumor to be treated, we can continuously monitor the delivered light dose to that tissue, and provide on-line measurements of the total accumulated light dose at each monitor point. Several fibers can be implanted into the tumor, or they can also be used as surface monitors of the diffusely reflected light.

Transport of light within tissue is amenable to modeling and analysis by mathematical methods. A frequency-domain measurement system can be used to make multifrequency measurements of the type developed by Madsen et al. [69], and fit to diffusion theory analytic calculations. These measurements are robust and independent of many tissue-air boundary artifacts, assuming that the distance between the source and detector fiber optics is smaller than the dimension of the tissue itself. Numerical models for light transport in tissue consist of Monte Carlo simulations and Finite Element-based diffusion theory programs which have been implemented on complex and realistic tissue geometries [70–72]. These calculations will be useful to predict light fluence in specific geometries and to give a first order estimate of the dosimetry light fields expected in vivo, to aid in treatment planning for each specific tumor type and organ location.

22.5 New indications

22.5.1 Oncology

22.5.1.1 Melanoma

While BCC and SCC have been covered in some detail in earlier chapters, some pre-clinical and clinical work has been carried out on cutaneous pigmented melanoma. It has been assumed that because of the absorption spectrum of melanin that PS with absorption maxima in the near-infrared, i.e. between 700 and 900 nm would be particular suited for this application, although it is not apparent that this assumption has been rigorously tested. It may be that powerful PS with good pharmacokinetics and tumor-localizing properties would perform well despite the need to illuminate at wavelengths at which melanin has an appreciable absorption or scattering co-efficient. There have been a few clinical studies reported in which pigmented melanoma has been treated with PDT [73,74] in addition to several reports of PDT being used in animal models [75–79]. Another reason for considering PDT for treatment of cutaneous melanoma is that this particular tumor is well known to be immunogenic and PDT is known to be particularly adept at inducing an immune response aginst treated tumors (see later). Busetti et al. have obtained interesting results [80,81] from the photothermal targeting of pigmented melanoma with short pulsed lasers in combination with PDT.

22.5.1.2 Kaposi's sarcoma

Targeting of endothelial cells and of neovasculature by PDT are viable approaches for this tumor [82]. PDT has been tested for treatment of the cutaneous and oral lesions typical of this disease [83]. Twenty-five patients received 1.0 mg/kg of Photofrin 48 h before exposure to 100–400 J/cm^2 of 630 nm light. Of the 348 lesions treated, 289 were evaluable: 32.5% had complete clinical response, 63.3% had partial clinical response and 4.2% were clinical failures. There was a strong correlation between response and light dose: 54% of lesions achieved a complete clinical response at optimum light dose (>250 J/cm^2). There was no correlation of response with CD4 cell count nor was there a change in CD4 cell count post-treatment. At 400 J/cm^2 full field scabbing and necrosis occurred in 90% of the treated fields. Thus, the maximum tolerated dose was determined to be 300 J/cm^2. At light doses of 250 J/cm^2 and below the toxicities were limited to erythema and edema in the treatment field. Forty-three biopsies were taken 0.5 h to 4 months post-treatment. These showed little change in the B and T cell infiltrates identified. Kaposi's sarcoma cells disappeared post-treatment in certain lesions. However the need for therapy of HIV-associated Kaposi's sarcoma has been dramatically reduced [84] by highly effective therapies using anti-retroviral drugs and protease inhibitors [85,86].

22.5.1.3 Role of immune stimulation

There have been substantial advances in the understanding of the PDT-induced tumor-specific immune reaction. This effect may not be relevant to the initial tumor ablation, but may be decisive in attaining long-term tumor control. Anti-cancer immunity elicited by PDT has the attributes of an inflammation primed immune development process [87] and bears similarities to the immune reaction induced by tumor inflammation caused by bacterial vaccines or some cytokines. Macrophages phagocytose large numbers of cancer cells killed or damaged by the cytotoxic effects of PDT. Directed by powerful inflammation-associated signaling, the antigen presenting cells will process tumor-specific peptides and present them on their membranes in the context of major histocompatibility class II molecules. Presentation of tumor peptides, accompanied by intense accessory signals, creates conditions for the recognition of tumor antigens by helper T lymphocytes. These lymphocytes become activated and in turn sensitize cytotoxic T cells to tumor specific epitopes. The activity of tumor sensitized lymphocytes is not limited to the original PDT treated site but can include disseminated and metastatic lesions of the same cancer. Thus, although the PDT treatment is localized to the tumor site, its effect can have systemic attributes due to the induction of an immune reaction. PDT generated tumor sensitized lymphocytes can be recovered from distant lymphoid tissues (spleen, lymph nodes) at protracted times after light treatment [88].

22.5.2 Non-oncological

PDT was developed for the treatment of malignant tumors, but non-malignant diseases, vascular and inflammatory disorders have become targets of photosensitization. This development has been supported by the improved understanding of PDT mechanisms at the molecular, cellular and tissue level, and the appreciation that PDT is more versatile than just to be used to destroy cells and tumors.

22.5.2.1 Vascular lesions

Kaposi's sarcoma is a malignant vascular lesion that is discussed above (5.1.2). Other vascular lesions, and specifically inborn vascular malformation such as portwine stains will remain targets of selective photothermolysis (PTL) [89]. PTL is based on selective absorption by hemoglobin of the laser pulses of 577 nm light. This wavelength range has a limited tissue penetration depth but is sufficient for the early, more superficial lesions in infancy [90]. For older portwine stains or more deeply situated vascular spaces, longer wavelengths are more desirable. Photosensitizers that absorb in the red or far-red region of the visible spectrum may make these lesions accessible to PDT. Benzoporphyrin derivative (BPD-MA, Verteporphin®) has an absorption maximum at 690 nm. BPD-MA has been approved for the treatment of age-related macular degeneration of the retina, which in essence is an acquired vascular lesion [91]. Light exposures are done at an early time after systemic BPD-MA administration when the vast majority of the photosensitizer is intravascularly localized. The same approach has also been successful in treating benign intraocular vascular tumors in humans [92]. These data indicate that similar therapeutic strategies may be useful in the management of cutaneous vascular lesions. One water soluble PS, lutetium texaphyrin may have great potential for thick lesions, because it efficiently absorbs at 730 nm [93,94]. This wavelength provides increased penetration into the tissue and may make deep-seated hemangiomas an indication for PDT.

22.5.2.2 Psoriasis

ALA-induced PpIX and exogenous photosensitizers have been used with some success in the treatment of psoriasis as described in an earlier chapter. The treatment of psoriasis and other benign disorders may utilize photosensitizers for conventional PDT but may also apply specific PS targeting strategies. Recent clinical studies have demonstrated that treatment with antibodies binding to lymphocyte surface antigens may result in improvement of psoriatic skin disease. Humanized antibodies against several leukocyte antigens such as CD25 [95], CD11a [96], and CD152 [97] have shown promise in clinical studies. The conjugation of one or several PS to these antibodies may offer several advantages for clinical regimens. Firstly, the therapeutic activity of the antibody may be enhanced, allowing the use of smaller or less frequent doses of antibodies, which reduces both costs and side effects. Secondly, the choice of irradiations (e.g. use of violet or blue instead of red light) allows the phototoxic reactions to be confined to the skin.

It has been shown that several therapies that are potent in inducing apoptosis are also very efficient in clearing psoriasis [98–100]. For example, PUVA induces apoptosis in a large portion of lymphocytes at doses that barely affect survival of epidermal keratinocytes [101]. Treatment of cells with certain PDT regimens leads rapidly to apoptosis (see chapter on molecular mechanisms of PDT), and antibody targeting of photosensitizers may help to selectively inactivate infiltrating lymphocytes or lymphocyte subsets. The great efficacy of PDT to induce apoptosis also indicates, that even without the use of conjugated antibodies, PDT may be quite powerful in depleting inflammatory cells in the infiltrate of psoriatic skin.

22.5.2.3 Wound healing

Dysfunctional wound healing can involve a slowing of the process, which can lead to indolent and chronic wounds. Such slowing can be due to factors such as diabetes [102], vascular disease [103], infection [104], immune suppression [105], or advanced age [106]. Alternatively, dysfunctional wound healing can involve an accentuation or loss of control of the healing response, which can lead to undesirable sequelae such as hyperproliferative scars and keloids.

Macrophages are central to the complex process of wound healing, which involves removal of dead tissue, formulation of granulation tissue, neovascularization, stimulation of locomotion and proliferation of fibroblasts and keratinocytes, and production of collagen types I and III. Photodynamic therapy can destroy large amounts of tissue with a good healing response and good cosmetic result [107]. Photodynamic therapy can be used either to stimulate or suppress cellular responses such as cytokine release and immune function. Whether the photodynamic therapy causes stimulation or suppresion depends on the dosage. Low dose photodynamic therapy stimulates cytokine release and immune function, while high dose photodynamic therapy suppresses those processes [108,109]. Photodynamic therapy has major effects on macrophages. Low dose photodynamic therapy activates macrophages. This enhances their cytotoxicity against tumor cells [110,111]. High dose photodynamic therapy leads to production of TNF alpha, and eventually, macrophage death [112]. Fibrosis is a response to injury in which new extracellular matrix is rapidly laid down producing dense bands of collagen that are the microscopic hallmark of scarring. The extent and duration of fibrosis often far exceeds the apparent need for wound healing, causing hypertrophic scars and contractures that limit function or distort anatomy. In the peritoneum, joints, tendon sheaths, or essentially any body space with an epithelial lining, injury and fibrosis can lead to adhesions which bind tissues together. Scars remain metabolically hyperactive long after injury, both producing and degrading extracellular matrix at a rate many times that of uninjured tissue. The initiation and control of many concerted processes responsible for wound healing are governed by molecules which direct cell activity such as cytokines, growth factors, and adhesion molecules. In particular, the extracellular matrix growth factors transforming growth factor β (TGF-β), platelet derived growth factor (PDGF), and basic fibroblast growth factor (bFGF) appear to initiate and/or sustain fibrosis. Specifically TGF-β. appears to be the dominant cytokine governing the aggressiveness of the scarring response. TGF-β has been implicated in hepatic fibrosis, pulmonary fibrosis, scleroderma, and keloids. It stimulates collagen and fibronectin formation, suppresses collagenase and induces production of collagenase inhibitors. Increased TGF-β levels, increasing scarring, and more rapid healing responses are associated with disorientation and thinning of type I collagen fibers with abnormal production of proteoglycans and glycosaminoglycans in wound extracellular matrix. We have proposed that by delivering the appropiate dose of PDT (perhaps targeted to macrophages) it may be possible to avoid or lessen the fibrosis that occurs after wounds, burns, and abdominal surgery [113].

Hypertrophic scar is marked by excess collagen accumulation secondary to an increased vascularization response in the scar and an increase in fibroblast cell density. It is currently the most debilitating long-term complication of the surviving burn patient, and at present, there is no routinely effective form of therapy. Reiken et al. [114]

investigated the potential use of PIT to treat hypertrophic scar. They used an antibody-targeted PS (Sn-chlorin e6 linked to a Mab that binds to human myofibroblasts which they prepared, and tested in a model consisting of 1 mm^3 human hypertrophic scar tissue implants in athymic mice. These implants increase approximately 20-fold in volume over a period of 15 days. Four days after implantation immunoconjugate was injected directly into scar implants, and allowed to diffuse throughout for 24 hr before implants were illuminated with laser light at 630 nm (120 J/cm^2). Treatment caused a significant reduction in total growth compared to the untreated controls (P < 0.05). No effect was observed when an irrelevant conjugate (anti-Pseudomonas aeruginosa) was used.

As previously mentioned, PDT can up- or down-regulate the expression of inflammatory mediators, cytokines and growth factors depending on the dose and the cellular target [113,115,116]. Therefore we proposed that low-dose PDT (especially when targeted to macrophages) may stimulate wound healing by stimulating expression of growth factors [116]. Growth factors can be grouped functionally, according to the role they play in initiating and controlling the various phases of wound healing [117]. A first group consists of chemotactic growth factors, i.e. those that attract inflammatory cells such as monocyte/macrophage and fibroblasts to the cell site. A second group consists of growth factors that act as mitogens to stimulate cellular proliferation. A third group consists of growth factors that stimulate angiogenesis [118]. A fourth group consists of growth factors that affect the production and degradation of the extracellular matrix. A fifth group of growth factors consists of those that influence the synthesis of cytokines and growth factors of neighboring cells. Growth factors are distinguished from other cytokine subclasses by their ability to act as mitogens, chemoattractants and proliferation inducers on cells of epithelial, endothelial, and mesenchymal origins. Exogenously applied growth factors (particularly bFGF, PDGF, EGF, and TGF-β) have been used to stimulate wound healing [119–121]. A difficulty in using this approach has been in formulating the growth factors in such a way as to ensure their sustained slow release in a biologically active form [122–124].

We have been able to demonstrate using targeted PS conjugates and low levels of red light that certain growth factors that are very important in the wound healing response are induced both at the mRNA and protein levels [125]. Specifically keratinocytes and macrophages can be stimulated to produce TGF β that is responsible for attracting fibroblasts and synthesizing new collagen, and vascular endothelial growth factor responsible for the growth of new blood vessels needed for wounds to heal [118].

22.5.3 Infections

22.5.3.1 Herpes simplex lesions
The only real clinical application of PDT for an infectious disease was investigated some time ago for treating recurrent herpes simplex lesions. In the 1970s there was numerous reports both in clinical trials and clinical practice of applying a PS as a topical solution to the lesions followed by illumination [126,127]. These lesions were frequently herpes genitalis [128] but also peri-oral and corneal lesions were treated [129]. A popular treatmnent regimen was to topically apply a 1% aqueous solution of

neutral red followed by a 15 minute exposure to a 40 W bulb filtered to transmit 440–550 nm light. However the treatment was discontinued after reports of marginal effectiveness [130] and possible carcinogenicity [131].

22.5.3.2 Infected wounds and burns

Berthiaume et al. [132] evaluated the efficacy of antibody-targeted photolysis to kill bacteria in vivo using specific antibacterial photosensitizer (PS) immunconjugates. After infecting the dorsal skin in mice with *Pseudomonas aeruginosa*, both specific and non-specific tin (IV) chlorin e6-monoclonal antibody conjugates were injected at the infection site. After a 15 min incubation period, the site was exposed to 630 nm light with a power density of 100 mW/cm^2 for 1600 seconds. Irradiation resulted in a greater then 75% decrease in the number of viable bacteria at sites treated with a specific conjugate, whereas normal bacterial growth was observed in animals that were untreated or treated with a non-specific conjugate. Antibody-targeted photolysis may be a selective and versatile tool for treating a variety of infections.

Bacterial contamination of wounds is a significant cause of morbidity, delayed wound healing and increased hospital stay. In accidents and combat contamination of wounds is sometimes unavoidable, and by the time the wound is treated the bacteria may have penetrated sufficiently into the tissue to be resistant to topically applied antiseptics. Although wound infections are treated with topical and systemic antibiotics, the rapid emergence of multi-antibiotic resistant strains of bacteria is of considerable concern. Photodynamic destruction of bacteria in wounds may be an effective means of killing the bacteria while simultaneously stimulating the host immune system. In order for a photodynamic treatment to be effective it is necessary to establish the factors which govern the susceptibility of various bacterial strains to photodynamic inactivation.

Gram-negative bacteria have developed sophisticated defenses against many toxic insults [133,134]. They have a complex many layered outer barrier structure consisting of a glycocalyx, lipolysaccharide, outer membrane lipid bilayer, periplasm, peptidoglycan cell wall, and plasma membrane lipid bilayer. This barrier keeps out most PS, and is the reason why cationic PS must be used [135] or other measures taken to disturb the outer membrane [136]. However there are other manipulations available which may still further increase the permeability of Gram negative bacteria. EDTA can chelate calcium and magnesium ions which are pivotal in binding lipolysaccharide to the outer membrane [137].

Staphylococcus aureus is a Gram-positive organism which can develop multiple resistance to antibiotics. Its primary habitat is skin which it can either infect or colonize. *S aureus* infections are often nocosomial in origin, and can be major problems in hospitals dealing with burns and immunocompromised patients in intensive care units. MRSA expresses a novel penicillin binding protein which confers on it resistance to all β-lactam antibiotics. Coincidentally it is usually resistant to many other groups of antibiotics such as aminoglycosides, chloramphenicol, clindamycin, fluoroquinolones and macrolides. Only vancomycin and minocycline for systemic infections and mupirocin for topical infections have been found to be widely effective, but now resistance to both these agents is beginning to be reported [138–140]. Removal of MRSA colonization from nasal and other skin surfaces of patients and other carriers has proved a major problem [141]. Due to fears that MRSA will shortly become resistant to all known antibiotics, research into alternative ways of killing this bacterium seems timely [142].

It has been shown that Gram-positive bacteria are relatively easy to kill by PDT, while Gram-negative bacteria are resistant unless specific methods are adopted to ensure the photosensitizer can penetrate the bacterium [143]. We have shown that polycationic chlorin e6 conjugates are specific and efficient photoinactivators of both Gram-positive and Gram-negative bacteria [144]. Gram-negative bacteria however prove significantly harder to kill than Gram-positive species. We have shown that it is absolutely necessary to have a minimum degree of polycationic charge (defined as number of primary amino groups per ce6 molecule) in order for the conjugate to penetrate the outer membrane typical of Gram negative bacteria [145,146]. *S aureus* is a Gram-positive organism which is very easily killed by polycationic chlorin e6 conjugates. It is killed approximately 100 times more than *Escherichia coli* and 1000 times more than *Pseudomonas aeruginosa*. We have demonstrated in vivo photodynamic killing of luminescent *E coli* and *P aeruginosa* in infected excisional wounds on the mouse skin [147,148]. Because the luminescence emitted from the bacteria was able to be imaged by a sensitive CCD camera [149], we could show that after topical application of PS conjugate and subsequent illumination with red light, there was a light dose dependant loss of luminescence from the infecting bacteria.

22.5.3.3 Dermatophytoses: yeasts and fungi

Although much in vitro work has been carried out on PDT mediated killing of yeasts [150,151] and fungi [152], as yet there have been no in vivo tests reported. In principal PDT with topically delivered PS could be applied against cutaneous manifestations of candidiasis [153] and dermatophytosis [154]. In addition oral [155] and vulval [156] lesions could also be treated. These diseases are prevalent among HIV-infected individuals [157] and others suffering from immune suppression.

22.5.3.4 Periodontal disease

Largely due to a considerable body of work by Wilson and collaborators the use of PDT has been proposed to combat the bacteria that are the cause of epriodontal disease [158–160]. A complex mixture of Gram-positive and Gram-negative species grow as a biofilm known as subgingival plaque that derives from supragingival plaque [161]. Early supragingival biofilm occurs by the formation of columnar microcolonies that coalesce and grow within the colony in a direction from the tooth surface to the outer surface of the plaque. The subgingival biofilm consists of a zone of Gram-negative and flagellated cells, as well as spirochetes, located adjacent to the epithelial lining of the pocket while Gram-positive rods and cocci appear to be forming a tightly adherent band of organisms on the enamel or root surface [162]. It is thought that the cause of tissue destruction is a combination of the host reaction to the chronic infection leading to release of metalloproteinases and elastase from neutrophils [163,164] and the realease of bacterial derived enzymes that destroy the periodontal ligament [165] and lead to tooth loss. Current treatment regime for chronic periodontitis involves the mechanical removal of the microbial biofilms accumulating at the gingival margin and/or subgingivally. The procedure involves root surface debridement, a labor, intensive, unpleasant, scraping away of biofilm from within the periodontal pocket. The use of light and PS as a bactericidal agent in vivo is proposed as an attractive method of eliminating oral bacteria in periodontal microbial biofilms [144]. The PS may be injected into the periodontal pocket followed by the delivery of light via fiber optics into the periodontal pocket. The technique would offer the following advantages over the use of traditional antimicrobial agents: (a) Relatively rapid application of the drug in the dental pocket (antiseptics and antibiotics are difficult to maintain at high concentrations within the lesion [166]) and rapid bacterial killing after a short time of exposure of the dental pockets to light. (b) As the killing is mediated by reactive oxygen species, the development of resistance would be unlikely. (c) Killing could easily be confined to the disease lesion by restricting irradiation to this region, so that microflora at other sites would remain intact.

22.6 Conclusion

The outlook for PDT as a part of the cancer treatment armamentarium appears to be promising both for dematological indications and for tumors in other locations. When sufficient progress has been made in PS targeting, this may allow PDT to be advantageously used for widespread multifocal disease that can be reached by light. Research concerning the immunological consequences of PDT will allow combination therapies with the precise cytokine or growth factor necessary for full potentiation of the immune response. When fundamental research has elucidated which factors enable one patient to perform well while another performs badly, the treatment can be tailored to the individual's disease. Research into new PS is gathering momentum and will allow the particular optimum compound to be chosen for particular applications. For the non-cancer applications the progress to the clinic is somewhat behind that of the cancer treatment, but there may be many applications to autoimmune diseases and other hyperproliferative disorders.

Acknowledgment

Support of the authors was provided by US Department of Defense Contract N00014-94-1-0927.

References

1. L.A. Hansen, C.C. Sigman, F. Andreola, S.A. Ross, G.J. Kelloff, L.M. De Luca (2000). Retinoids in chemoprevention and differentiation therapy. *Carcinogenesis*, **21**, 1271–1279.
2. T.R. Randolph (2000). Acute promyelocytic leukemia (AML-M3) – Part 1: Pathophysiology, clinical diagnosis, and differentiation therapy. *Clin. Lab. Sci.*, **13**, 98–105.
3. O.W. Petersen, L. Ronnov-Jessen, V.M. Weaver, M.J. Bissell (1998). Differentiation and cancer in the mammary gland: shedding light on an old dichotomy. *Adv. Cancer Res.*, **75**, 135–161.

4. B. Ortel, N. Chen, J. Brissette, G.P. Dotto, E. Maytin, T. Hasan (1998). Differentiation-specific increase in ALA-induced protoporphyrin IX accumulation in primary mouse keratinocytes. *Br. J. Cancer*, **77**, 1744–1751.

5. R.L. Lipson, E.J. Baldes (1960). The photodynamic properties of a particular hematoporphyin derivative. *Arch. Dermatol.*, **82**, 508.

6. R.W. Boyle, D. Dolphin (1996). Structure and biodistribution relationships of photodynamic sensitizers. *Photochem. Photobiol.*, **64**, 469–485.

7. M.R. Hamblin, E.L. Newman (1994). On the mechanism of the tumour-localising effect in photodynamic therapy. *J. Photochem. Photobiol. B*, **23**, 3–8.

8. G. Jori, E. Reddi (1993). The role of lipoproteins in the delivery of tumour-targeting photosensitizers. *Int. J. Biochem.*, **25**, 1369–1375.

9. J.C. Maziere, R. Santus, P. Morliere, J.P. Reyftmann, C. Candide, L. Mora, S. Salmon, C. Maziere, S. Gatt, L. Dubertret (1990). Cellular uptake and photosensitizing properties of anticancer porphyrins in cell membranes and low and high density lipoproteins. *J. Photochem. Photobiol. B*, **6**, 61–68.

10. D. Kessel, A. Morgan, G.M. Garbo (1991). Sites and efficacy of photodamage by tin etiopurpurin in vitro using different delivery systems. *Photochem. Photobiol.*, **54**, 193–196.

11. M. Kongshaug, J. Moan, S.B. Brown (1989). The distribution of porphyrins with different tumour localising ability among human plasma proteins. *Br. J. Cancer*, **59**, 184–188.

12. M. Korbelik (1992). Low density lipoprotein receptor pathway in the delivery of Photofrin: how much is it relevant for selective accumulation of the photosensitizer in tumors? *J. Photochem. Photobiol. B*, **12**, 107–109.

13. F. Yuan, M. Leunig, D.A. Berk, R.K. Jain (1993). Microvascular permeability of albumin, vascular surface area, and vascular volume measured in human adenocarcinoma LS174T using dorsal chamber in SCID mice. *Microvasc. Res.*, **45**, 269–289.

14. B.A. Allison, P.H. Pritchard, J.G. Levy (1994). Evidence for low-density lipoprotein receptor-mediated uptake of benzoporphyrin derivative [published erratum appears in: *Br. J. Cancer*, Jan. 1995, **71**(1), 214]. *Br. J. Cancer*, **69**, 833–839.

15. W.G. Roberts, T. Hasan (1992). Role of neovasculature and vascular permeability on the tumor retention of photodynamic agents. *Cancer Res.*, **52**, 924–930.

16. M. Korbelik, G. Krosl (1995). Photofrin accumulation in malignant and host cell populations of a murine fibrosarcoma. *Photochem. Photobiol.*, **62**, 162–168.

17. M. Korbelik, G. Krosl, D.J. Chaplin (1991). Photofrin uptake by murine macrophages. *Cancer Res.*, **51**, 2251–2255.

18. R. Pottier, J.C. Kennedy (1990). The possible role of ionic species in selective biodistribution of photochemotherapeutic agents toward neoplastic tissue. *J. Photochem. Photobiol. B*, **8**, 1–16.

19. E. Reddi (1997). Role of delivery vehicles for photosensitizers in the photodynamic therapy of tumours. *J. Photochem. Photobiol. B*, **37**, 189–195.

20. K. Woodburn, E. Sykes, D. Kessel (1995). Interactions of Solutol HS 15 and Cremophor EL with plasma lipoproteins. *Int. J. Biochem. Cell Biol.*, **27**, 693–699.

21. K. Woodburn, C.K. Chang, S. Lee, B. Henderson, D. Kessel (1994). Biodistribution and PDT efficacy of a ketochlorin photosensitizer as a function of the delivery vehicle. *Photochem. Photobiol.*, **60**, 154–159.

22. F. Ginevra, S. Biffanti, A. Pagnan, R. Biolo, E. Reddi, G. Jori (1990). Delivery of the tumour photosensitizer zinc(II)-phthalocyanine to serum proteins by different liposomes: studies in vitro and in vivo. *Cancer Lett.*, **49**, 59–65.

23. F. Jiang, L. Lilge, B. Logie, Y. Li, M. Chopp (1997). Photodynamic therapy of 9L gliosarcoma with liposome-delivered photofrin. *Photochem. Photobiol.*, **65**, 701–706.

24. A.M. Richter, E. Waterfield, A.K. Jain, A.J. Canaan, B.A. Allison, J.G. Levy (1993). Liposomal delivery of a photosensitizer, benzoporphyrin derivative monoacid ring A (BPD), to tumor tissue in a mouse tumor model. *Photochem. Photobiol.*, **57**, 1000–1006.

25. L. Polo, A. Segalla, G. Jori, G. Bocchiotti, G. Verna, R. Franceschini, R. Mosca, P.G. De Filippi (1996). Liposome-delivered 131I-labelled Zn(II)-phthalocyanine as a radio-diagnostic agent for tumours. *Cancer Lett.*, **109**, 57–61.

26. E. Allemann, J. Rousseau, N. Brasseur, S.V. Kudrevich, K. Lewis, J.E. van Lier (1996). Photodynamic therapy of tumours with hexadecafluoro zinc phthalocyanine formulated in PEG-coated poly(lactic acid) nanoparticles. *Int. J. Cancer*, **66**, 821–824.

27. R. Margalit, E. Silbiger (1985). Albumin microspheres as delivery systems for photodynamic drugs: physico-chemical studies and their implications for in vivo situations. *J. Microencapsul.*, **2**, 183–196.

28. U. Schmidt-Erfurth, W. Bauman, E. Gragoudas, T.J. Flotte, N.A. Michaud, R. Birngruber, T. Hasan (1994). Photodynamic therapy of experimental choroidal melanoma using lipoprotein-delivered benzoporphyrin. *Ophthalmology*, **101**, 89–99.

29. C. Larroque, A. Pelegrin, J.E. Van Lier (1996). Serum albumin as a vehicle for zinc phthalocyanine: photodynamic activities in solid tumour models. *Br. J. Cancer*, **74**, 1886–1890.

30. R. Duncan, F. Spreafico (1994). Polymer conjugates. Pharmacokinetic considerations for design and development. *Clin. Pharmacokinet.*, **27**, 290–306.

31. N.S. Soukos, M.R. Hamblin, T. Hasan (1997). The effect of charge on cellular uptake and phototoxicity of polylysine chlorin e6-conjugates. *Photochem. Photobiol.*, **65**, 723–729.

32. B.A. Goff, M. Bamberg, T. Hasan (1991). Photoimmunotherapy of human ovarian carcinoma cells ex vivo. *Cancer Res.*, **51**, 4762–4767.

33. N.L. Krinick, Y. Sun, D. Joyner, J.D. Spikes, R.C. Straight, J. Kopecek (1994). A polymeric drug delivery system for the simultaneous delivery of drugs activatable by enzymes and/or light. *J. Biomater. Sci. Polym. Ed.*, **5**, 303–324.

34. N. Davis, D. Liu, A.K. Jain, S.Y. Jiang, F. Jiang, A. Richter, J.G. Levy (1993). Modified polyvinyl alcohol-benzoporphyrin derivative conjugates as phototoxic agents. *Photochem. Photobiol.*, **57**, 641–647.

35. M.R. Hamblin, M. Rajadhyaksha, T. Momma, N.S. Soukos, T. Hasan (1999). In vivo fluorescence imaging of the transport of charged chlorin e6 conjugates in a rat orthotopic prostate tumour. *Br. J. Cancer*, **81**, 261–268.

36. H.B. Ris, V. Im Hof, C.M. Stewart, D. Mettler, H.J. Altermatt (1998). Endobronchial photodynamic therapy: comparison of mTHPC and polyethylene glycol-derived mTHPC on human tumor xenografts and tumor-free bronchi of minipigs. *Lasers Surg. Med.*, **23**, 25–32.

37. H.B. Ris, T. Krueger, A. Giger, C.K. Lim, J.C. Stewart, U. Althaus, H.J. Altermatt (1999). Photodynamic therapy with mTHPC and polyethylene glycol-derived mTHPC: a comparative study on human tumour xenografts. *Br. J. Cancer*, **79**, 1061–1066.

38. N. Brasseur, R. Ouellet, C. La Madeleine, J.E. van Lier (1999). Water-soluble aluminium phthalocyanine-polymer conjugates for PDT: photodynamic activities and pharmacokinetics in tumour-bearing mice. *Br. J. Cancer*, **80**, 1533–1541.

39. R. Bachor, C.R. Shea, R. Gillies, T. Hasan (1991). Photosensitized destruction of human bladder carcinoma cells treated with chlorin e6-conjugated microspheres. *Proc. Natl. Acad. Sci. USA*, **88**, 1580–1584.

40. M.B. Vrouenraets, G.W. Visser, F.A. Stewart, M. Stigter, H. Oppelaar, P.E. Postmus, G.B. Snow, G.A. van Dongen (1999). Development of meta-tetrahydroxyphenylchlorin-monoclonal antibody conjugates for photoimmunotherapy. *Cancer Res.*, **59**, 1505–1513.

41. A.R. Oseroff, D. Ohuoha, T. Hasan, J.C. Bommer, M.L. Yarmush (1986). Antibody-targeted photolysis: selective photodestruction of human T-cell leukemia cells using monoclonal antibody-chlorin e6 conjugates. *Proc. Natl. Acad. Sci. USA*, **83**, 8744–8748.

42. T. Hasan, A. Lin, D. Yarmush, A. Oseroff, M. Yarmush (1989). Monoclonal antibody-chromophore conjugates as selective phototoxins. *J. Controlled Release*, **10**, 107–117.

43. F.N. Jiang, S. Jiang, D. Liu, A. Richter, J.G. Levy (1990). Development of technology for linking photosensitizers to a model monoclonal antibody. *J. Immunol. Methods*, **134**, 139–149.

44. V. Omelyanenko, P. Kopeckova, C. Gentry, J.G. Shiah, J. Kopecek (1996). HPMA copolymer-anticancer drug-OV-TL16 antibody conjugates. 1. influence of the method of synthesis on the binding affinity to OVCAR-3 ovarian carcinoma cells in vitro. *J. Drug Target*, **3**, 357–373.

45. M.R. Hamblin, J.L. Miller, T. Hasan (1996). Effect of charge on the interaction of site-specific photoimmunoconjugates with human ovarian cancer cells. *Cancer Res.*, **56**, 5205–5210.

46. D. Mew, C.K. Wat, G.H. Towers, J.G. Levy (1983). Photoimmunotherapy: treatment of animal tumors with tumor-specific monoclonal antibody-hematoporphyrin conjugates. *J. Immunol.*, **130**, 1473–1477.

47. J.K. Steele, D. Liu, A.T. Stammers, S. Whitney, J.G. Levy (1988). Suppressor deletion therapy: selective elimination of T suppressor cells in vivo using a hematoporphyrin conjugated monoclonal antibody permits animals to reject syngeneic tumor cells. *Cancer Immunol. Immunother.*, **26**, 125–131.

48. B.A. Goff, U. Hermanto, J. Rumbaugh, J. Blake, M. Bamberg, T. Hasan (1994). Photoimmunotherapy and biodistribution with an OC125-chlorin immunoconjugate in an in vivo murine ovarian cancer model. *Br. J. Cancer*, **70**, 474–480.

49. B.A. Goff, J. Blake, M.P. Bamberg, T. Hasan (1996). Treatment of ovarian cancer with photodynamic therapy and immunoconjugates in a murine ovarian cancer model. *Br. J. Cancer*, **74**, 1194–1198.

50. K.L. Molpus, M.R. Hamblin, I. Rizvi, T. Hasan (2000). Intraperitoneal photo-immunotherapy of ovarian carcinoma xenografts in nude mice using charged photoimmunoconjugates. *Gynecol. Oncol.*, **76**, 397–404.

51. L.R. Duska, M.R. Hamblin, M.P. Bamberg, T. Hasan (1997). Biodistribution of charged F(ab')2 photoimmunoconjugates in a xenograft model of ovarian cancer. *Br. J. Cancer*, **75**, 837–844.

52. L. Lilge, K. Molpus, T. Hasan, B.C. Wilson (1998). Light dosimetry for intraperitoneal photodynamic therapy in a murine xenograft model of human epithelial ovarian carcinoma. *Photochem. Photobiol.*, **68**, 281–288.

53. S. Schmidt, U. Wagner, P. Oehr, D. Krebs (1992). Clinical use of photodynamic therapy in gynecologic tumor patients – antibody-targeted photodynamic laser therapy as a new oncologic treatment procedure. *Zentralbl. Gynakol.*, **114**, 307–311.

54. S. Schmidt, U. Wagner, B. Schultes, P. Oehr, W. Decleer, W. Ertmer, H. Lubaschowski, H.J. Biersack, D. Krebs (1992). Photodynamic laser therapy with antibody-bound dyes. A new procedure in therapy of gynecologic malignancies. *Fortschr. Med.*, **110**, 298–301.

55. L.R. Duska, M.R. Hamblin, J.L. Miller, T. Hasan (1999). Combination photo-immunotherapy and cisplatin: effects on human ovarian cancer ex vivo. *J. Natl. Cancer Inst.*, **91**, 1557–1563.

56. U. Schmidt-Erfurth, H. Diddens, R. Birngruber, T. Hasan (1997). Photodynamic targeting of human retinoblastoma cells using covalent low-density lipoprotein conjugates. *Br. J. Cancer*, **75**, 54–61.

57. M.R. Hamblin, E.L. Newman (1994). Photosensitizer targeting in photodynamic therapy. II. Conjugates of haematoporphyrin with serum lipoproteins. *J. Photochem. Photobiol. B*, **26**, 147–157.

58. M.R. Hamblin, E.L. Newman (1994). Photosensitizer targeting in photodynamic therapy.I. Conjugates of haematoporphyrin with albumin and transferrin. *J. Photochem. Photobiol. B.*, **26**, 45–56.

59. A. Gijsens, P. De Witte (1998). Photocytotoxic action of EGF-PVA-Sn(IV)chlorin e6 and EGF-dextran-Sn(IV)chlorin e6 internalizable conjugates on A431 cells. *Int. J. Oncol.*, **13**, 1171–1177.

60. T.V. Akhlynina, A.A. Rosenkranz, D.A. Jans, A.S. Sobolev (1995). Insulin-mediated intracellular targeting enhances the photodynamic activity of chlorin (e6). *Cancer Res.*, **55**, 1014–1019.

61. T.V. Akhlynina, D.A. Jans, A.A. Rosenkranz, N.V. Statsyuk, I.Y. Balashova, G. Toth, I. Pavo, A.B. Rubin, A.S. Sobolev (1997). Nuclear targeting of chlorin (e6) enhances its photosensitizing activity. *J. Biol. Chem.*, **272**, 20328–20331.

62. L.E. Rhodes, M.M. Tsoukas, R.R. Anderson, N. Kollias (1997). Iontophoretic delivery of ALA provides a quantitative model for ALA pharmacokinetics and PpIX phototoxicity in human skin. *J. Invest. Dermatol.*, **108**, 87–91.

63. S. Lee, D.J. McAuliffe, T.J. Flotte, N. Kollias, A.G. Doukas (1998). Photomechanical transcutaneous delivery of macromolecules. *J. Invest Dermatol.*, **111**, 925–929.

64. S. Lee, N. Kollias, D.J. McAuliffe, T.J. Flotte, A.G. Doukas (1999). Topical drug delivery in humans with a single photomechanical wave. *Pharm. Res.*, **16**, 1717–1721.

65. R. Vanbever, V. Preat (1999). In vivo efficacy and safety of skin electroporation. *Adv. Drug Deliv. Rev.*, **35**, 77–88.

66. S. Mitragotri, J. Kost (2000). Low-frequency sonophoresis: a noninvasive method of drug delivery and diagnostics. *Biotechnol. Prog.*, **16**, 488–492.

67. W. Beyer (1996). Systems for light application and dosimetry in photodynamic therapy. *J. Photochem. Photobiol. B*, **36**, 153–156.

68. L. Lilge, N. Pomerleau, B. McIlroy, L. Vesselov (2000). Multi-tasking optical fiber probes for fluence-rate and fluorescent drug monitoring in vivo. *Proc. Soc. Photo-Opt. Instr. Eng.*, **3909**, 66–75.

69. S.J. Madsen, E.R. Anderson, R.C. Haskell, B.J. Tromberg (1994). Portable, high-bandwidth frequency-domain photon migration instrument for tissue spectroscopy. *Opt. Lett.*, **19**, 1934–1936.

70. B.W. Pogue, L. Lilge, M.S. Patterson, B.C. Wilson, T. Hasan (1997). The absorbed photodynamic dose examined from pulsed and cw light using tissue-simulating dosimeters. *Applied Optics.*, **36**, 7257–7269.

71. B.W. Pogue, G.C. Burke (1998). Fiber optic bundle design for quantitative fluorescence measurement from tissue. *Appl. Opt.*, **37**, 7429–7436.

72. B.W. Pogue, T. McBride, U. Osterberg, K. Paulsen (1999). Comparison of imaging geometries for diffuse optical tomography of tissue. *Opt. Exp.*, **4**, 270–286.

73. J.S. McCaughan, Jr., J.T. Guy, W. Hicks, L. Laufman, T.A. Nims, J. Walker (1989). Photodynamic therapy for cutaneous and subcutaneous malignant neoplasms. *Arch. Surg.*, **124**, 211–216.

74. M.F. Renschler, A.R. Yuen, T.J. Panella, T.J. Wieman, S. Dougherty, L. Esserman, M. Panjehpour, S.W. Taber, V.H. Fingar, E. Lowe, J.S. Engel, B. Lum, K.W. Woodburn, W. Cheong, R.A. Miller (1998). Photodynamic therapy trials with lutetium texaphyrin (Lu-Tex) in patients with locally recurrent breast cancer. *Proc. SPIE.*, **3247**, 35–39.

75. A. Busetti, M. Soncin, G. Jori, M.A. Rodgers (1999). High efficiency of benzoporphyrin derivative in the photodynamic therapy of pigmented malignant melanoma. *Br. J. Cancer*, **79**, 821–824.
76. N. Michailov, M. Peeva, I. Angelov, D. Wohrle, S. Muller, G. Jori, F. Ricchelli, M. Shopova (1997). Fluence rate effects on photodynamic therapy of B16 pigmented melanoma. *J. Photochem. Photobiol. B*, **37**, 154–157.
77. L.H. Young, M.A. Howard, L.K. Hu, R.Y. Kim, E.S. Gragoudas (1996). Photodynamic therapy of pigmented choroidal melanomas using a liposomal preparation of benzoporphyrin derivative. *Arch. Ophthalmol.*, **114**, 186–192.
78. R. Biolo, G. Jori, M. Soncin, B. Rihter, M.E. Kenney, M.A. Rodgers (1996). Effect of photosensitizer delivery system and irradiation parameters on the efficiency of photodynamic therapy of B16 pigmented melanoma in mice. *Photochem. Photobiol.*, **63**, 224–228.
79. K.W. Woodburn, Q. Fan, D. Kessel, Y. Luo, S.W. Young (1998). Photodynamic therapy of B16F10 murine melanoma with lutetium texaphyrin. *J. Invest Dermatol.*, **110**, 746–751.
80. A. Busetti, M. Soncin, G. Jori, M.E. Kenney, M.A. Rodgers (1998). Treatment of malignant melanoma by high-peak-power 1064 nm irradiation followed by photodynamic therapy. *Photochem. Photobiol.*, **68**, 377–381.
81. A. Busetti, M. Soncin, E. Reddi, M.A. Rodgers, M.E. Kenney, G. Jori (1999). Photothermal sensitization of amelanotic melanoma cells by Ni(II)-octabutoxy-naphthalocyanine. *J. Photochem. Photobiol. B*, **53**, 103–109.
82. K.M. Hebeda, M.T. Huizing, P.A. Brouwer, F.W. van der Meulen, H.J. Hulsebosch, P. Reiss, J. Oosting, C.H. Veenhof, P.J. Bakker (1995). Photodynamic therapy in AIDS-related cutaneous Kaposi's sarcoma. *J. Acquir. Immune Defic. Syndr. Hum. Retrovirol.*, **10**, 61–70.
83. Z.P. Bernstein, B.D. Wilson, A.R. Oseroff, C.M. Jones, S.E. Dozier, J.S. Brooks, R. Cheney, L. Foulke, T.S. Mang, D.A. Bellnier, T.J. Dougherty (1999). Photofrin photodynamic therapy for treatment of AIDS-related cutaneous Kaposi's sarcoma. *Aids*, **13**, 1697–1704.
84. J.L. Jones, D.L. Hanson, M.S. Dworkin, H.W. Jaffe (2000). Incidence and trends in Kaposi's sarcoma in the era of effective antiretroviral therapy. *J. Acquir. Immune Defic. Syndr.*, **24**, 270–274.
85. M. Bower, P. Fox, K. Fife, J. Gill, M. Nelson, B. Gazzard (1999). Highly active antiretroviral therapy (HAART) prolongs time to treatment failure in Kaposi's sarcoma. *Aids*, **13**, 2105–2111.
86. A.M. Cattelan, M.L. Calabro, S.M. Aversa, M. Zanchetta, F. Meneghetti, A. De Rossi, L. Chieco-Bianchi (1999). Regression of AIDS-related Kaposi's sarcoma following antiretroviral therapy with protease inhibitors: biological correlates of clinical outcome. *Eur. J. Cancer*, **35**, 1809–1815.
87. M. Korbelik (1996). Induction of tumor immunity by photodynamic therapy. *J. Clin. Laser Med. Surg.*, **14**, 329–334.
88. M. Korbelik, G.J. Dougherty (1999). Photodynamic therapy-mediated immune response against subcutaneous mouse tumors. *Cancer Res.*, **59**, 1941–1946.
89. R.R. Anderson, J.A. Parrish (1983). Selective photothermolysis: precise microsurgery by selective absorption of pulsed radiation. *Science*, **220**, 524–527.
90. L.L. Polla, S.L. Jacques, R.J. Margolis, M.R. Prince, R.R. Anderson, J.A. Parrish, O.T. Tan (1987). Selective photothermolysis: contribution to the treatment of flat angiomas (port wine stains) by laser. *Ann. Dermatol. Venereol.*, **114**, 497–505.
91. U. Schmidt-Erfurth, J.W. Miller, M. Sickenberg, H. Laqua, I. Barbazetto, E.S. Gragoudas, L. Zografos, B. Piguet, C.J. Pournaras, G. Donati, A.M. Lane, R. Birngruber, H. van den

Berg, H.A. Strong, U. Manjuris, T. Gray, M. Fsadni, N.M. Bressler (1999). Photodynamic therapy with verteporfin for choroidal neovascularization caused by age-related macular degeneration: results of retreatments in a phase 1 and 2 study. *Arch. Ophthalmol.*, **117**, 1177–1187.

92. I. Barbazetto, U. Schmidt-Erfurth (2000). Photodynamic therapy of choroidal hemangioma: two case reports. *Graefes Arch. Clin. Exp. Ophthalmol.*, **238**, 214–221.

93. S.W. Young, K.W. Woodburn, M. Wright, T.D. Mody, Q. Fan, J.L. Sessler, W.C. Dow, R.A. Miller (1996). Lutetium texaphyrin (PCI-0123): a near-infrared, water-soluble photosensitizer. *Photochem. Photobiol.*, **63**, 892–897.

94. G. Kostenich, A. Orenstein, L. Roitman, Z. Malik, B. Ehrenberg (1997). In vivo photodynamic therapy with the new near-IR absorbing water soluble photosensitizer lutetium texaphyrin and a high intensity pulsed light delivery system. *J. Photochem. Photobiol. B*, **39**, 36–42.

95. J.G. Krueger, I.B. Walters, M. Miyazawa, P. Gilleaudeau, J. Hakimi, S. Light, A. Sherr, A.B. Gottlieb (2000). Successful in vivo blockade of CD25 (high-affinity interleukin 2 receptor) on T cells by administration of humanized anti-Tac antibody to patients with psoriasis. *J. Am. Acad. Dermatol.*, **43**, 448–458.

96. A. Gottlieb, J.G. Krueger, R. Bright, M. Ling, M. Lebwohl, S. Kang, S. Feldman, M. Spellman, K. Wittkowski, H.D. Ochs, P. Jardieu, R. Bauer, M. White, R. Dedrick, M. Garovoy (2000). Effects of administration of a single dose of a humanized monoclonal antibody to CD11a on the immunobiology and clinical activity of psoriasis. *J. Am. Acad. Dermatol.*, **42**, 428–435.

97. J.R. Abrams, S.L. Kelley, E. Hayes, T. Kikuchi, M.J. Brown, S. Kang, M.G. Lebwohl, C.A. Guzzo, B.V. Jegasothy, P.S. Linsley, J.G. Krueger (2000). Blockade of T lymphocyte costimulation with cytotoxic T lymphocyte-associated antigen 4-immunoglobulin (CTLA4Ig) reverses the cellular pathology of psoriatic plaques, including the activation of keratinocytes, dendritic cells, and endothelial cells. *J. Exp. Med.*, **192**. 681–694.

98. J.G. Krueger, J.T. Wolfe, R.T. Nabeya, V.P. Vallat, P. Gilleaudeau, N.S. Heftler, L.M. Austin, A.B. Gottlieb (1995). Successful ultraviolet B treatment of psoriasis is accompanied by a reversal of keratinocyte pathology and by selective depletion of intraepidermal T cells. *J. Exp. Med.*, **182**, 2057–2068.

99. T.R. Coven, I.B. Walters, I. Cardinale, J.G. Krueger (1999). PUVA-induced lymphocyte apoptosis: mechanism of action in psoriasis. *Photodermatol. Photoimmunol. Photomed.*, **15**, 22–27.

100. F.P. Murphy, T.R. Coven, L.H. Burack, P. Gilleaudeau, I. Cardinale, R. Auerbach, J.G. Krueger (1999). Clinical clearing of psoriasis by 6-thioguanine correlates with cutaneous T-cell depletion via apoptosis: evidence for selective effects on activated T lymphocytes. *Arch. Dermatol.*, **135**, 1495–1502.

101. V.P. Vallat, P. Gilleaudeau, L. Battat, J. Wolfe, R. Nabeya, N. Heftler, E. Hodak, A.B. Gottlieb, J.G. Krueger (1994). PUVA bath therapy strongly suppresses immunological and epidermal activation in psoriasis: a possible cellular basis for remittive therapy. *J. Exp. Med.*, **180**, 283–296.

102. C.S. Rosenberg (1990). Wound healing in the patient with diabetes mellitus. *Nurs. Clin. North Am.*, **25**, 247–261.

103. M. Kamler, H.A. Lehr, R.K. Saetzler, T.J. Galla, K. Messmer (1992). Impact of ischemia on tissue oxygenation and wound healing: improvement by vasoactive medication, *Adv. Exp. Med. Biol.*, **316**, 419–424.

104. L. Dillin, P. Slabaugh (1986). Delayed wound healing, infection, and nonunion following open reduction and internal fixation of tibial plafond fractures. *J. Trauma*, **26**, 1116–1119.

105. J. Casey, W.R. Flinn, J.S. Yao, V. Fahey, J. Pawlowski, J.J. Bergan (1983). Correlation of immune and nutritional status with wound complications in patients undergoing vascular operations. *Surgery*, **93**, 822–827.

106. M.J. Reed (1998). Wound repair in older patients: preventing problems and managing the healing. *Geriatrics*, **53**, 88–94.

107. H. Koren, G. Alth, G.M. Schenk, R.H. Jindra (1994). Photodynamic therapy – an alternative pathway in the treatment of recurrent breast cancer. *Int. J. Radiat. Oncol. Biol. Phys.*, **28**, 463–466.

108. M.O. Obochi, A.J. Canaan, A.K. Jain, A.M. Richter, J.G. Levy (1995). Targeting activated lymphocytes with photodynamic therapy: susceptibility of mitogen-stimulated splenic lymphocytes to benzoporphyrin derivative (BPD) photosensitization. *Photochem. Photobiol.*, **62**, 169–175.

109. N. Yamamoto, T.W. Sery, J.K. Hoober, N.P. Willett, D.D. Lindsay (1994). Effectiveness of photofrin II in activation of macrophages and in vitro killing of retinoblastoma cells. *Photochem. Photobiol.*, **60**, 160–164.

110. N. Yamamoto, S. Homma, Y. Nakagawa, M. Hayami, H. Imanaga, M. Kurimoto, M. Mitsuhashi, T. Kimoto (1992). Activation of mouse macrophages by in vivo and in vitro treatment with a cyanine dye, lumin. *J. Photochem. Photobiol. B*, **13**, 295–306.

111. N. Yamamoto, J.K. Hoober, N. Yamamoto, S. Yamamoto (1992). Tumoricidal capacities of macrophages photodynamically activated with hematoporphyrin derivative. *Photochem. Photobiol.*, **56**, 245–250.

112. S. Evans, W. Matthews, R. Perry, D. Fraker, J. Norton, H.I. Pass (1990). Effect of photodynamic therapy on tumor necrosis factor production by murine macrophages. *J. Natl. Cancer Inst.*, **82**, 34–39.

113. K.B. Trauner, T. Hasan, M.R. Hamblin (1999). Inhibition of fibrosis by photodynamic therapy. USA Patent Number 5,913,884.

114. S.R. Reiken, S.F. Wolfort, F. Berthiaume, C. Compton, R.G. Tompkins, M.L. Yarmush (1997). Control of hypertrophic scar growth using selective photothermolysis. *Lasers Surg. Med.*, **21**, 7–12.

115. M. Obochi, J.-S. Tao, D. Hunt, J. Levy (1996). Photodynamic therapy as a biological modifier. *SPIE Proc.*, **2675**, 122–131.

116. T. Hasan, M.R. Hamblin, K.B. Trauner (2000). Acceleration of wound healing by photodynamic therapy. USA Patent Number 6,107,466.

117. N. Bennett, G. Schultz (1993). Growth factors and wound healing: biochemical properties of growth factors and their receptors. *Am. J. Surg.*, **165**, 728–737.

118. S. Frank, G. Hubner, G. Breier, M.T. Longaker, D.G. Greenhalgh, S. Werner (1995). Regulation of vascular endothelial growth factor expression in cultured keratinocytes. Implications for normal and impaired wound healing. *J. Biol. Chem.*, **270**, 12607–12613.

119. P.A. Puolakkainen, D.R. Twardzik, J.E. Ranchalis, S.C. Pankey, M.J. Reed, W.R. Gombotz (1995). The enhancement in wound healing by transforming growth factor-beta 1 (TGF-beta 1) depends on the topical delivery system. *J. Surg. Res.*, **58**, 321–329.

120. R.L. Brown, M.P. Breeden, D.G. Greenhalgh (1994). PDGF and TGF-alpha act synergistically to improve wound healing in the genetically diabetic mouse. *J. Surg. Res.*, **56**, 562–570.

121. G.L. Brown, L.J. Curtsinger, M. White, R.O. Mitchell, J. Pietsch, R. Nordquist, A. von Fraunhofer, G.S. Schultz (1988). Acceleration of tensile strength of incisions treated with EGF and TGF-beta. *Ann. Surg.*, **208**, 788–794.

122. K. Fukunaga, S. Hijikata, K. Ishimura, R. Sonoda, T. Irie, K. Uekama (1994). Aluminium beta-cyclodextrin sulphate as a stabilizer and sustained-release carrier for basic fibroblast growth factor. *J. Pharm. Pharmacol.*, **46**, 168–171.

123. M.J. Davies, C.A. Mitchell, M.A. Maley, M.D. Grounds, A.R. Harvey, G.W. Plant, D.J. Wood, Y. Hong, T.V. Chirila (1997). In vitro assessment of the biological activity of basic fibroblast growth factor released from various polymers and biomatrices. *J. Biomater. Appl.*, **12**, 31–56.

124. S.B. Nicoll, S. Radin, E.M. Santos, R.S. Tuan, P. Ducheyne (1997). In vitro release kinetics of biologically active transforming growth factor-beta 1 from a novel porous glass carrier. *Biomaterials*, **18**, 853–859.

125. T. Hasan, M.R. Hamblin (1997). Light mediated modulation of wound healing. *Proceedings of the Twentieth Biennial Cornea Research Conference* (abstract, p. 34). Boston, MA.

126. R.S. Bartholomew, M. Clarke, C.I. Phillips (1977). "Dye/light" Dye-induced photosensitization of herpes virus. A clinical trial on humans. *Trans. Ophthalmol. Soc. U.K.*, **97**, 508–509.

127. R.H. Kaufman, H.L. Gardner, D. Brown, C. Wallis, W.E. Rawls, J.L. Melnick (1973). Herpes genitalis treated by photodynamic inactivation of virus. *Am. J. Obstet. Gynecol.*, **117**, 1144–1146.

128. E.G. Friedrich, Jr., R.H. Kaufman, P.J. Lynch, D. Woodruff (1976). Vulvar histology after neutral red photoinactivation of herpes simplex virus. *Obstet. Gynecol.*, **48**, 564–570.

129. C. Moore, C. Wallis, J.L. Melnick, M.D. Kuns (1972). Photodynamic treatment of herpes keratitis. *Infect Immun.*, **5**, 169–171.

130. M.G. Myers, M.N. Oxman, J.E. Clark, K.A. Arndt (1975). Failure of neutral-red photodynamic inactivation in recurrent herpes simplex virus infections. *N. Engl. J. Med.*, **293**, 945–949.

131. L.S. Kucera, J.P. Gusdon (1976). Transformation of human embryonic fibroblasts by photodynamically inactivated herpes simplex virus, type 2 at supra-optimal temperature. *J. Gen. Virol.*, **30**, 257–261.

132. F. Berthiaume, S. Reiken, M. Toner, R. Tompkins, M. Yarmush (1994). Antibody-targeted photolysis of bacteria in vivo. *Biotechnology*, **12**, 703–706.

133. M. Vaara (1992). Agents that increase the permeability of the outer membrane. *MicroBiol. Rev.*, **56**, 395–411.

134. R.E. Hancock (1997). The bacterial outer membrane as a drug barrier. *Trends Microbiol.*, **5**, 37–42.

135. M. Merchat, G. Bertolini, P. Giacomini, A. Villanueva, G. Jori (1996). Meso-substituted cationic porphyrins as efficient photosensitizers of gram-positive and gram-negative bacteria. *J. Photochem. Photobiol. B*, **32**, 153–157.

136. Y. Nitzan, M. Gutterman, Z. Malik, B. Ehrenberg (1992). Inactivation of gram-negative bacteria by photosensitized porphyrins. *Photochem. Photobiol.*, **55**, 89–96.

137. G. Bertoloni, F. Rossi, G. Valduga, G. Jori, J. van Lier (1990). Photosensitizing activity of water- and lipid-soluble phthalocyanines on Escherichia coli. *FEMS Microbiol. Lett.*, **59**, 149–155.

138. I. Eltringham (1997). Mupirocin resistance and methicillin-resistant Staphylococcus aureus (MRSA). *J. Hosp. Infect.*, **35**, 1–8.

139. K. Hiramatsu (1998). The emergence of Staphylococcus aureus with reduced susceptibility to vancomycin in Japan. *Am. J. Med.*, **104**, 29.

140. I.A. Holder, A.N. Neely (1998). Fear of MRSA – potential for future disaster. *Burns*, **24**, 99–103.

141. R. Fekety (1987). The management of the carrier of methicillin-resistant Staphylococcus aureus. *Curr. Clin. Top. Infect. Dis.*, **8**, 169–180.

142. M. Rahman (1998). Alternatives to vancomycin in treating methicillin-resistant Staphylococcus aureus infections. *J. Antimicrob. Chemother.*, **41**, 325–328.

143. Z. Malik, H. Ladan, Y. Nitzan (1992). Photodynamic inactivation of Gram-negative bacteria: problems and possible solutions. *J. Photochem. Photobiol. B*, **14**, 262–266.

144. N.S. Soukos, L.A. Ximenez-Fyvie, M.R. Hamblin, S.S. Socransky, T. Hasan (1998). Targeted antimicrobial photochemotherapy. *Antimicrob. Agents Chemother.*, **42**, 2595–2601.

145. M.R. Hamblin, D.A. O'Donnell, N. Murthy, K. Rajagopalan, M.E. Sherwood, T. Hasan (2000). Photoinactivation of antibiotic-resistant bacteria using polycationic photosensitizer conjugates. *Proceedings of the13th International Congress on Photobiology* (abstract 392, p. 131). San Francisco, CA.

146. M.R. Hamblin, N.S. Soukos, T. Hasan (1997). Selective photoinactivation of gram. positive and gram. negative bacteria while sparing mammalian cells. *Proceedings of the 7th Congress of the European Society for Photobiology* (abstract O99, p. 79). Stresa, Italy.

147. M.R. Hamblin, D.A. O'Donnell, N. Murthy, C.H. Contag, A.T. McManus, T. Hasan (2000). Photodynamic inactivation of luminescent bacteria in contaminated and infected wounds in the mouse. *Proceedings of the13th International Congress on Photobiology* (abstract 393, p. 131). San Francisco, CA.

148. M.R. Hamblin, D.A. O'Donnell, N. Murthy, I. Rizvi, T. Hasan (1999). Targeted photoinactivation of gram. negative bacterial contamination in wounds. *Proceedings of the 8th Congress of the European Society for Photobiology*. Granada, Spain.

149. C.H. Contag, P.R. Contag, J.I. Mullins, S.D. Spilman, D.K. Stevenson, D.A. Benaron (1995). Photonic detection of bacterial pathogens in living hosts. *Mol. Microbiol.*, **18**, 593–603.

150. T. Zoladek, B.N. Nguyen, I. Jagiello, A. Graczyk, J. Rytka (1997). Diamino acid derivatives of porphyrins penetrate into yeast cells, induce photodamage, but have no mutagenic effect. *Photochem. Photobiol.*, **66**, 253–259.

151. M. Paardekooper, A.W. De Bruijne, J. Van Steveninck, P.J. Van den Broek (1995). Intracellular damage in yeast cells caused by photodynamic treatment with toluidine blue. *Photochem. Photobiol.*, **61**, 84–89.

152. V. Carre, O. Gaud, I. Sylvain, O. Bourdon, M. Spiro, J. Blais, R. Granet, P. Krausz, M. Guilloton (1999). Fungicidal properties of meso-arylglycosylporphyrins: influence of sugar substituents on photoinduced damage in the yeast Saccharomyces cerevisiae. *J. Photochem. Photobiol. B*, **48**, 57–62.

153. F.C. Odds (1994). Pathogenesis of Candida infections. *J. Am. Acad. Dermatol.*, **31**, S2–5.

154. S.J. Rupke (2000). Fungal skin disorders. *Prim. Care*, **27**, 407–421.

155. J.E. Bouquot, S.J. Fenton (1988). Juvenile juxtavermilion candidiasis: yet another form of an old disease? *J. Am. Dent. Assoc.*, **116**, 187–192.

156. J.A. Kamarashev, S.G. Vassileva (1997). Dermatologic diseases of the vulva. *Clin. Dermatol.*, **15**, 53–65.

157. T.G. Berger (1993). Treatment of bacterial, fungal, and parasitic infections in the HIV-infected host. *Semin. Dermatol.*, **12**, 296–300.

158. M. Wilson, J. Dobson (1993). Lethal photosensitization of oral anaerobic bacteria. *Clin. Infect. Dis.*, **16**, (Suppl. 4) S414–415.

159. M. Wilson (1996). Susceptibility of oral bacterial biofilms to antimicrobial agents. *J. Med. Microbiol.*, **44**, 79–87.

160. M. Wilson, J. Dobson, S. Sarkar (1993). Sensitization of periodontopathogenic bacteria to killing by light from a low-power laser. *Oral Microbiol. Immunol.*, **8**, 182–187.

161. M.A. Listgarten (2000). The structure of dental plaque. *Periodontol.*, **5**, 52–65.

162. M.A. Listgarten (1986). Pathogenesis of periodontitis. *J. Clin. Periodontol.*, **13**, 418–430.

163. H.Y. Chen, S.W. Cox, B.M. Eley, P. Mantyla, H. Ronka, T. Sorsa (2000). Matrix metalloproteinase-8 levels and elastase activities in gingival crevicular fluid from chronic adult periodontitis patients. *J. Clin. Periodontol.*, **27**, 366–369.
164. J. Liebana, A. Castillo (1994). Physiopathology of primary periodontitis associated with plaque. Microbial and host factors. A review. Part 1. *Aust. Dent. J.*, **39**, 228–232.
165. J. Potempa, J. Travis (1996). Porphyromonas gingivalis proteinases in periodontitis, a review. *Acta. Biochim. Pol.*, **43**, 455–465.
166. S.P. Vyas, V. Sihorkar, V. Mishra (2000). Controlled and targeted drug delivery strategies towards intraperiodontal pocket diseases. *J. Clin. Pharm. Ther.*, **25**, 21–42.

Index